AF509462

Air Quality Characterisation and Modelling

Air Quality Characterisation and Modelling

Editors

José Carlos Magalhães Pires
Álvaro Gómez-Losada

MDPI • Basel • Beijing • Wuhan • Barcelona • Belgrade • Manchester • Tokyo • Cluj • Tianjin

Editors
José Carlos Magalhães Pires
Department of Chemical
Engineering
University of Porto
Porto
Portugal

Álvaro Gómez-Losada
Departament of Quantitative
Methods
Universidad Loyola Andalucía
Sevilla
Spain

Editorial Office
MDPI
St. Alban-Anlage 66
4052 Basel, Switzerland

This is a reprint of articles from the Special Issue published online in the open access journal *Sustainability* (ISSN 2071-1050) (available at: www.mdpi.com/journal/sustainability/special_issues/air_quality_characterisation).

For citation purposes, cite each article independently as indicated on the article page online and as indicated below:

LastName, A.A.; LastName, B.B.; LastName, C.C. Article Title. *Journal Name* **Year**, *Volume Number*, Page Range.

ISBN 978-3-0365-7821-7 (Hbk)
ISBN 978-3-0365-7820-0 (PDF)

Contents

About the Editors

José Carlos Magalhães Pires

JCM Pires graduated in Chemical Engineering from the Faculty of Engineering of Porto University (FEUP) in 2004, with a final classification of 16. He worked in two chemical companies (SunChemical and Tagol) and then started his PhD in Environmental Engineering in 2006 at FEUP. J.C.M. Pires is an Assistant Researcher at LEPABE. He has strong expertise in environmental applications of microalgae (CO_2 capture and wastewater treatment), microalgal technology, process modelling, and integration. His teaching activity comprises several multidisciplinary topics, such as the Project of Engineering (for Chemical Engineering, Environmental Engineering, and Bioengineering), and Air Pollution, among others. He is/was the Principal Investigator of five research projects (total funding 0.80 M€): (i) FCT Investigator Grant IF/01341/2015 - Design configurations of photobioreactors for cultivation of microalgae: bioprocess modelling and sustainability assessment (2017–2021); (ii) project PIV4Algae - Process Intensification for microalgal production and Valorisation (2018–2022); (iii) Power2Biofuel –Storing Renewable Energy in Liquid Biofuels through Microalgae (2021–2022, working with Galp Energy company); (iv) Environmental Management at IMPETUS (2021–2022); and (v) PhotoBioValue - Light effect on photobioreactor design for microalgae cultivation: enhancement of photosynthetic efficiency and biomass value (2022–2024). He is/was also involved in (11) other research projects. He is (co)-author of 76 papers (>4100 citations, h-index=29), 6 books, and 27 book chapters. He was on the list of the world's most cited scientists (career-long impact and single-year impact) released by Stanford University in 2020–2022. J.C.M. Pires (co)-supervises/(co)-supervised more than 40 students, including Post-Doc Fellows, and PhD and MSc students. He is a member of the Editorial Board of seven international journals, including Bioengineering and Biotechnology (Frontiers) and Algal Research (Elsevier).

Álvaro Gómez-Losada

Álvaro Gómez-Losada has been a data scientist (non-tenure track position) at the Joint Research Centre, the in-house science service of the European Commission, since 2017. Previously, he worked as a data scientist at Banco Santander in Madrid. He holds a PhD, MSc, and BSc in Statistics from the University of Seville, and a MSc and BSc in Biology from the University of Córdoba. He specializes in the design, implementation, evaluation, and application of machine learning algorithms to different fields (e.g., competition in e-commerce platforms, the economics of climate change, and atmospheric pollution). He has co-authored more than 15 papers in international peer-reviewed journals dealing with the application of data science. Since 2023, he has been a lecturer in Statistics at University Loyola (Seville, Spain).

Preface to "Air Quality Characterisation and Modelling"

Ambient air pollution is the primary environmental health risk concern worldwide that causes seven million preventable deaths per year and the loss of healthy years of life; it threatens the sustaining of the environment through acidification and eutrophication. Almost all of the global population breathes air that exceeds the latest health-based guideline levels set by the World Health Organization, with low- and middle-income countries experiencing the highest exposure. Key sources of air pollution are road transport vehicles, domestic heating and industrial installations, and transboundary emissions. These sources produce the main ambient air pollutants of concern to which the population is exposed: particulate matter, ground-level ozone, nitrogen dioxide, carbon monoxide, and sulphur dioxide.

Air quality models constitute a complementary approach to monitoring and characterising air pollution. Spatial and temporal variability of air pollution is mainly investigated across urban areas or those hosting industrial activities, especially in developed countries, where the highest concentration of air pollutants is expected. Understanding these spatial and temporal variabilities is essential for both the implementation of air quality policies and the definition of effective measures to mitigate air pollution and its effects.

This Special Issue aims to showcase selected and original research articles concerning air quality characterisation and modelling. It includes 16 manuscripts covering a wide range of modelling topics, geographical scopes, and the characterisation of different fractions of air pollution.

Among modelling approaches are emission inventories, simulation scenarios, dispersion models, source contribution techniques, and, more particularly, cluster analysis and linear and non-linear regressions, to cite a few. In some manuscripts, air quality indexes or coefficients are used to characterise air quality; others, such as the thermal sensation index or a particular decomposition of the Gini coefficient, are not as frequently used. The geographical scope was diverse, both in urban agglomerations and rural areas, with air quality studies in several countries. Among the studied facets of air pollution were outdoor and indoor pollution, thermal comfort, or the quantification of suspendable road dust. Other included studies examined the role of administrations in air quality, evaluating the effectiveness of local emission reduction plans or carbon trading policies.

We hope the air quality characterisation and modelling community will find this special issue to be an informative and useful collection of articles and serve as an impetus to spur much more research on the topic.

Acknowledgements

This work was financially supported by (i) LA/P/0045/2020 (ALiCE) and UIDB/00511/2020-UIDP/00511/2020 (LEPABE), funded by national funds through FCT/MCTES (PIDDAC).

José Carlos Magalhães Pires and Álvaro Gómez-Losada
Editors

 *sustainability*

Review

The History of Air Quality in Utah: A Narrative Review

Logan E. Mitchell [1],* and Chris A. B. Zajchowski [2]

[1] Department of Atmospheric Sciences, The University of Utah, Salt Lake City, UT 84112, USA
[2] Department of Human Movement Sciences, Old Dominion University, Norfolk, VA 23529, USA
* Correspondence: logan.mitchell@utah.edu

Abstract: Utah has a rich history related to air pollution; however, it is not widely known or documented. This is despite air quality being a top issue of public concern for the state's urban residents and acute episodes that feature some of the world's worst short-term particulate matter exposure. As we discuss in this narrative review, the relationship between air pollution and the state's residents has changed over time, as fuel sources shifted from wood to coal to petroleum and natural gas. Air pollution rose in prominence as a public issue in the 1880s as Utah's urban areas grew. Since then, scientific advances have increased the understanding of air quality impacts on human health, groups of concerned citizens worked to raise public awareness, policy makers enacted legislation to improve air quality, and courts upheld rights to clean air. Utah's air quality future holds challenges and opportunities and can serve as useful case for other urbanizing regions struggling with air quality concerns. Population growth and changing climate will exacerbate current air quality trends, but economically viable clean energy technologies can be deployed to reduce air pollution, bringing substantial public health and economic benefits to the state's residents and other settings with similar public health concerns.

Keywords: air pollution; air quality; Utah; environmental history; national ambient air quality standards

Citation: Mitchell, L.E.; Zajchowski, C.A.B. The History of Air Quality in Utah: A Narrative Review. *Sustainability* 2022, 14, 9653. https://doi.org/10.3390/su14159653

Academic Editors: José Carlos Magalhães Pires and Álvaro Gómez-Losada

Received: 31 May 2022
Accepted: 20 July 2022
Published: 5 August 2022

Publisher's Note: MDPI stays neutral with regard to jurisdictional claims in published maps and institutional affiliations.

1. Introduction

The city's atmosphere can be cleared of smoke and grime, but not in a single day or year, not by a single group or group of persons, not by a single invention nor without efforts or price. There is nothing magical going to happen. It will take a properly guided, united and continued effort to solve the problem.

—George W. Snow, Chief of Salt Lake Bureau of Mechanical Inspections, 8 February 1917 [1].

In recent decades, degraded air quality ranked among the top concerns of Utahns, particularly the state's residents who live along the rapidly urbanizing Wasatch Front (Figure 1) (e.g., [2]). Both social and ecological factors affect the state's ability to mitigate the impacts of anthropogenic air pollution. Utah is situated in the arid, intermountain, western United States (U.S.) with major population centers located in northern topographic basins. Two-thirds of Utah's population, roughly 2 million residents, live adjacent to the Wasatch Mountain range and the Great Salt Lake. Air pollutants are emitted almost entirely by local sources within Utah [3] and these topographic features exacerbate air quality events, leading this urbanizing region to regularly feature some of the worst short-term air pollution episodes in the world (e.g., [4]).

With air quality a dominant health risk of the 21st century (e.g., [5,6]), the prevalence of acute air pollution events in Utah presents significant challenges to public health, quality of life, and economic vitality. The impacts from poor air quality episodes on human health (e.g., respiratory, circulatory, cancer, mortality, etc.) are well documented [4,6,7], as are cascading economic impacts (e.g., health care costs, decreased worker productivity, etc.) [8,9]

and environmental impacts (e.g., O_3 injury to plants, viewshed impacts from haze, etc.) [10]. Moreover, recent scholarship documenting the impact of degraded air quality in Utahn's quality of life, ranging from recess closures for children (e.g., [11]) to substituted outdoor recreation for adults (e.g., [12]), demonstrates the holistic impacts of air pollution on the wellness of state residents.

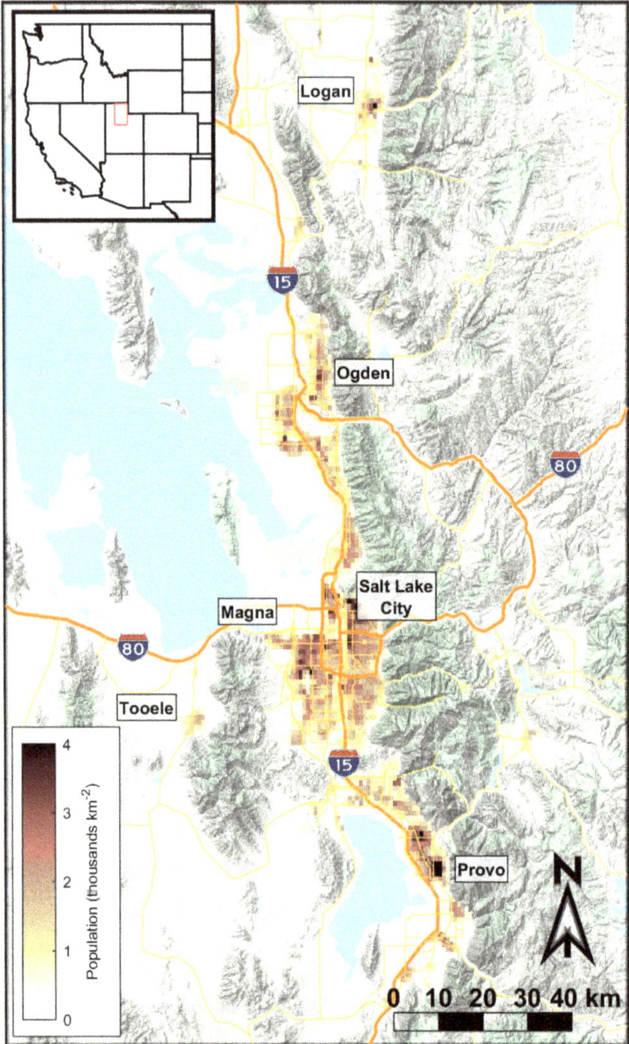

Figure 1. Map of the Wasatch Front and major Utah population centers. Note. The mountainous topography surrounding metropolitan areas exacerbates air quality challenges.

As we document in this narrative review, Utah's struggles with poor air quality are not new. Recent digitization of newspaper articles and scientific reports increased the availability of historical accounts of how Utahns historically grappled with air pollution issues. The dominant narrative in U.S. air quality history focuses on eastern cities, such as Pittsburgh, Philadelphia, and Chicago, where coal combustion led to severe air quality issues, as well as Californian cities such as Los Angeles, where photochemical smog was the primary air quality issue [13]. However, drawing from resources recently made available, we situate Utah's important role in the national and international history of air pollution, focusing on scientific advances, public opinion, and policy responses from the mid-19th century to present day. In doing so, we discuss how this history informs the present understanding of air resource management in Utah, the success of regulatory and scientific advances in reducing air pollution events, and potential future efforts to mitigate poor air quality conditions in the rapidly urbanizing state. Our findings help forward not only technical solutions for other urban areas plagued by poor air quality, but also present structural and behavioral challenges and opportunities to pursue clean air and improve quality of life.

2. Materials and Methods

Here, we deploy a narrative review to understand the history of air quality in Utah, with a particular focus on northern urban centers along the Wasatch Front. Narrative reviews differ from other systematic review types, in part, as they are a qualitative synthesis of literatures from diverse fields without statistical analyses [14]. Additionally, they are useful in understanding the "historical account" of a topic (p. 755, [14]), such as the coupled human and natural system of air quality in Utah. Narrative reviews are becoming increasingly common in interdisciplinary scholarship (e.g., [15,16]), making them relevant to our use of various forms of data (i.e., newspaper articles, scientific manuscripts, policy statements, etc.) to illuminate the historical arc of the human relationship with air quality in Utah.

We conducted a keyword search-based literature review using the Utah Digital Newspaper Archive (https://digitalnewspapers.org/, accessed on 1 May 2020) for articles containing terms such as "air quality", "smoke", "nuisance", "inversion", "smog", and "abatement". We applied a snowballing search methodology using the references in the newspaper articles to identify and include peer-reviewed articles (e.g., primary research, reviews, commentaries), theses, government reports, online reports, and press releases. Each identified item was assessed for relevance by a member of the study team. This review is intended to cover the major events and public issues related to the history of air quality in Utah but is not meant to be an exhaustive account.

3. Results

3.1. 1800s to 1910s

Air quality has been an issue in Utah for as long as human residents have been combusting fuel. Early non-indigenous explorers in the 1800s noted how blue smoke from wood fires would hang in the Salt Lake Valley for extended periods of time [17]. Mormon pioneers settled in the Salt Lake Valley in 1847, and as they developed Salt Lake City, air pollution quickly became a persistent problem. One of the earliest notable comments about air pollution in Utah surfaced in 1860 from Brigham Young, the second president of The Church of Jesus Christ of Latter-day Saints. In those days, homes were heated primarily with wood burning stoves, making adequate ventilation essential. As Young said:

> *What constitutes health, wealth, joy, and peace? In the first place, good, pure air is the greatest sustainer of animal life. Other elements of life we can dispense with for a time, but this seems to be essential every moment; hence the necessity of well-ventilated dwelling houses—especially the rooms occupied for sleeping.* [18]

Young's focus on the importance of indoor air quality preceded the science that would corroborate it. Even today, there is a continued international focus on the indoor air quality impacts from burning natural gas for home cooking and the need for adequate ventilation [19].

By the late 1800s, coal became the dominant fuel for factories and home heating both in Utah and nationally (Figure 2). Although Utah's population of 200,000 was smaller than the Wasatch Front's approximately 2 million residents today, unregulated emissions and inefficient combustion created a persistent "smoke nuisance" [20]. The direct health impacts of air quality were not well known at the time, but there was an understanding that the smoke nuisance degraded the cleanliness of the air and caused other socioeconomic impacts, particularly from soot particles falling from the air, soiling clothing and affecting outdoor markets that sold furniture, fruit, groceries, and other products. There was even an understanding that ambient air pollution from outside the home penetrated into homes and affected indoor air quality [20,21].

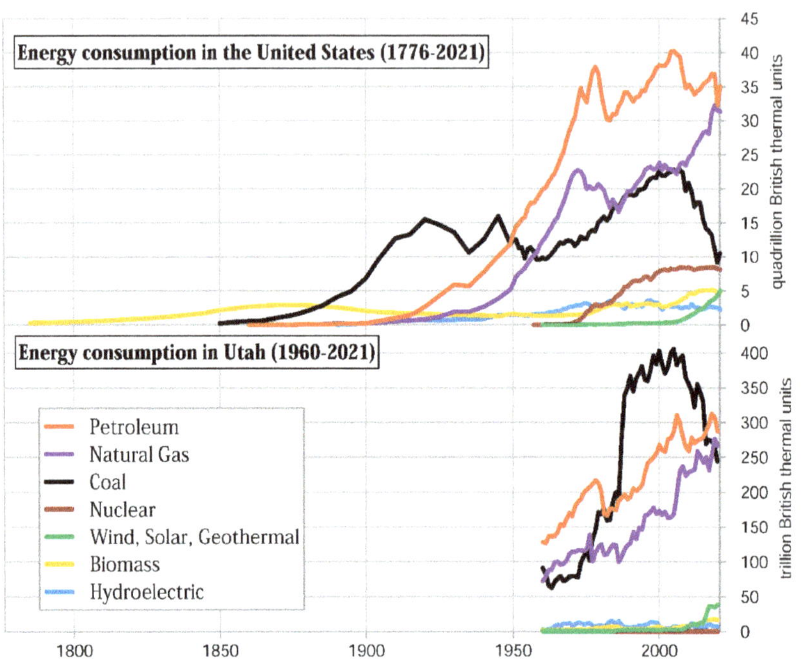

Figure 2. Energy consumption in the United States (**top**) and in Utah (**bottom**) by fuel type [22,23]. Utah energy consumption data are only available after 1960.

Published commentary advocating how to address air pollution in the late 1800s is not wholly dissimilar from certain attitudes existent in the state today [24]. One preference was for private industry to take initiative to reduce emissions and avoid regulation. For example, a Deseret Evening News editorial in 1881 stated:

> *Salt Lake is beginning to suffer in some degree from the effects of the smoke nuisance, and judging from the dense clouds which arise especially in the evening in the central part of the city, we are of the opinion that the time is not far distant when some municipal regulation will have to be adopted and enforced here—as elsewhere—to abate it. But would it not just as well for those whose business is such as to require the use of coal in large quantities, to look about for some remedy for excessive smoke, of their own volition, and thus avoid the necessity of declaring their works a public nuisance? We think so and throw out the suggestion in kindness. [25]*

However, at that time, low-emission technologies were not available, and efforts to address inefficient combustion of coal were nascent. As the population continued to grow and the air pollution problem became worse, focus turned to municipal ordinances to control emissions.

The first legislative effort focused on improving air quality in Utah was a Salt Lake City municipal ordinance mandating the installation of emission control devices ("smoke consumers") on large furnaces in 1891 [26,27]. These devices were designed to control the rate that coal was added to the boiler to optimize combustion efficiency since incomplete combustion caused excess "smoke" air pollution. Since this ordinance was introduced five years before Utah became the 45th state in the United States of America, in 1896, and Utah continues to have air pollution challenges today, with several counties in nonattainment for National Ambient Air Quality Standards, it is fair to say poor air quality has been a permanent fixture in Utah for the entire history of the state.

Today, there is an understanding that wintertime air pollution is exacerbated by multi-day temperature inversions that last several days to weeks and trap emissions near the surface [28,29]; however, in the late 1800s, that phenomenon was not understood. Instead, there was a basic understanding of the interaction between topography and meteorology. Urban planners of the time recognized that if a factory was located at the mouth of one of the canyons adjacent to Salt Lake City, the emissions from the factory would be transported through the city and worsen air pollution issues; thus, factories were built west of the Jordan River that defines the topographic low point within the Salt Lake Valley [30]. These urban planning decisions in the 1890s reverberate to the present day since the commercial and industrial infrastructure generally remain in the same location over a century later. Areas of the city near industrial facilities were then the focus of redlining practices in the 1930s, dictating where racial minorities lived. These demographic patterns persist today contribute to modern environmental justice challenges [31].

The same 1893 article that discussed the location of factories also articulated an important value system that appears throughout Utah's history, stating:

> *Factories that blacken the city with smoke can be as much a detriment as they are an advantage, for Salt Lake has as much to expect from the increase she will receive from persons who will select it as their residence on account of its pure air and cleanliness as it has to gain from factories.* [30]

This desire to balance economic development with environmental stewardship is present throughout Utah's history across all aspects of society, including activists, regulators, academics, stakeholders, and policy makers. A recent example is Governor Spencer Cox's OneUtah Roadmap that incorporates sustainable growth that addresses air quality and climate change as part of the state's overall economic advancement strategy [32].

Smelting and refining of ores in the Salt Lake Valley from nearby mines began in the 1900s, with the first copper smelter opening in Murray in 1899, and several more opening in the following years. These early smelters featured no pollution controls and emitted toxic smoke that included lead, arsenic, and sulfur dioxide. In the years that followed, crops near the smelters began to die and farmers blamed the "smelter smut" for causing these crop failures. Farmers eventually sued the smelter owners and federal courts forced the smelters to process ore with less than 10% sulfur content and prevented them from emitting arsenic [33]. In the aftermath of the court decision, several of the smelters moved to the less-populous Tooele Valley, and efforts to address toxic air emissions led to several innovations [34]: specifically, filters installed at the smelting factories collected lead, arsenic, and other solid particles that were then processed into chemical products and sold. Smelter owners also funded research into farming practices that examined how crops responded to sunlight, fertilizer, watering, etc. This latter effort, in part, was design to discredit farmers, who they accused of "smoke farming", i.e., suing smelters instead of growing crops [35] (Figure 3).

Since the Mormon Farmer Can No Longer Farm the Smelters, He Asks Counsel From the Authorities.

Figure 3. Anti-farmer sentiment related to suit of smelting operations depicted in this cartoon, Salt Lake Telegram, 28 December 1906, digitalnewspapers.org [35].

By 1912, dozens of cities across the country were passing ordinances targeted at improving air quality [13,36]. Salt Lake City was no exception, and by that point passed several ordinances that imposed fines or hired inspectors to suggest efficiency improvements at factories [37–39]. There was also a growing recognition that air pollution would affect tourism [40]. The Salt Lake Telegram published an in-depth six-part series about air quality that discussed impacts on health, economics, and ecosystems:

> *Aside from the immense cost of smoke to the owners of the plants which make it–a cost represented in needless coal bills–the national smoke bill represents millions of dollars of economic waste to the public at large. The blighting influence of a municipal smoke pall on the health of a community alone might justify its abatement, for it has been shown at home and abroad that smoky cities are cities with high death rates from all bronchial and pulmonary diseases.* [41]

In the following years, efforts to improve air quality saw sustained community interest. Several women's groups banded together to form a "Smoke Abatement League" to assist in public education efforts and lobby for regulations [42–44]. The Smoke Abatement League gathered information about how cities were reducing emissions from across the country and printed 7000 leaflets that they distributed across the city [45,46].

3.2. Air Quality Research (1910s–1920s)

In 1914, the University of Utah began to officially study air pollution. The 1914 Utah Legislature established the Metallurgical department and began a collaboration with the U.S. Bureau of Mines to investigate how low-grade coal contributed to air pollution. The research partnership investigated methods of processing soft coal into a "fool proof" fuel that did not produce smoke when it was burned [47]. The Bureau of Mines also became involved in air quality abatement campaigns and studies with an overall objective to increase efficiency and prevent waste in the utilization of mineral resources throughout the U.S. Salt Lake City was chosen as a demonstration city for this effort. Osborne Monnett, the head of the project, noted after he arrived that "in no other city of the country has the smoke problem been attacked with the vigor shown in this campaign" [48].

One aspect of this effort was to determine if smelters located 8–16 miles away from Salt Lake City contributed a disproportionate amount to the poor air quality in the city. Samples of sulfur dioxide (SO_2) were collected around the city and at various times of day. The highest concentrations of SO_2 were located in the business district of the city, not downwind of the nearby smelters [49]. Another popular hypothesis at the time was that SO_2 emitted from smelters created a "heavy blanket which prevents the smoke from rising". That a gas could create a blanket above the city to prevent atmospheric mixing "seemed untenable" to St. John Perrot, but he tested this hypothesis by collecting SO_2 samples from several altitudes with a biplane [50] (Figure 4). Only three samples were collected, but each demonstrated that there was not a high concentration layer of SO_2 over the city that prevented vertical mixing. While there was an understanding at the time that wintertime pollution episodes were associated with low wind speeds, it would be several decades before it was widely understood that the vertical temperature profile of the atmosphere during inversions was responsible for the lack of mixing in the atmosphere and buildup of pollutants [51–53].

The study concluded in the following year and several reports were issued. Total suspended particles (TSP) as high as 2500 $\mu g/m^3$ were measured in the early 1920s, which would be considered "hazardous" by today's federal air quality standards [54]. Additionally, enamel pails were used as passive samplers to measure soot deposition of 300 to 1000 t/mi^2 in the residential and industrial districts, respectively, during the winter heating season. About 30–50 percent of the soot fall was combustible, indicating that much of it resulted from incomplete combustion and represented wasted energy resources. In December 1919 and January 1920, SO_2 concentrations in Salt Lake City averaged 150 ppb, with the highest measurement being 800 ppb, which is far higher than today's federal air quality SO_2 standard of 75 ppb and typical observations of approximately 1 ppb.

At that time, there was also a growing understanding of the ways that air pollution was affecting human health, particularly in connection with tuberculosis and pneumonia [55]. The first systematic inventory of large point sources concluded that the large heating and industrial plants in the business district were a key contributor to the air quality issues [56]. The interventions and technological approaches to improving air quality at this time were still focused on burning fuel without visible smoke. A scientific concept that was not understand at the time was that pollutants such as particulate matter could be formed from secondary chemical reactions of gaseous precursor species and that coal combustion, even if performed efficiently, was still contributing to air pollution [57]. Despite this lack of understanding, Monnett and his colleague's effort concluded that air quality problems would persist while soft coal was used for domestic heating. So, when natural gas was introduced into the Salt Lake Valley in 1930, there was optimism that its usage would improve air quality conditions, and the natural gas industry began promoting its usage to improve local air quality as compared to coal combustion [54,58] (Figure 5).

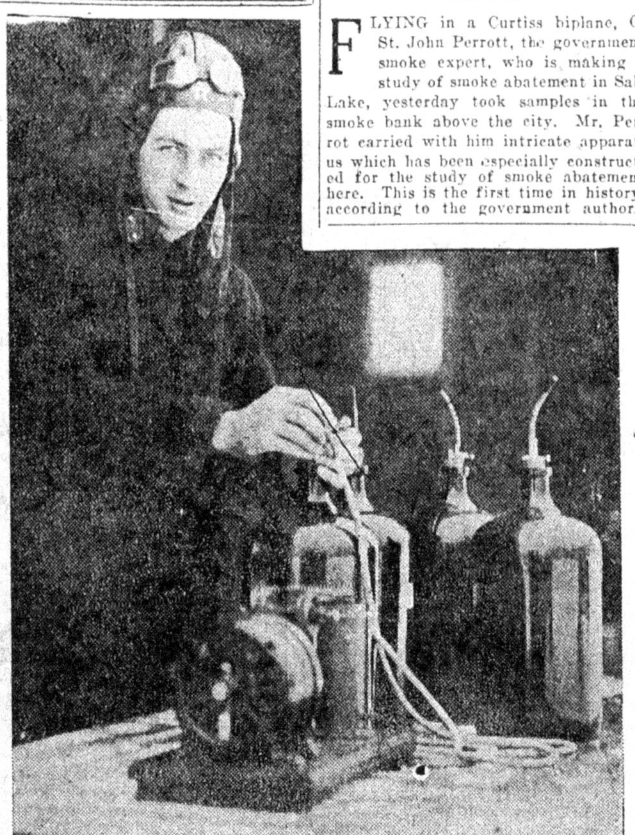

Figure 4. Photo of G. St. John Perrot and the sampling flasks used in the first aircraft sampling campaign to study SLC's air pollution, Salt Lake Tribune, 10 November 1919, digitalnewspapers.org [50].

Figure 5. Natural gas advertisement, Salt Lake Tribune, 24 October 1930, digitalnewspapers.org [58].

3.3. 1930s–1960s

During the economic challenges of the Great Depression, public attention on air quality issues declined, but the problem persisted. Jobs for air quality monitoring of point sources were created as part of the New Deal Works Projects Administration (WPA) program in Salt Lake City [59]. Interest was renewed in 1941 when Salt Lake City again considered new smoke ordinances modeled after a program in St. Louis, Missouri. For 30 days, the Salt Lake Telegram published daily air quality reports under the title "Today's Smoke" [60] (Figure 6). In addition, the Telegram published the time series of 13 years of daily Ringelmann chart observations from the Smoke Abatement Division that began after the major studies of the 1920s [61] (Figure 7). After some debate, regulations were passed that went into effect on 1 October 1941. Shortly thereafter, on 7 December 1941, the attack on Pearl Harbor occurred and the country mobilized for WWII.

During the war, air quality regulations were relaxed in favor of increasing production capacity and public attention to air quality once again receded. Throughout this period, visibility was tracked at the Salt Lake Airport and used as a proxy for visible air pollution in the winter months (November–February). A retrospective analysis of winter visibility conditions suggested that air quality worsened significantly during the war years in the 1940s but improved markedly after WWII was over [53] (Figure 8). For example, soot deposition from coal combustion was 83 t/mi^2 in 1942–1943 and decreased to 22 t/mi^2 in 1957 [62].

Figure 6. "Today's Smoke" column, Salt Lake Telegram, 7 March 1941, digitalnewspapers.org [60] (**left**) and the present-day equivalent (**right**).

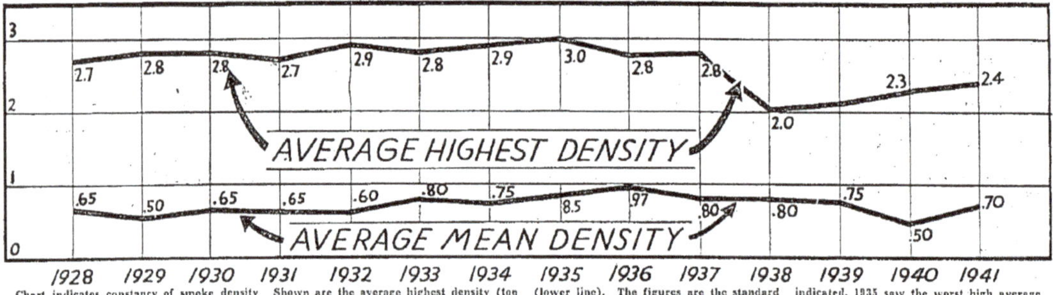

Figure 7. Smoke density from Ringelmann smoke charts collected by the Salt Lake City Smoke Abatement Division, Salt Lake Telegram, 1 March 1941, digitalnewspapers.org [61].

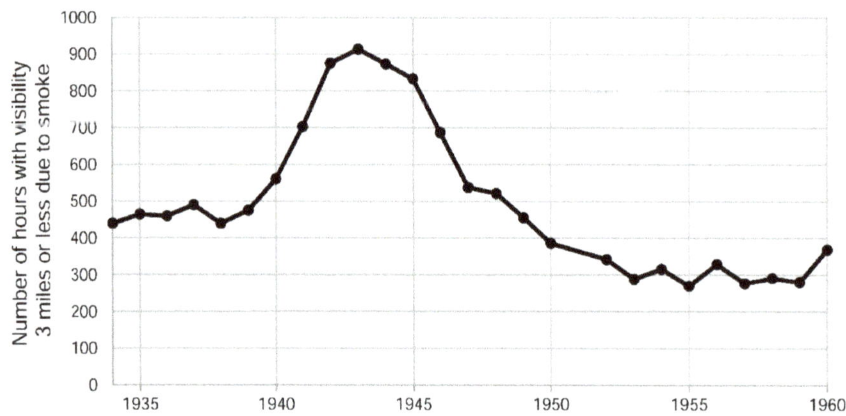

Figure 8. Five-year average of the number of hours with visibility three miles or less at the Salt Lake Airport in winter (November–February), which was used as a proxy for air pollution [53].

After WWII concluded and through the 1950s, there was a substantial decrease in coal consumption as homes and industries transitioned from coal usage to natural gas

(Figure 2). However, while air pollution from coal combustion was declining, a new air quality problem was emerging: vehicle emissions were unregulated. In large population centers in the western U.S. that featured abundant sunshine and reliance on personal vehicles for transportation (e.g., Los Angeles), severe levels of photochemical smog became a serious issue. This led to several U.S. states forming air quality control agencies. This time period also saw several high-profile, high-mortality air pollution episodes associated with wintertime inversions and industrial pollution such as in Meuse Valley, Belgium in 1930 [63]; Donora, Pennsylvania in 1948 [64,65]; and the London "fog" event in 1952 [66]. These high-profile events and action across several states led the federal government to pass the Air Pollution Control Act of 1955. This act provided technical assistance for state air pollution control boards and funding for the U.S. Public Health Service to study the health impacts of air pollution, but did not impose regulations on emissions. In 1959, the California Motor Vehicle Control Board set the first vehicle emissions standards to control nitrogen oxide (NO_x) emissions that took effect in 1963 model year vehicles.

Back in Utah, the role of persistent wintertime atmospheric stratification events ("inversions") that exacerbated air quality issues along the Wasatch Front began to be studied and understood. This was first explored in 1947 [52], but it was not until daily radiosondes began launching in 1956 that it was extensively studied [51,53]. It was noted that local inversions were often accompanied by high pressure aloft, which could persist for 2–3 weeks at a time and during these extended time periods, air pollutants would accumulate. Research on how inversions play a key role in exacerbating air quality conditions has continued through the present day and is the focus of upcoming projects [28,29,67].

In the early 1960s, Representative David King called for a major assessment of the air resources in the Utah that was submitted to the Legislature in 1962 [62]. The report summarized the history of air pollution to date, described the latest understanding of how topography and meteorology contributed to air pollution, how population growth was contributing to the issue, specific pollutants, and what efforts were being made to further study and address the issue. Some of the focus at the time was related to exposure to nuclear fallout, and monitoring stations were being installed accordingly. The report concluded by finding a lack of continuous measurements of air pollutants taken with standardized procedures and recommended that additional systematic measurements take place in Utah. In response to this report, the Utah Legislature codified air pollution as a public nuisance and made it a misdemeanor to create air pollution injurious to human life, plants, animals, or property [68]. The air pollution control law was updated significantly in 1967 with the passage of the Air Conservation Act that created the Air Conservation Committee within the State Department of Health and empowering it in control, abatement, and prevention of air pollution [69], which formed the foundation of what today is the Division of Air Quality.

Throughout the 1960s, air quality and meteorology research continued by, for example, deploying new air quality measurement sites [70,71], studying impacts on health and ecosystems [72,73], and developing better understanding of inversions [74]. Private companies were also working to install pollution control devices [75], and discussion began about how electric cars could help to mitigate air pollution [76] (Figure 9). A conference on Air Quality was held in 1969 at the University of Utah, sponsored by the Utah Air Conservation Committee, women's civic groups, researchers, and legislators [77]. Throughout this time, however, problems with air pollution persisted.

3.4. The Federal Regulatory Framework (1970s)

The dawn of the federal air quality regulatory era began in 1970 with an amendment of the Clean Air Act, which created the Environmental Protection Agency (EPA). The EPA established the first National Ambient Air Quality Standards (NAAQS) to protect public health and welfare in 1971. Around the same time, the EPA conducted a series of epidemiologic studies, known as the Community Health and Environmental Surveillance System (CHESS), and the Wasatch Front was one of six urban regions selected for the study [78]. The CHESS studies had several methodological and operational problems, which were discovered in part by comparing the EPA CHESS air quality data against data collected by the Utah Division of Health. These issues spurred a U.S. congressional investigation and after several years led to major improvements in EPA measurement and quality control procedures [79].

The electric automobile--for clean, quiet, efficient urban transportation--is the subject of sharply increased interest these days, as a means of combatting air pollution, noise and traffic congestion.

BOUNTIFUL Light and Power said this week that pressures carry out provisions of the bill.

Sen. Edmund Muski (D., Maine), is chairman of a Senate Public Works subcommittee which recently issued a report urging creation of a federal task force to develop alternatives to the internal combustion engine.

DEVELOPMENT of the elec-

an electric car known as the "Electrovair" and an electric truck. Such vehicles are capable of speeds up to 80 miles per hour and will have a range of 100 miles, sufficient for commuting and shopping.

"The electric vehicle can prove to be at least a partial answer to many of our urban problems today," Mayor Swapp said. "Air pollution, traffic and parking congestion and noise all can be reduced by development of a modern electric vehicle. This would also be an excellent off-peak load builder for our city owned utility."

Figure 9. Editorial discussing how electric vehicles could improve air quality along the Wasatch Front, Davis County Clipper, 18 November 1966, digitalnewspapers.org [76].

While the CHESS saga unfolded on the federal stage, the Utah Division of Health used their own monitoring data to evaluate compliance with the new NAAQS in a report on "Utah Health Facts" [80]. This report found widespread exceedances for total suspended particles (TSP), sulfur dioxide (SO_2), carbon monoxide (CO), oxidants (primarily ozone, O_3), and nitrogen oxides (NO_x) (Table 1). Although the NAAQS thresholds for each pollutant became more stringent over time as air quality impacts became better understood, the number of exceedances in 1970 demonstrate that air quality conditions were far worse than they are today. The report discussed ways in which Utah could achieve compliance with air quality standards but noted that several factors were beyond Utah's control; some of the TSP exceedances were caused by natural dust storms, and exceedances of CO and NO_x were due to vehicle emissions regulated at the federal level. Air quality research in Utah continued to increase in the 1970s, with at least 26 active research projects across a wide range of disciplines being active in 1972 [81].

The potential influence of dust storms on TSP was particularly important for Utah. At that time, there were several competing methods to measure particulate pollution. In the United Kingdom, the "smoke shade" method collected particles on filter paper and then examined their optical properties to establish the level of "black smoke" (BS). A similar technique used in the U.S. measured the "coefficient of haze" (CoH). These devices utilized pumps with a low flow rate that biased the measurements towards smaller particles [82,83] and are analogous to modern black carbon measurements [84,85]. These instruments were originally devised as a way to quantify subjective smoke density observations, but in the U.S. there was a desire to use the TSP method since it was thought to be a more objective technique. The TSP measurement method used a high-volume sampler with much higher flow rates that collected nearly all suspended particles, including a large range of particle sizes up to 60 μm in diameter, or about the same diameter as human hair [79]. However, these large particles can be filtered out by the body's defense mechanisms in the nose, mouth, and upper respiratory tracts and, while irritable, do not cause the same kinds of substantial health effects that smaller particles cause. These deficiencies and differing approaches led to disagreements within the epidemiology community, with some scientists supporting strong standards and other scientists supported by the American Iron and Steel Institute arguing that the research was inconclusive about the health risks of low concentrations of particulates [82,86,87]. Utah was caught in the middle of this controversy since many of the TSP exceedances were from natural dust storms and, since there was uncertainty about the health impacts, there was not a strong focus on addressing particulate pollution [80].

Table 1. Performance on federal air quality standards in Utah, 1970 [80].

Pollutant	Avg. Time	NAAQS of 1970	No. of Exceedances in 1970
TSP [1]	24 h Primary	$260\ \mu g\ m^{-3}$	5
TSP [1]	24 h Secondary	$150\ \mu g\ m^{-3}$	32
SO_2	24 h Primary	140 ppb	11
SO_2	24 h Secondary	100 ppb	32
SO_2	3 h	500 ppb	36
CO	8 h	9 ppm	641
CO	1 h	35 ppm	2
Oxidants [2]	1 h	80 ppb	42
NO_x [3]	Annual	50 ppb	1

[1] TSP = total suspended particles, [2] oxidants primarily refer to ozone, [3] NO_x = nitrogen oxides ($NO + NO_2$).

To investigate the health impacts of smaller particles, the EPA created a new monitoring network called the Inhalable Particulate Network (IPN) in 1979 that eventually spread to 157 cities [88,89]. The IPN deployed several new instruments, one of them being a dichotomous sampler that measured the mass of aerosols in two size fractions: fine particulate matter with a diameter smaller than 2.5 μm ($PM_{2.5}$), thought to be derived primarily from combustion processes, and coarse particulate matter with a diameter of 2.5–15 μm. When added together, they yielded the mass of all particles smaller than 15 μm (PM_{15}). In 1981, based on recommendations from the scientific community, the size of the coarse fraction was changed to 2.5–10 μm to obtain measurements of PM_{10}.

The IPN initially hosted two sites in Utah, one at the Salt Lake City Health Department (6 South, 200 East), and one in Magna, UT (Brockbank Jr. high school). Only the Salt Lake City site was equipped with the new dichotomous sampler, so these were the first measurements of $PM_{2.5}$ collected in Utah [90] (Figure 10). These 24 h average measurements were laborious, so they were only collected every sixth day. Based on the current 24 h $PM_{2.5}$ NAAQS of $35\ \mu g\ m^{-3}$, the winter of 1981–1982 exceeded this level on at least four days. If we assume that the sampling strategy of once every sixth day was representative of conditions that winter, it indicates that concentrations exceeded that level on about 24 days. This number of days is approximately equivalent to the 18 exceedances per winter that were found in the decade after regular measurements of $PM_{2.5}$ began in Salt Lake City in 1998 [29].

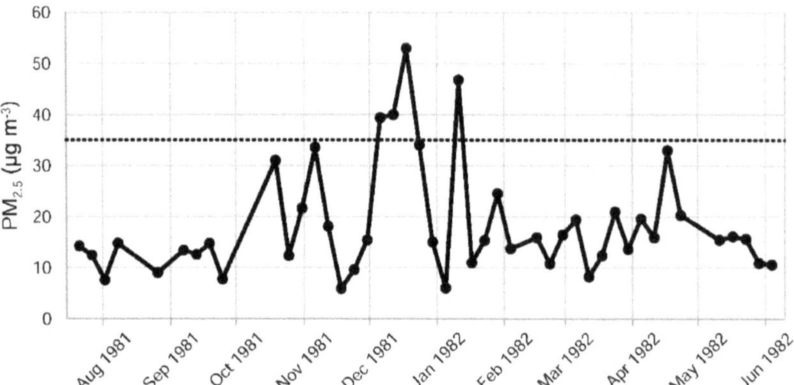

Figure 10. The first measurements of fine particulate matter (PM$_{2.5}$) in Utah in the Inhalable Particulate Network (IPN) [90]. Measurements are 24 h averages made every sixth day. Dashed line indicates the current 24 h PM$_{2.5}$ National Ambient Air Quality Standard.

3.5. Science, Regulations, Disinformation and State Action (1980s–1990s)

The 1980s–1990s were a critical time for air quality in Utah. Data from the IPN were leading to an increased awareness of the impact of particulates on health, but it was the Geneva Steel case study that shocked the scientific community and once again thrust Utah onto the global scientific stage. These scientific advances then led to increased regulatory scrutiny, and industrial polluters sought to spread disinformation about the scientific findings to delay the implementation of regulations. Meanwhile, sustained public dissatisfaction with air quality and exceedances of federal air quality standards led the Utah state government to explore state-level efforts to address air pollution. All these events occurred simultaneously, but since each of them have their own historical narrative arc, we discuss each in turn.

3.5.1. Scientific Advances

In the 1980s, a natural experiment at the Geneva Steel plant in Provo, UT, demonstrated the significant health impacts of particulate air pollution [91,92]. The Geneva plant was a large industrial source of air pollution, which was the focus of air quality concerns for some time (e.g., [62,93,94]). Nearby monitoring of Total Suspended Particles began in the early 1960s, and PM$_{10}$ was added in April 1985 [95]. On 1 August 1986, a labor strike forced the steel plant to shut down for 13 months. When it reopened on 1 September 1987, there was a resumption of emissions from the smokestack [96] (Figure 11). In the following months, there were near daily complaints in the Provo Daily Herald about the pollution from the plant, contrasted with plant workers and industry lobbyists who claimed that the pollution was not that bad and the jobs provided by the plant were more important than any potential health impacts from pollution (e.g., [97–99]). The strike and plant closure allowed for a comparison of health outcomes during the winter season when the plant was shut down and the winter seasons before and after. The design of this natural experiment controlled for many of the typical confounding variables in health studies, such as smoking, weight, and physical activity, by using the same population of the same city in a different year as comparison. The results showed that bronchitis and asthma hospital admissions for preschool-age children in Provo were approximately twice as frequent in the winters when the steel mill was operating versus the winter when it was idled [95,100]. This startling result demonstrated the considerable health impacts from particulate pollution and set the stage for a series of influential scientific studies.

Figure 11. Restarting the Geneva steel mill after a 13-month closure due to a labor strike caused an increase in pollution, Provo Daily Herald, 13 September 1987, digitalnewspapers.org. Note: Reprinted with permission from [96]. 1987, Ryan Christner.

Building on the Geneva Steel case study, Pope joined a team of researchers at the EPA and Harvard University examining health impacts of particulate pollution across multiple cities. In what became known as the "Harvard Six Cities" study, the team examined the measurements from the IPN network in six cities with sufficiently long records (5–8 years) and found a robust linear relationship between $PM_{2.5}$ and mortality after controlling for confounding variables [101] (Salt Lake City was not included in this study because it only collected one year of $PM_{2.5}$ data from the IPN (Figure 10)). This was a groundbreaking result because the relationship between mortality and $PM_{2.5}$ was so clear, even at low concentrations typically found in cities. Even the investigators were skeptical of the strength of the relationship between $PM_{2.5}$ and mortality, so Pope sought a way to replicate the results with a broader population data set. He collaborated with investigators working on the American Cancer Society (ACS) Cancer Prevention Study II (CPS-II) that had long patient history records to compare the IPN data to health outcomes for 500,000 people across 151 cities. The results of this study confirmed the relationship between fine particulate and mortality in the six cities study [102]. This pair of studies demonstrated how important minimizing fine particulates was for public health and they have since become some of the most cited publications in the field of air pollution research [103]. Additionally, this pair of studies set off protracted struggles over particulate matter regulations between industry, public health advocates, scientists, and regulators that continue to the present day.

3.5.2. Regulatory Advances

The Clean Air Act requires the EPA to re-evaluate the NAAQS for each pollutant every five years based on the latest scientific knowledge. The original 1970 particulate standard was based on TSP, but given the methodological and epidemiological questions about TSP, it was not until 1987 that the TSP standard was changed to PM_{10}. In 1992, despite the new research on particulates being published by Pope and others, the EPA declined to conduct its five-year review of the standard and the American Lung Association thus sued the EPA in 1994. The American Lung Association won their lawsuit, and the court ordered the EPA to review the standard by 1997. Based on the new scientific findings of the importance of fine particulate matter ($PM_{2.5}$) on public health, the EPA created a new standard for $PM_{2.5}$ in addition to the existing PM_{10} standard. Industrial trade organizations then sued to prevent the standards from being implemented; the cases worked their way up through U.S. courts until, in 2001, the Supreme Court unanimously decided that the EPA had the authority to establish the standards and the agency had a suitable approach for determining the standards. The Supreme Court remanded the remaining issues to the D.C. Circuit Court of Appeal, which rejected all remaining challenges to the $PM_{2.5}$ standards in 2002 and determined that the standards were not "arbitrary and capricious". A more comprehensive account of the legal conflict at the national level can be found elsewhere [13,91].

3.5.3. Disinformation

Faced with tighter regulations, industry trade groups mounted disinformation campaigns questioning the health impacts of particulate pollution by using techniques similar to those employed by tobacco and fossil fuel companies [104,105]. These efforts are well documented at the national level [91,106], but events in Utah are not as widely documented. A few months after the publication of the Geneva Steel case study in the summer of 1989, the Utah Air Conservation Committee met to discuss the State Implementation Plan to gain compliance with the PM_{10} NAAQS. To cast doubt about the scientific findings, Geneva Steel hired Dr. Steven Lamm, an epidemiologist associated with John Hopkins and Georgetown University, to reanalyze the Geneva Steel case study data [107]. Lamm examined the data for a few weeks, and instead of submitting his analysis for peer review, he held a press conference where he presented his claims publicly to create the appearance of scientific disagreement and inject Geneva's desired narrative into newspaper articles [108,109]. Lamm claimed that a respiratory syncytial virus (RSV) caused the variation in pediatric respiratory hospital admissions, not variations in PM_{10} from the Geneva plant. This was despite the fact that Pope's analysis controlled for effects like this by examining hospital admissions in several cities across Utah that would have had similar levels of RSV [100]. Lamm did eventually submit his analysis for peer review several months later, but the very short paper was not published until 1994, five years after Lamm held his press conference [110]. In the intervening years, a large body of research was published linking fine particulates to a wide range of health impacts, including hospitalizations, lung function, respiratory symptoms, school absences, and mortality [111–117]. That said, the disinformation effort to create misleading news coverage had the desired effect of creating an artificial controversy that muddled public understanding of the health impacts of air quality in Utah for years (e.g., [118,119]).

3.5.4. State Initiatives in Utah

While the science of air pollution was advancing and conflicts between regulators and industries were playing out in Utah, the state government was working on its own air quality initiatives in response to sustained public pressure and the continued violation of federal air quality standards. In 1989, Governor Norman Bangerter created the Governor's Clean Air Commission (GCAC) composed of community leaders, elected officials, and industry leaders along the Wasatch Front to develop a plan for Utah to address its air quality concerns. The GCAC was organized into five working groups (energy utilization, industry, socioeconomic, technology assessment, and transportation) that each released a report in 1990, and then the final report of 146 legislative and budget recommendations was released in 1991 [120,121]. The Energy Utilization working group report noted that Utah was growing rapidly in terms of population, households, jobs, and vehicles, all of which would create additional air pollution in the coming decades. Given the underlying growth assumptions, the working group outlined three overall policy options for Utah to address air pollution: (1) allow air pollution to increase, (2) meet federal air quality standards, or (3) reduce pollution levels significantly below federal standards.

The first option was not considered viable because of significant public concern about air pollution and because Utah was already exceeding the NAAQS, violating federal law. The second option would take substantial work to achieve and was viewed as the most feasible path; however, the committee also explored what it would take to achieve the third option. In their exploration of approaches to reduce pollution levels significantly below federal standards, they noted that there were economic, technical, legal/regulatory, socioeconomic, and political barriers, and significant tradeoffs would be necessary to significantly improve air quality. It was determined by the group that, at the time, there were no technical or economically viable solutions available to dramatically improve air quality. Governor Bangerter discussed the need to strike a balance between environmental issues and economic development [122]. In the final report, the priority policy recommendations from the GCAC to meet federal air quality standards included recommendations to reduce wood burning, incentivize clean vehicles, improve vehicle emissions inspections, fund public education programs, and improve emission inventories [121]. Three decades later, despite substantial progress, these same policies continue to be state priorities to address air pollution and meet federal air quality standards.

3.6. Air Quality and the New Millennium

In the 30 years since the GCAC report was written, meteorological modeling has advanced dramatically and new air quality concerns, such as in Utah's Uinta basin, surfaced [123–125]. One ongoing scientific question concerns the formation of fine particulates during wintertime inversions resulting from complex interactions of meteorology, emission sources, and atmospheric chemistry. A future field campaign using an aircraft to examine the atmospheric chemistry during inversions is planned for the coming years [67], https://atmos.utah.edu/aquarius/index.php (accessed on 1 May 2022). Incidentally, this field campaign would come ~100 years after the St. John Perrot's first aircraft field campaign to study air pollution during inversion conditions.

Meanwhile, growing scientific evidence of the health impacts from air quality led to tightening federal air quality standards and to increased public awareness. Nationally, the initial $PM_{2.5}$ NAAQS of 65 µg m^{-3}, averaged over 24 h, was introduced in 1997. Upon further data collection and analysis, the 24 h standard was lowered to 35 µg m^{-3} in 2006. After a delay and change in administration, the standard was implemented in 2009, and Salt Lake and Davis counties as well as portions of Cache, Weber, Box Elder, Tooele, and Utah counties were found to be out of compliance with the 24 h $PM_{2.5}$ NAAQS [126].

This official designation of "moderate" nonattainment triggered the State Implementation Planning (SIP) process requiring Utah to plan for reducing emissions to meet federal $PM_{2.5}$ standards. It also raised public awareness of the issue; in 2014, a survey of public opinion found that air quality was the third highest priority for Utahns but ranked last in terms of the state's performance on priorities [2]. In the winter during these years, there were frequent, large rallies for clean air at the Utah state capitol (e.g., [127] (Figure 12). By 2015, Utah still did not meet the 24 h standard and several areas were reclassified as "serious" nonattainment, triggering further action to reduce emissions. Finally, in 2019, Utah had a three-year period with measurements just below the 24 h $PM_{2.5}$ NAAQS [128]. This is likely due to reduced emissions that contribute to $PM_{2.5}$ but might also be due to frequent winter storms and a lack of extended inversion conditions over several years. The EPA is currently reevaluating the $PM_{2.5}$ NAAQS based on extensive studies showing the health impacts of particulate pollution (e.g., [129]) and, if the standards are lowered, Utah will once again be out of compliance with federal air quality standards.

Figure 12. Protesters at the Utah State Capitol in 2014 demanding government action to address poor air quality. Note: Reprinted with permission from [126]. 2014, Scott Sommerdorf.

The other major air pollutant of concern in Utah in recent decades is ozone. Ozone is a secondary pollutant that forms when VOCs (volatile organic compounds) and nitrogen oxides (NO and NO_2, a.k.a. NO_x) combine in the presence of ultraviolet sunlight. Similar to $PM_{2.5}$, the NAAQS for ozone tightened as the scientific evidence for health impacts grew. The standards were 80 ppb (averaged over 8 h) in 1997 and were lowered to 75 ppb in 2008 and 70 ppb in 2015. At the current standard, Salt Lake and Davis counties as well as portions of Weber, Tooele, Utah, Uintah, and Duchesne counties are out of compliance and the state is designated as "moderate" for compliance with federal ozone standards [130].

There are several factors contributing to high ozone concentrations in Utah today. Local emissions of NO_x and VOCs from fossil fuel combustion constitute a primary factor affecting urban and rural areas downwind of metropolitan regions [131]. In addition, in recent decades, atmospheric methane has increased globally, and the long-range transport of ozone precursors from fossil fuel combustion in Asia is contributing to higher levels of background ozone [132,133]. As temperatures warm from the changing climate, the chemical reactions forming ozone will also speed up, leading to higher ozone concentrations. Lastly, precursors in wildfire smoke can form ozone, and increasing wildfires in the western U.S. will also contribute to increasing ozone trends in Utah [134,135].

Utah's unique topography also contributes to elevated ozone in two relatively unusual ways. First, lake breezes along the Great Salt Lake lead to elevated ozone in the urban areas bordering the lake [136,137]. This process has been documented by field campaigns for discrete time periods but the frequency and climatology of elevated ozone lake breezes contributing to exceeding NAAQS have not been investigated along the Wasatch Front. Second, unusually elevated ozone was discovered in 2009 to occur in the Uinta basin in the wintertime [125]. Ozone was typically only monitored in the summertime when ultraviolet radiation and warm temperatures contribute to typical ozone formation conditions. However, a station in the Uinta basin was left on over the 2009–2010 winter and recorded 8 h average ozone concentrations of ~120 ppb, far higher than those typically observed along the Wasatch Front in the summertime of ~75 ppb. The Utah Division of Air Quality and teams of researchers studied this phenomenon in subsequent years, finding that exceptionally high VOC concentrations from oil and gas operations are trapped near the surface by strong thermal inversions, and then reflective snow on the ground increases UV radiation that then produces elevated ozone concentrations [123]. Studies of this phenomenon are ongoing.

3.7. The Future of Air Quality in Utah

The future of air quality in Utah will be dictated by how the state responds to several challenges and opportunities. Utah has one of the fastest growing populations in the U.S. and the population is projected to increase by 66% by 2060, putting upward pressure on emissions with more vehicles, buildings, and commercial or industrial activities. In addition, the changing climate is expected to exacerbate air quality conditions by increasing ozone concentrations, particulates from wildfire smoke, and windblown dust from drying lake beds (as well as many other climate impacts unrelated to air quality) [138]. Despite these challenges, there is enormous potential to address air pollution as we are currently in the middle of a rapid energy and economic transition away from fossil fuel-based energy towards renewable and zero-emission energy sources. This transition is being driven by energy innovation and a rapid decline in prices over the past decade that is expected to continue (e.g., [139–143]).

In addition to the technical and environmental aspects, there is growing interest among policy makers to structurally tackle these challenges. In 2018, the Utah Legislature adopted a resolution on environmental and economic stewardship, committing to use sound science to reduce emissions that are contributing to the changing climate and air pollution [144]. The Legislature then requested a report from the University of Utah's Kem Gardner Policy Institute outlining a roadmap for the state to reduce emissions. The "Utah Roadmap: Positive Solutions on Climate and Air Quality" recommended seven mileposts for the state government to take to address these challenges in the coming decades [145]. This approach has gained broad support among leaders from business, government, faith, and civic institutions across Utah [146].

It is notable how different the situation is from any time since the 1880s, when there were not technically, economically, or politically viable solutions available to significantly improve air quality. Today, the balance between environmental stewardship and economic development has shifted and technically, economically, and politically viable solutions are available that present a historic opportunity to lower air pollution to near background levels in the coming decades. Doing this would have enormous public health and economic benefits and would protect Utah from any future tightening of federal air quality standards.

4. Conclusions

Air quality has had an important and storied legacy in Utah, affecting many aspects of society. Scientific studies in Utah and by Utahn scientists have substantially contributed to the scientific understanding of air quality. The dominant pollution sources changed over time with changes in the production and combustion of fuels used for energy and the state's unique topography of high mountain basins and lakes that exacerbate air quality conditions. In response to regulations and public pressure (Figure 12), policies to improve air quality were implemented based on the best available scientific information and the economically and technically viable solutions available. These policies have improved air quality over time, and although Utah still exceeds federal air quality standards, air quality today is likely better than at any time since the late 19th century. The coming decades present a unique opportunity to substantially improve air quality since clean energy technologies are economically viable and are being rapidly deployed. At the same time, many scientific advances will be made and the societal benefits of clean air will become clearer. If Utah harnesses innovation and leadership, the future of air quality will be bright.

Author Contributions: Conceptualization, L.E.M.; methodology, L.E.M. and C.A.B.Z.; software, L.E.M.; validation, L.E.M.; investigation, L.E.M.; resources, L.E.M.; data curation, L.E.M.; writing—original draft preparation, L.E.M.; writing—review and editing, L.E.M. and C.A.B.Z.; visualization, L.E.M. and C.A.B.Z.; funding acquisition, L.E.M. All authors have read and agreed to the published version of the manuscript.

Funding: This research was funded by NOAA grant number NA19OAR4310078, NSF grant number 1912664, and the Uinta Institute.

Data Availability Statement: Not applicable.

Acknowledgments: We are grateful for the Utah Digital Newspapers (https://digitalnewspapers. org/, accessed on 1 May 2020), Hathi Trust, and Google Books for digitizing historical documents and making them freely available, which made this research possible. We also thank Alfred Mowdood for research assistance.

Conflicts of Interest: The authors declare no conflict of interest. The funders had no role in the design of the study; in the collection, analyses, or interpretation of data; in the writing of the manuscript, or in the decision to publish the results.

References

1. Check Smoke Evil by Educational Campaign. *Salt Lake Telegram*. 8 February 1917. Available online: https://newspapers.lib.utah.edu/details?id=20017332 (accessed on 18 April 2020).
2. Envision Utah. 2014 Values Study Results. 2014. Available online: https://yourutahyourfuture.org/images/final_values_study_report.pdf (accessed on 8 June 2020).
3. Utah Division of Air Quality. *Utah's Air Quality 2021 Annual Report*; Utah Division of Air Quality: Salt Lake City, UT, USA, 2022. Available online: https://documents.deq.utah.gov/air-quality/planning/air-quality-policy/DAQ-2022-000342.pdf (accessed on 13 July 2022).
4. Ou, J.; Pirozzi, C.S.; Horne, B.D.; Hanson, H.A.; Kirchhoff, A.C.; Mitchell, L.E.; Coleman, N.C.; Arden Pope, C. Historic and Modern Air Pollution Studies Conducted in Utah. *Atmosphere* **2020**, *11*, 1094. [CrossRef]
5. Lelieveld, J.; Evans, J.S.; Fnais, M.; Giannadaki, D.; Pozzer, A. The contribution of outdoor air pollution sources to premature mortality on a global scale. *Nature* **2015**, *525*, 367–371. [CrossRef] [PubMed]
6. Landrigan, P.J.; Fuller, R.; Acosta, N.J.R.; Adeyi, O.; Arnold, R.; Basu, N.; Baldé, A.B.; Bertollini, R.; Bose-O'Reilly, S.; Boufford, J.I.; et al. The Lancet Commission on pollution and health. *Lancet* **2017**, *391*, 462–512. [CrossRef]

7. Di, Q.; Wang, Y.; Zanobetti, A.; Wang, Y.; Koutrakis, P.; Choirat, C.; Dominici, F.; Schwartz, J.D. Air Pollution and Mortality in the Medicare Population. *N. Engl. J. Med.* **2017**, *376*, 2513–2522. [CrossRef] [PubMed]
8. Errigo, I.M.; Abbott, B.W.; Mendoza, D.L.; Mitchell, L.; Sayedi, S.S.; Glenn, J.; Kelly, K.E.; Beard, J.D.; Bratsman, S.; Carter, T.; et al. Human Health and Economic Costs of Air Pollution in Utah: An Expert Assessment. *Atmosphere* **2020**, *11*, 1238. [CrossRef]
9. Zivin, J.G.; Neidell, M. Air pollution's hidden impacts. *Science* **2018**, *359*, 39–40. [CrossRef]
10. U.S. EPA. *Integrated Science Assessment (ISA) of Ozone and Related Photochemical Oxidants*; U.S. EPA: Washington, DC, USA, 2013. Available online: https://cfpub.epa.gov/ncea/isa/recordisplay.cfm?deid=247492 (accessed on 1 December 2017).
11. Mendoza, D.L.; Pirozzi, C.S.; Crosman, E.T.; Liou, T.G.; Zhang, Y.; Cleeves, J.J.; Bannister, S.C.; Anderegg, W.R.L.; Robert, P.I. Impact of low-level fine particulate matter and ozone exposure on absences in K-12 students and economic consequences. *Environ. Res. Lett.* **2020**, *15*, 114052. [CrossRef]
12. Zajchowski, C.A.B.; Brownlee, M.T.J.; Blacketer, M.; Rose, J.; Rumore, D.L.; Watson, J.; Dustin, D.L. "Can you take me higher?": Normative thresholds for air quality in the Salt Lake City Metropolitan area. *J. Leis. Res.* **2019**, *50*, 157–180. [CrossRef]
13. Bachmann, J. Will the Circle Be Unbroken: A History of the U.S. National Ambient Air Quality Standards. *J. Air Waste Manag. Assoc.* **2007**, *57*, 652–697. [CrossRef]
14. Siddaway, A.P.; Wood, A.M.; Hedges, L.V. How to Do a Systematic Review: A Best Practice Guide for Conducting and Reporting Narrative Reviews, Meta-Analyses, and Meta-Syntheses. *Annu. Rev. Psychol.* **2019**, *70*, 747–770. [CrossRef]
15. Franchini, M.; Mannucci, P.M. Mitigation of air pollution by greenness: A narrative review. *Eur. J. Intern. Med.* **2018**, *55*, 1–5. [CrossRef] [PubMed]
16. Holland, I.; DeVille, N.V.; Browning, M.H.E.M.; Buehler, R.M.; Hart, J.E.; Hipp, J.A.; Mitchell, R.; Rakow, D.A.; Schiff, J.E.; White, M.P.; et al. Measuring Nature Contact: A Narrative Review. *Int. J. Environ. Res. Public. Health* **2021**, *18*, 4092. [CrossRef]
17. Fairbanks, J.L. City Planning. *Munic. Rec.* **1920**, *9*, 4–7. Available online: https://babel.hathitrust.org/cgi/pt?id=nyp.33433015315769&view=1up&seq=253 (accessed on 27 April 2020).
18. Deseret News. Remarks by President Brigham Young. 26 September 1860. Available online: https://newspapers.lib.utah.edu/details?id=2585051 (accessed on 22 October 2021).
19. Lebel, E.D.; Finnegan, C.J.; Ouyang, Z.; Jackson, R.B. Methane and Nox Emissions from Natural Gas Stoves, Cooktops, and Ovens in Residential Homes. *Environ. Sci. Technol.* **2022**, *56*, 2529–2539. [CrossRef] [PubMed]
20. The Smoke Nuisance. *Deseret News*. 9 May 1883. Available online: https://newspapers.lib.utah.edu/details?id=2644770 (accessed on 30 March 2020).
21. The Smoke Nuisance. *Salt Lake Tribune*. 26 June 1893. Available online: https://newspapers.lib.utah.edu/details?id=12517223 (accessed on 30 March 2020).
22. U.S. Energy Information Administration Total Energy Monthly Data. Available online: https://www.eia.gov/totalenergy/data/monthly/index.php (accessed on 21 May 2022).
23. U.S. Energy Information Administration State Energy Data System (SEDS): 1960–2020. Available online: https://www.eia.gov/state/seds/seds-data-complete.php?sid=UT (accessed on 18 July 2022).
24. Johnston, F.R. Small Business Owners on Compliance with Environmental Regulations in Utah's Manufacturing Industry: A Case Study. Ph.D. Thesis, Northcentral University, Prescott Valley, AZ, USA, 2016.
25. The Nuisance of Smoke. *Deseret News*. 21 December 1881. Available online: https://newspapers.lib.utah.edu/details?id=2634016 (accessed on 17 April 2020).
26. Salt Lake City Corporation. *Smoke Emitted from Chimneys*; Salt Lake City Corporation: Salt Lake City, UT, USA, 1891; pp. 294–295.
27. To Abate the Smoke Nuisance. *Deseret Evening News*. 3 September 1891. Available online: https://newspapers.lib.utah.edu/details?id=1588551 (accessed on 30 March 2020).
28. Lareau, N.P.; Crosman, E.; Whiteman, C.D.; Horel, J.D.; Hoch, S.W.; Brown, W.O.J.; Horst, T.W. The Persistent Cold-Air Pool Study. *Bull. Am. Meteorol. Soc.* **2012**, *94*, 51–63. [CrossRef]
29. Whiteman, C.D.; Hoch, S.W.; Horel, J.D.; Charland, A. Relationship between particulate air pollution and meteorological variables in Utah's Salt Lake Valley. *Atmos. Environ.* **2014**, *94*, 742–753. [CrossRef]
30. Location of Factories. *Salt Lake Herald-Republican*. 22 February 1893. Available online: https://newspapers.lib.utah.edu/details?id=11049267 (accessed on 4 April 2020).
31. Lane, H.M.; Morello-Frosch, R.; Marshall, J.D.; Apte, J.S. Historical Redlining Is Associated with Present-Day Air Pollution Disparities in U.S. Cities. *Environ. Sci. Technol. Lett.* **2022**, *9*, 345–350. [CrossRef]
32. Governor Spencer Cox Gov. Cox and Lt. Gov. Henderson Release One Utah Roadmap Update. Available online: https://governor.utah.gov/2021/10/20/gov-cox-and-lt-gov-henderson-release-one-utah-roadmap-update/ (accessed on 29 May 2022).
33. American Smelting & Refining Co. v. Godfrey, 158 F. 225. 1907. Available online: https://cite.case.law/f/158/225/ (accessed on 16 April 2020).
34. Church, M.A. Smoke Farming: Smelting and Agricultural Reform in Utah, 1900–1945. *Utah Hist. Q.* **2004**, *72*, 196–218.
35. Looking for Dividends. *Salt Lake Tribune*. 28 December 1906. Available online: https://newspapers.lib.utah.edu/details?id=13893263 (accessed on 19 July 2022).
36. Flagg, S.B. *City Smoke Ordinances and Smoke Abatement*; United States Department of the Interior, Bureau of Mines; Government Printing Office: Washington, DC, USA, 1912; Available online: https://digital.library.unt.edu/ark:/67531/metadc12845/ (accessed on 13 April 2020).

37. Smoke Trouble Can be Stopped. *Salt Lake Herald-Republican*. 3 February 1908. Available online: https://newspapers.lib.utah.edu/details?id=11872203 (accessed on 9 March 2022).

38. Council Aids in Smoke Abatement. *Salt Lake Herald-Republican*. 7 November 1911. Available online: https://newspapers.lib.utah.edu/details?id=9937330 (accessed on 9 March 2022).

39. New Anti-Smoke Ordnance Will Be Strictly Enforced, Says the Mayor. *Salt Lake Telegram*. 30 April 1903. Available online: https://newspapers.lib.utah.edu/details?id=18371733 (accessed on 9 March 2022).

40. Smoke Clouds Greet Tourists in Salt Lake. *Salt Lake Herald-Republican*. 19 December 1912. Available online: https://newspapers.lib.utah.edu/details?id=10092789 (accessed on 17 April 2020).

41. The Smoke Nuisance II-The Economic Problem. *Salt Lake Telegram*. 18 November 1912. Available online: https://newspapers.lib.utah.edu/details?id=19570261 (accessed on 17 April 2020).

42. Women Will Assist in Fight on Smoke. *Salt Lake Tribune*. 7 October 1913. Available online: https://newspapers.lib.utah.edu/details?id=14397839 (accessed on 17 April 2020).

43. Smoke Talks Engage Ears of Officials. *Salt Lake Tribune*. 22 January 1914. Available online: https://newspapers.lib.utah.edu/details?id=14493134 (accessed on 17 April 2020).

44. Form League for Smoke Abatement. *Salt Lake Tribune* . 7 April 1914. Available online: https://newspapers.lib.utah.edu/details?id=14478448 (accessed on 17 April 2020).

45. Women Gather Data to Wage War on Smoke. *Salt Lake Telegram*. 31 January 1914. Available online: https://newspapers.lib.utah.edu/details?id=15979425 (accessed on 17 April 2020).

46. Push Campaign to Lessen Smoke. *Salt Lake Tribune*. 25 August 1914. Available online: https://newspapers.lib.utah.edu/details?id=14610418 (accessed on 17 April 2020).

47. Metallurgists to Attact Zion Smoke. *Salt Lake Tribune*. 28 September 1914, pp. A1–A2. Available online: https://newspapers.lib.utah.edu/details?id=14578784 (accessed on 17 April 2020).

48. Airplane to Aid in Smoke Probe. *Salt Lake Tribune*. 2 November 1919. Available online: https://newspapers.lib.utah.edu/details?id=15076993 (accessed on 18 April 2020).

49. Perrott, G.S.J. Smoke Problem at Salt Lake City. *Power Plant Eng.* **1920**, *24*, 784–785.

50. Smoke Expert Makes Flight Studies Currents Over City. *Salt Lake Tribune*. 10 November 1919. Available online: https://newspapers.lib.utah.edu/details?id=15043354 (accessed on 18 April 2020).

51. Dickson, C.R. A synoptic climatology of diurnal inversions in the Jordan Valley. Master's Thesis, University of Utah, Salt Lake City, UT, USA, 1957.

52. Schmalz, W.M. Some Notes on Visibilities at Salt Lake Airport. *Bull. Am. Meteorol. Soc.* **1947**, *28*, 179–186. [CrossRef]

53. Williams, P. Air Pollution Potential over the Salt Lake Valley of Utah as Related to Stability and Wind Speed. *J. Appl. Meteorol.* **1964**, *3*, 92–97. [CrossRef]

54. Gudmundsen, A. *Nine Years of Smoke-Abatement Work at Salt Lake City*; U.S. Department of Commerce, Bureau of Mines: Washington, DC, USA, 1930; Available online: https://catalog.hathitrust.org/Record/006865386 (accessed on 13 April 2020).

55. Monnett, O. *Smoke Abatement*; U.S. Department of Commerce, Bureau of Mines: Washington, DC, USA, 1923; Available online: http://hdl.handle.net/2027/mdp.39015077560111 (accessed on 13 April 2020).

56. Monnett, O.; Perrott, G.S.J.; Clark, H.W. *Smoke-Abatement Investigation at Salt Lake City, Utah*; U.S. Bureau of Mines: Washington, DC, USA, 1926; Available online: https://babel.hathitrust.org/cgi/pt?id=mdp.39015024579750&view=1up&seq=5 (accessed on 26 April 2020).

57. Kuprov, R.; Eatough, D.J.; Cruickshank, T.; Olson, N.; Cropper, P.M.; Hansen, J.C. Composition and secondary formation of fine particulate matter in the Salt Lake Valley: Winter 2009. *J. Air Waste Manag. Assoc.* **2014**, *64*, 957–969. [CrossRef] [PubMed]

58. Be a Better Neighbor Remove the Spot. *Salt Lake Telegram*. 24 October 1930. Available online: https://newspapers.lib.utah.edu/details?id=15680318 (accessed on 30 March 2022).

59. Assistant Smoke Abatement Engineer Appointed by City. *Salt Lake Telegram* . 24 December 1937. Available online: https://newspapers.lib.utah.edu/details?id=18727234 (accessed on 20 April 2020).

60. Today's Smoke. *Salt Lake Telegram*. 7 March 1941. Available online: https://newspapers.lib.utah.edu/details?id=16906496 (accessed on 19 July 2022).

61. Diagram Reveals That Unrelenting Black Menace Cloaked City Most Heavily in 1935. *Salt Lake Telegram*. 1 March 1941. Available online: https://newspapers.lib.utah.edu/details?id=16899625 (accessed on 6 April 2020).

62. Heaney, R.J.; Winn, G.S.; Thorne, W.; Lloyd, L.H. *Air Resources of Utah*; Utah Legislative Council; Air Pollution Advisory Committee: Salt Lake City, UT, USA, 1962. Available online: https://digitallibrary.utah.gov/awweb/guest.jsp?smd=1&cl=all_lib&lb_document_id=62246 (accessed on 14 April 2020).

63. Nemery, B.; Hoet, P.H.; Nemmar, A. The Meuse Valley fog of 1930: An air pollution disaster. *Lancet* **2001**, *357*, 704–708. [CrossRef]

64. Jacobs, E.T.; Burgess, J.L.; Abbott, M.B. The Donora Smog Revisited: 70 Years after the Event That Inspired the Clean Air Act. *Am. J. Public Health* **2018**, *108*, S85–S88. [CrossRef] [PubMed]

65. Schrenk, H.H.; Heimann, H.; Clayton, G.D.; Gafafer, W.M.; Wexler, H. *Air Pollution in Donora, Pa. Epidemiology of the Unusual Smog Episode of October 1948*; Public Health Bulletin; Public Health Service, U.S. Government Printing Office: Washington, DC, USA, 1949; Available online: https://hdl.handle.net/2027/ucl.c060945791 (accessed on 27 April 2020).

66. Anderson, H.R. Air pollution and mortality: A history. *Atmos. Environ.* **2009**, *43*, 142–152. [CrossRef]

67. Hallar, A.G.; Brown, S.S.; Crosman, E.; Barsanti, K.C.; Cappa, C.D.; Faloona, I.; Fast, J.; Holmes, H.A.; Horel, J.; Lin, J.; et al. Coupled Air Quality and Boundary-Layer Meteorology in Western U.S. Basins during Winter: Design and Rationale for a Comprehensive Study. *Bull. Am. Meteorol. Soc.* **2021**, *102*, E2012–E2033. [CrossRef]
68. Utah Legislature. Air Pollution. 1963. Available online: https://digitallibrary.utah.gov/aw-server/rest/product/purl/USL/i/48 ef689a-2211-4b03-b0b3-921eefa674a9 (accessed on 8 April 2022).
69. Utah Legislature. Air Conservation Act. 1967. Available online: https://digitallibrary.utah.gov/aw-server/rest/product/purl/ USL/i/ea9963f4-565f-42d7-8d8b-9c5dfe7fe2cd (accessed on 8 April 2022).
70. Air Pollution Studies Planned by Utah State Health Department. *Salt Lake Times* . 13 April 1962. Available online: https://newspapers.lib.utah.edu/details?id=13317874 (accessed on 7 April 2022).
71. Air Pollution Equipment Being Installed. *Springville Herald*. 9 July 1964. Available online: https://newspapers.lib.utah.edu/details?id=22363878 (accessed on 30 March 2022).
72. Red Butte Canyon Goes to Service. *Davis County Clipper*. 9 May 1969. Available online: https://newspapers.lib.utah.edu/details?id=823736 (accessed on 8 April 2022).
73. Salt Lake City; Emphysema Capital. *The Daily Utah Chronicle* . 19 May 1969. Available online: https://newspapers.lib.utah.edu/details?id=22473861 (accessed on 8 April 2022).
74. Jackman, D.N. A Study of Meteorological Effect on Air Pollution in the Salt Lake Valley. Master's Thesis, University of Utah, Salt Lake City, UT, USA, 1968.
75. PSCIP Company Cooperating with City on Smog Problem. *Springville Herald*. 19 March 1964. Available online: https://newspapers.lib.utah.edu/details?id=22362037 (accessed on 13 April 2022).
76. Bountiful Mayor Looks to Electric Car. *Davis County Clipper*. 18 November 1966. Available online: https://newspapers.lib.utah.edu/details?id=791722 (accessed on 13 April 2022).
77. Air Pollution Conference Planned in SLC. *Springville Herald*. 30 January 1969. Available online: https://newspapers.lib.utah.edu/details?id=22394024 (accessed on 13 April 2022).
78. Hertz, M.; Truppi, L.; English, T.; Sovocool, G.W.; Burton, R.; Heiderscheit, T.; Hinton, D. Human Exposure to Air Pollutants in Salt Lake Basin Communities, 1940–1971. 1974. Available online: https://nepis.epa.gov/Exe/ZyPURL.cgi?Dockey=20015S4I.txt (accessed on 27 April 2022).
79. U.S. Government. *The Environmental Protection Agency's Research Program with Primary Emphasis on the Community Health and Environmental Surveillance System (CHESS), an Investigative Report*; U.S. Government Printing Office: Washington, DC, USA, 1976; Available online: https://www.google.com/books/edition/The_Environmental_Protection_Agency_s_Re/_MhCiLTlSbQC?hl=en&gbpv=0 (accessed on 22 May 2022).
80. Utah Center for Health Statistics. *Utah Health Facts*; Utah State Division of Health: Salt Lake City, UT, USA, 1972. Available online: https://digitallibrary.utah.gov/aw-server/rest/product/purl/USL/f/b2f9ab69-027a-4762-8838-52c922954c4d (accessed on 11 May 2020).
81. Center for Air Environment Studies. *Guide to Research in Air Pollution*; U.S. EPA, Research Triangle Park: Durham, NC, USA, 1972. Available online: https://nepis.epa.gov/Exe/ZyPURL.cgi?Dockey=20013P97.txt (accessed on 11 May 2020).
82. Holland, W.W.; Bennett, A.E.; Cameron, I.R.; Florey, C.D.V.; Leeder, S.R.; Schilling, R.S.F.; Swan, A.V.; Waller, R.E. Health effects of particulate pollution: Reappraising the evidence. *Am. J. Epidemiol.* **1979**, *110*, 527. [CrossRef]
83. Sullivan, J.L. The Calibration of Smoke Density. *J. Air Pollut. Control Assoc.* **1962**, *12*, 474–478. [CrossRef]
84. Kirchstetter, T.W.; Preble, C.V.; Hadley, O.L.; Bond, T.C.; Apte, J.S. Large reductions in urban black carbon concentrations in the United States between 1965 and 2000. *Atmos. Environ.* **2017**, *151*, 17–23. [CrossRef]
85. Kirchstetter, T.W.; Aguiar, J.; Tonse, S.; Fairley, D.; Novakov, T. Black carbon concentrations and diesel vehicle emission factors derived from coefficient of haze measurements in California: 1967–2003. *Atmos. Environ.* **2008**, *42*, 480–491. [CrossRef]
86. Shy, C.M. Epidemiologic evidence and the United States air quality standards. *Am. J. Epidemiol.* **1979**, *110*, 661–671. [CrossRef] [PubMed]
87. Ware, J.H.; Thibodeau, L.A.; Speizer, F.E.; Colome, S.; Ferris, B.G. Assessment of the health effects of atmospheric sulfur oxides and particulate matter: Evidence from observational studies. *Environ. Health Perspect.* **1981**, *41*, 255–276. [CrossRef] [PubMed]
88. Hinton, D.; Sune, J.; Suggs, J.; Barnard, W. *Inhalable Particulate Network Report: Operation and Data Summary (Mass Concentrations Only)*; U.S. EPA: Research Triangle Park, NC, USA, 1985. Available online: https://nepis.epa.gov/Exe/ZyPURL.cgi?Dockey=50 0024O4.txt (accessed on 6 June 2020).
89. Miller, F.J.; Gardner, D.E.; Graham, J.A.; Lee, R.E.; Wilson, W.E.; Bachmann, J.D. Size Considerations for Establishing a Standard for Inhalable Particles. *J. Air Pollut. Control Assoc.* **1979**, *29*, 610–615. [CrossRef]
90. Hinton, D.; Sune, J.; Suggs, J.; Barnard, W. *Inhalable Particulate Network Report: Data Listing (Mass Concentrations Only) Vol II April 1979-December 1982*; U.S. EPA: Research Triangle Park, NC, USA, 1984. Available online: https://nepis.epa.gov/Exe/ZyPURL. cgi?Dockey=20015OZX.txt (accessed on 17 September 2020).
91. Dockery, D.W. Health Effects of Particulate Air Pollution. *Ann. Epidemiol.* **2009**, *19*, 257–263. [CrossRef] [PubMed]
92. Smart, M.D. Clearing the Air. *Y Magazine*. Spring. 2007. Available online: https://magazine.byu.edu/article/clearing-the-air/ (accessed on 11 November 2021).
93. Air Quality Plan Opposed in Part. *Provo Daily Herald* . 14 September 1972. Available online: https://newspapers.lib.utah.edu/details?id=23795843 (accessed on 29 May 2022).

94. Christian, P. U.S. Steel Wins OK for Pollution, Cash-Saving Idea. *The Daily Herald*. 23 August 1982. Available online: https://newspapers.lib.utah.edu/details?id=23901151 (accessed on 29 May 2022).
95. Pope, C.A. Respiratory disease associated with community air pollution and a steel mill, Utah Valley. *Am. J. Public Health* **1989**, *79*, 623–628. [CrossRef]
96. Geneva Belches Back to Life with Plume of Smoke. *Provo Daily Herald*. 13 September 1987. Available online: https://newspapers.lib.utah.edu/details?id=24048164 (accessed on 19 July 2022).
97. Feedback: Pollution in Air Is Bread on the Table. *Provo Daily Herald*. 10 February 1988. Available online: https://newspapers.lib.utah.edu/details?id=24064701 (accessed on 21 May 2022).
98. Feedback: Geneva Pollution Takes Its Toll. *Provo Daily Herald*. 6 March 1988. Available online: https://newspapers.lib.utah.edu/details?id=24063610 (accessed on 21 May 2022).
99. Evans, M. Dirty Air Angers Residents. *Provo Daily Herald*. 25 February 1988. Available online: https://newspapers.lib.utah.edu/details?id=24065124 (accessed on 21 May 2022).
100. Pope, C.A. Respiratory Hospital Admissions Associated with PM10 Pollution in Utah, Salt Lake, and Cache Valleys. *Arch. Environ. Health Int. J.* **1991**, *46*, 90–97. [CrossRef]
101. Dockery, D.W.; Pope, C.A.; Xu, X.; Spengler, J.D.; Ware, J.H.; Fay, M.E.; Ferris, B.G.; Speizer, F.E. An Association between Air Pollution and Mortality in Six U.S. Cities. *N. Engl. J. Med.* **1993**, *329*, 1753–1759. [CrossRef]
102. Pope, C.A.; Thun, M.J.; Namboodiri, M.M.; Dockery, D.W.; Evans, J.S.; Speizer, F.E.; Heath, C.W. Particulate Air Pollution as a Predictor of Mortality in a Prospective Study of U.S. Adults. *Am. J. Respir. Crit. Care Med.* **1995**, *151*, 669–674. [CrossRef]
103. Zell, H.; Quarcoo, D.; Scutaru, C.; Vitzthum, K.; Uibel, S.; Schöffel, N.; Mache, S.; Groneberg, D.A.; Spallek, M.F. Air pollution research: Visualization of research activity using density-equalizing mapping and scientometric benchmarking procedures. *J. Occup. Med. Toxicol. Lond. Engl.* **2010**, *5*, 5. [CrossRef]
104. Oreskes, N.; Conway, E.M. *Merchants of Doubt: How a Handful of Scientists Obscured the Truth on Issues from Tobacco Smoke to Global Warming*, 1st ed.; Bloomsbury Press: New York, NY, USA, 2010; ISBN 978-1-59691-610-4.
105. Supran, G.; Oreskes, N. Assessing ExxonMobil's climate change communications (1977–2014). *Environ. Res. Lett.* **2017**, *12*, 084019. [CrossRef]
106. Kaiser, J. Showdown Over Clean Air Science. *Science* **1997**, *277*, 466–469. [CrossRef] [PubMed]
107. Morrey, S. Geneva to Counter Study by Pope. *Provo Daily Herald*. 30 August 1989. Available online: https://newspapers.lib.utah.edu/details?id=24081733 (accessed on 16 November 2021).
108. Adams, B. Illness Blamed on Virus, Not Pollution: Doctor Hired by Geneva to Analyze Data Disputes Findings of BYU Professor. *Deseret News*. 2 September 1989. Available online: https://www.deseret.com/1989/9/2/18822398/illnesses-blamed-on-virus-not-pollution-br-doctor-hired-by-geneva-to-analyze-data-disputes-findings (accessed on 16 November 2021).
109. Morrey, S. Expert: Virus, Not Pollution, Causes Illness. *Provo Daily Herald*. 1 September 1989. Available online: https://newspapers.lib.utah.edu/details?id=24081819 (accessed on 16 November 2021).
110. Lamm, S.H.; Hall, T.A.; Engel, A.; Rueter, F.H.; White, L.D. PM10 Particulates: Are They the Major Determinant of Pediatric Respiratory Admissions in Utah County, Utah (1985–1989). *Ann. Occup. Hyg.* **1994**, *38*, 969–972. [CrossRef]
111. Archer, V.E. Air Pollution and Fatal Lung Disease in Three Utah Counties. *Arch. Environ. Health Int. J.* **1990**, *45*, 325–334. [CrossRef] [PubMed]
112. Pope, C.A.; Schwartz, J.; Ransom, M.R. Daily Mortality and PM10 Pollution in Utah Valley. *Arch. Environ. Health Int. J.* **1992**, *47*, 211–217. [CrossRef] [PubMed]
113. Pope, C.A. Particulate pollution and health: A review of the Utah valley experience. *J. Expo. Anal. Environ. Epidemiol* **1996**, *6*, 23–34. [CrossRef] [PubMed]
114. Pope, C.A.; Dockery, D.W.; Spengler, J.D.; Raizenne, M.E. Respiratory Health and PM10 Pollution: A Daily Time Series Analysis. *Am. Rev. Respir. Dis.* **1991**, *144*, 668–674. [CrossRef] [PubMed]
115. Pope, C.A.; Dockery, D.W. Acute Health Effects of PM10 Pollution on Symptomatic and Asymptomatic Children. *Am. Rev. Respir. Dis.* **1992**, *145*, 1123–1128. [CrossRef] [PubMed]
116. Pope, C.A.; Kanner, R.E. Acute Effects of PM10 Pollution on Pulmonary Function of Smokers with Mild to Moderate Chronic Obstructive Pulmonary Disease. *Am. Rev. Respir. Dis.* **1993**, *147*, 1336–1340. [CrossRef]
117. Ransom, M.R.; Pope, C.A. Elementary school absences and PM10 pollution in Utah Valley. *Environ. Res.* **1992**, *58*, 204–219. [CrossRef]
118. Hicken, R. Geneva Officials Challenge PM10 Study's Results. *Provo Daily Herald*. 30 March 1991. Available online: https://newspapers.lib.utah.edu/details?id=24117199 (accessed on 16 November 2021).
119. Meyers, D. Blame It on Bugs, Not Air, Group says. *Provo Daily Herald*. 29 May 1997. Available online: https://newspapers.lib.utah.edu/details?id=24241709 (accessed on 16 November 2021).
120. Governor's Clean Air Commission. *Summary of Recommendations*; All Five Work Group Reports; Governor's Clean Air Commission: Salt Lake City, UT, USA, 1990. Available online: https://digitallibrary.utah.gov/awweb/guest.jsp?smd=1&cl=all_lib&lb_document_id=62223 (accessed on 28 April 2020).
121. Governor's Clean Air Commission. *Final Report*; Governor's Clean Air Commission: Salt Lake City, UT, USA, 1991. Available online: https://digitallibrary.utah.gov/awweb/guest.jsp?smd=1&cl=all_lib&lb_document_id=62036 (accessed on 28 April 2020).

122. Zimmerman, J.; Governor Urges Environmental Health Department. *Provo Daily Herald* . 12 January 1990. Available online: https://newspapers.lib.utah.edu/details?id=24086483 (accessed on 19 November 2021).

123. Edwards, P.M.; Brown, S.S.; Roberts, J.M.; Ahmadov, R.; Banta, R.M.; deGouw, J.A.; Dubé, W.P.; Field, R.A.; Flynn, J.H.; Gilman, J.B.; et al. High winter ozone pollution from carbonyl photolysis in an oil and gas basin. *Nature* **2014**, *514*, 351–354. [CrossRef]

124. Schnell, R.C.; Oltmans, S.J.; Neely, R.R.; Endres, M.S.; Molenar, J.V.; White, A.B. Rapid photochemical production of ozone at high concentrations in a rural site during winter. *Nat. Geosci.* **2009**, *2*, 120–122. [CrossRef]

125. Utah Division of Air Quality. *Utah Area Designation Recommendation for the 2015 8-Hour Ozone National Ambient Air Quality Standard*; Utah Division of Air Quality: Salt Lake City, UT, USA, 2016. Available online: https://www.epa.gov/sites/production/files/2016-11/documents/ut-rec-tsd.pdf (accessed on 14 October 2020).

126. U.S. EPA. *Air Quality Designations for the 2006 24-Hour Fine Particle (PM2.5)*; 40 CFR 81; U.S. EPA: Washington, DC, USA, 2009; pp. 58687–58781. Available online: https://www.federalregister.gov/documents/2009/11/13/E9-25711/air-quality-designations-for-the-2006-24-hour-fine-particle-pm25 (accessed on 25 April 2022).

127. Napier-Pearce, J. Utah Clean Air Rally Draws Thousands to Capitol. *The Salt Lake Tribune*. 26 January 2014. Available online: https://archive.sltrib.com/article.php?id=57447995&itype=CMSID#:~{}:text=In%20what%20organizers%20called%20the,gas%20masks%2C%20swarmed%20Capitol%20Hill (accessed on 25 April 2022).

128. U.S. EPA. *Clean Data Determination; Salt Lake City, Utah 2006 Fine Particulate Matter Standards Nonattainment Area*; 40 CFR 52; U.S. EPA: Denver, CO, USA, 2019; pp. 26053–26057. Available online: https://www.federalregister.gov/documents/2019/06/05/2019-11702/clean-data-determination-salt-lake-city-utah-2006-fine-particulate-matter-standards-nonattainment (accessed on 25 April 2022).

129. U.S. EPA. EPA to Reexamine Health Standards for Harmful Soot that Previous Administration Left Unchanged. Available online: https://www.epa.gov/newsreleases/epa-reexamine-health-standards-harmful-soot-previous-administration-left-unchanged (accessed on 25 April 2022).

130. U.S. EPA. *Determinations of Attainment by the Attainment Date, Extensions of the Attainment Date, and Reclassification of Areas Classified as Marginal for the 2015 Ozone National Ambient Air Quality Standards*; 40 CFR 52; U.S. EPA: Washington, DC, USA, 2022; pp. 21842–21858. Available online: https://www.federalregister.gov/documents/2022/04/13/2022-07513/determinations-of-attainment-by-the-attainment-date-extensions-of-the-attainment-date-and (accessed on 25 April 2022).

131. U.S. EPA. *Integrated Science Assessment for Ozone and Related Photochemical Oxidants*; U.S. EPA: Research Triangle Park, NC, USA, 2020. Available online: https://cfpub.epa.gov/ncea/isa/recordisplay.cfm?deid=348522 (accessed on 25 November 2020).

132. Langford, A.O.; Alvarez, R.J.; Brioude, J.; Fine, R.; Gustin, M.S.; Lin, M.Y.; Marchbanks, R.D.; Pierce, R.B.; Sandberg, S.P.; Senff, C.J.; et al. Entrainment of stratospheric air and Asian pollution by the convective boundary layer in the southwestern U.S. *J. Geophys. Res. Atmos.* **2017**, *122*, 1312–1337. [CrossRef]

133. Zhang, L.; Jacob, D.J.; Downey, N.V.; Wood, D.A.; Blewitt, D.; Carouge, C.C.; van Donkelaar, A.; Jones, D.B.A.; Murray, L.T.; Wang, Y. Improved estimate of the policy-relevant background ozone in the United States using the GEOS-Chem global model with $1/2 \times 2/3$ horizontal resolution over North America. *Atmos. Environ.* **2011**, *45*, 6769–6776. [CrossRef]

134. Abatzoglou, J.T.; Williams, A.P. Impact of anthropogenic climate change on wildfire across western US forests. *Proc. Natl. Acad. Sci. USA* **2016**, *113*, 11770–11775. [CrossRef] [PubMed]

135. Lin, M.; Horowitz, L.W.; Payton, R.; Fiore, A.M.; Tonnesen, G. US surface ozone trends and extremes from 1980 to 2014: Quantifying the roles of rising Asian emissions, domestic controls, wildfires, and climate. *Atmos. Chem. Phys.* **2017**, *17*, 2943–2970. [CrossRef]

136. Blaylock, B.; Horel, J.D.; Crosman, E.T. Impact of Lake Breezes on Summer Ozone Concentrations in the Salt Lake Valley. *J. Appl. Meteorol. Climatol.* **2016**, *56*, 353–370. [CrossRef]

137. Horel, J.; Crosman, E.; Jacques, A.; Blaylock, B.; Arens, S.; Long, A.; Sohl, J.; Martin, R. Summer ozone concentrations in the vicinity of the Great Salt Lake. *Atmos. Sci. Lett.* **2016**, *17*, 480–486. [CrossRef]

138. USGCRP. *Impacts, Risks, and Adaptation in the United States: Fourth National Climate Assessment, Volume II*; U.S. Global Change Research Program: Washington, DC, USA, 2018; p. 1515.

139. Abbott, B.W.; Bliss, A.; Barros, L.; Moyer, T.; Moore, F.; Rapp, M.; Gilbert, S.; Bekker, J.; Mitchell, L.; Hill, S.; et al. *Clean Electrification of the U.S. Economy*; Brigham Young University: Provo, UT, USA, 2021. [CrossRef]

140. Bloch, C.; Newcomb, J.; Shiledar, S.; Tyson, M. *Breakthrough Batteries: Powering the Era of Clean Electrification*; Rocky Mountain Institute: Basalt, CO, USA, 2019; p. 84. Available online: https://rmi.org/insight/breakthrough-batteries/ (accessed on 7 November 2019).

141. He, G.; Lin, J.; Sifuentes, F.; Liu, X.; Abhyankar, N.; Phadke, A. Rapid cost decrease of renewables and storage accelerates the decarbonization of China's power system. *Nat. Commun.* **2020**, *11*, 2486. [CrossRef]

142. Lazard. Lazard's Levelized Cost of Energy Analysis—Version 14.0. 2020, p. 21. Available online: https://www.lazard.com/media/451419/lazards-levelized-cost-of-energy-version-140.pdf (accessed on 21 October 2020).

143. Wilson, C.; Grubler, A.; Bento, N.; Healey, S.; Stercke, S.D.; Zimm, C. Granular technologies to accelerate decarbonization. *Science* **2020**, *368*, 36–39. [CrossRef]

144. Edwards, R.; Weiler, T. Concurrent Resolution on Environmental and Economic Stewardship. 2018. Available online: https://le.utah.gov/~{}2018/bills/static/HCR007.html (accessed on 20 October 2020).

145. Kem, C. *Gardner Policy Institute. The Utah Roadmap: Positive Solutions on Climate and Air Quality*; Kem C. Gardner Policy Institute: Salt Lake City, UT, USA, 2020; Available online: https://gardner.utah.edu/utahroadmap/ (accessed on 8 June 2020).
146. Utah Climate & Clean Air Compact. Available online: https://climateandcleanaircompact.org/ (accessed on 20 October 2020).

 *sustainability*

Article

Investigating the Relationship between Landscape Design Types and Human Thermal Comfort: Case Study of Beijing Olympic Forest Park

Lin Zhang , Haiyun Xu and Jianbin Pan *

Department of Landscape Architecture, College of Architecture and Urban Planning, Beijing University of Civil Engineering and Architecture, Beijing 100044, China
* Correspondence: panjianbin@bucea.edu.cn; Tel.: +86-134-0113-9795

Simple Summary: Beijing, China, is a megacity with a population of more than 20 million. The hot summer climate and environmental problems caused by urbanization affect the quality of human settlements. Based on the measured data in the Beijing Olympic Forest Park and the analysis of the human body comfort index model, the study concluded that the thermal comfort level of the double-layer plant community area composed of tall deciduous trees such as Sophora japonica and Ginkgo biloba and shrubs or grass native cover plants was higher than that of other areas.

Abstract: Urban green space can improve the local thermal environment and thus the quality of the urban residential environment. Taking the green space of Beijing Olympic Forest Park (BOFP) as an example, this study analysed sample points representing different plant community structures, plant community types, and landscape environments based on 15 years of continuous dynamic measurement and selected typical annual data (from 2020). The study analysed and explained the spatial differentiation characteristics of human thermal comfort (HTC) in green space areas of BOFP using the predicted mean vote (PMV)–predicted percentage dissatisfied (PPD) physical comfort index model, which comprehensively considers both the objective environment and people's subjective feelings and psychological states. The results showed that the level of HTC in the park's green space, across community types and across typical landscape environments, differed between areas with different community structures. PMV–PPD mathematical model fitting further verified the above results.

Keywords: landscape architecture; urban green space; human thermal comfort; spatial differentiation; evidence-based design

Citation: Zhang, L.; Xu, H.; Pan, J. Investigating the Relationship between Landscape Design Types and Human Thermal Comfort: Case Study of Beijing Olympic Forest Park. *Sustainability* **2023**, *15*, 2969. https://doi.org/10.3390/su15042969

Academic Editors: José Carlos Magalhães Pires and Álvaro Gómez-Losada

Received: 24 November 2022
Revised: 13 January 2023
Accepted: 23 January 2023
Published: 6 February 2023

1. Introduction

As urbanisation accelerates, changes in land cover type are reshaping the landscape patterns and physical environments of urban areas. Within built-up urban areas, hard paved surfaces such as roads, large-scale buildings and other structures, and gaseous pollutants and artificial heat discharged by human production and living activities have massively changed the urban thermal environment. In particular, the surface temperature in urban centres is significantly higher than that in the suburbs, which is known as the urban heat island (UHI) effect. The continuous intensification of the UHI effect is not only changing the process of atmospheric circulation between cities and suburbs but also increasing urban energy consumption and atmospheric pollution. This reduces human comfort in the urban environment, which is not conducive to the sustainable development of cities and other human settlements [1–3]. Studies of UHIs have focused on three main scales: the urban (including urban region), the landscape, and green space. One urban scale study analysed spatial changes in patterns of temperature in Changsha and their relationships with many related factors, such as natural and human factors, using composite methods. The study found that the UHI effect was consistent with the spatial development trajectory of the city and positively correlated with the intensity of urban construction. The UHI effect was also significantly correlated with urban landscape patterns and human factors, based on analysis of POI spatial big data [4]. Other studies have found significant correlations between urban surface temperature and factors such as surface coverage type and proportion [5], natural water area, and forestland plant community coverage [6–14]. In urban scale research, analysis has been conducted mainly through the interpretation and inversion of high-definition satellite image data [1,5–8,12,13,15,16]. At the landscape scale, Fu et al. (2020) analysed summer thermal environment factors in 183 key cities in China from 1990 to 2016 to identify the mechanisms driving the temporal and spatial evolution of the urban thermal environment in the process of rapid urbanization [17]. Research has shown that the characteristics of urban green space areas, such as community structure and landscape patterns [15,16,18–20], as well as buildings (floor area, size) and three-dimensional greening, play significant roles in regulating spatial patterns in the urban thermal environment [7,21–26]. It can be concluded that the surface temperature of urban green space areas is lower than that of general urban areas, resulting in an urban cold island (UCI) effect. In urban-scale studies, the most common data acquisition method is fixed-point measurement (as well as spatial interpolation and numerical simulation based on measured data) [20–22,27]. In cities that frequently experience a hot summer climate such as Beijing, China, UCI areas such as urban green space can not only significantly reduce and improve the scope and intensity of the UHI, but also increase the level of human comfort in this area and then offer urban residents recreation.

The human thermal comfort (HTC) index quantitatively describes people's subjective feelings about the local thermal environment. It does so digitally, in a landscape environment. A landscape environment is an objective environment created using natural landscape elements and based on people's subjective and objective needs. Research on human thermal comfort supports the fine-grained design, analysis, and evaluation of landscape architecture. Extensive research has investigated the regional climatic characteristics of case cities, different landscape/environmental scales, and influencing factors, using various research measures and methods. The discomfort index [27], the wet bulb globe temperature index [19], the synthesis index, physiological equivalent temperature [17,22], and the predicted mean vote–predicted percentage dissatisfied (PMV–PPD) thermal comfort index [21,25] have been used in related research. The above comfort indices have specific objects and ranges. The PMV–PPD thermal comfort index method was developed to analyse comfort in indoor spaces (e.g., train compartments) dependent on air conditioning in the 1970s. The index's composition includes not only temperature, relative humidity, and wind speed, which reflect objective environmental characteristics, but also human activities, clothing types, and psychological and physiological factors that affect human thermal comfort. In recent years, some researchers have applied this method to evaluate bodily comfort in open spaces in cities in different regions [20,21,25]. The current study used this method to explain the correlations between the characteristics of plant communities in urban green spaces and human thermal comfort. To summarise, urban scale research on the urban thermal environment has mainly described the spatial correlations between urban spatial morphological characteristics and UHI intensity from a quantitative perspective, aiming to serve urban economic and social development, material space planning, etc. Research on landscape patterns and green space has paid more attention to the factors affecting the intensity of UHIs, seeking to provide a scientific basis for the incremental planning, design, and stock renewal of urban space according to the desired intensity and scope of impact of UHI mitigation. As urban green spaces differ in their proportions of forestland, water bodies, grass, and hard pavements, the regional thermal environment shows some spatial differentiation. However, this differentiation has not been adequately reflected in quantitative research, which has thus far failed to examine the impact of different green space landscape types (including plant community structure, community type, and typical landscape environment) on human comfort.

To address the above shortcomings, the research team continuously and dynamically monitored the microenvironmental effects of green space plant communities (such as reducing air fungi, cooling and humidifying, and reducing negative ions in the air) in BOFP for 15 years, from 2005 to 2020. In a previous study, the team used the results to illustrate the spatial characteristics of regional thermal comfort in green space areas of BOFP [27]. The current study addressed the following research questions.

(1)   Which plant community structure(s) and type(s) of green space significantly affect human thermal comfort in BOFP?
(2)   What are the differences in human thermal comfort between typical landscape areas of green space in BOFP?

The research results provide a scientific basis for the planning, design, renewal, and optimisation of landscape green space in BOFP to improve the urban thermal environment and human thermal comfort.

2. Overview of the Study Area and Research Methods

2.1. Study Area

The main green space areas in BOFP are located in Chaoyang district, Beijing (Figure 1). They extend from Kehui Road in the south to Qinghe in the north (across the North Fifth Ring Road), east to Anli Road and west to Lincui Road. The park covers a total area of 680 hectares, of which the southern part covers about 380 hectares (including a venue area in the west) and the northern part covers about 300 hectares.

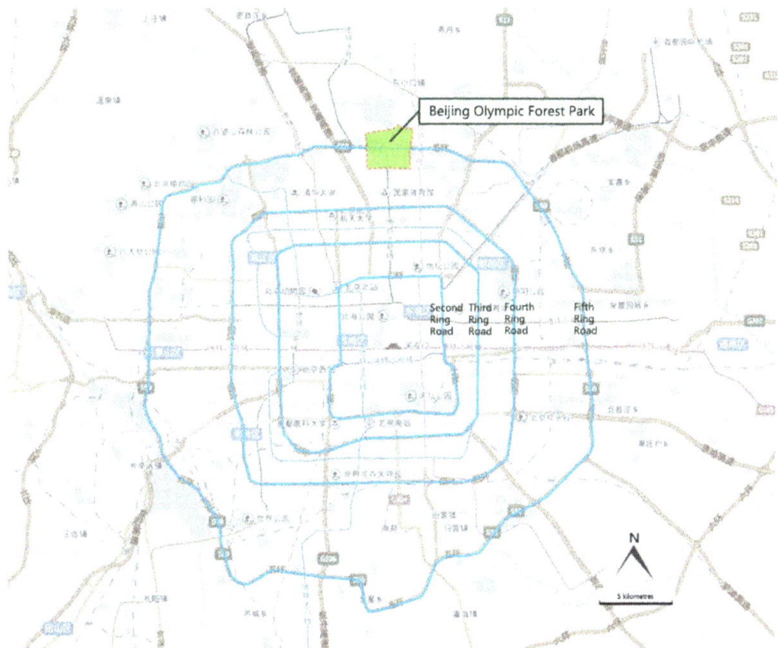

Figure 1. Location map of Beijing Olympic Forest Park.

2.2. Research Methods

2.2.1. Selection of Plant Community Index for Typical Green Space

Plant communities in the green space quadrat of BOFP were categorised as follows. Five types of plant community structure: tree–shrub–grass (TSG), tree-shrub (TS), tree-grass (TG), shrub-grass (SG), and grass/ground cover (G). Five plant community types:

coniferous plant community (CP), coniferous and broadleaved mixed plant community (CBP), deciduous broadleaved plant community (DBP), shrub (S) and grass/ground cover (G). Five typical environments: TSG multi-layer plant community (MPC), TS, TG, and SG double-layer plant community (DPC), G single-plant Community (SPC), waterfront plaza (WS), and waterfront plant community (WPC). CK denotes the comparison sample.

2.2.2. Data Collection

(1) Sample Setting

Using the chessboard sampling method, 17 experimental sample points were selected in the green space of BOFP (Figure 2 and Table 1). To ensure that the test sample points were far away from large-scale crowd activity areas (urban roads, squares, etc.), representative vegetation types were selected to fine-tune their positions [27]. The types and number of plant community structures covered more than three test sites. The two comparison samples were located in the Beijing Olympic Park paved square (near the underground business district), 1 km south of the south gate of BOFP; and in a paved square on the north side of the north fourth ring road of BOFP (near Beijing's Bird's Nest National Stadium). Compared with the former, the latter site has more hard pavement, less green space, less diverse plant communities, and more dense crowd activities.

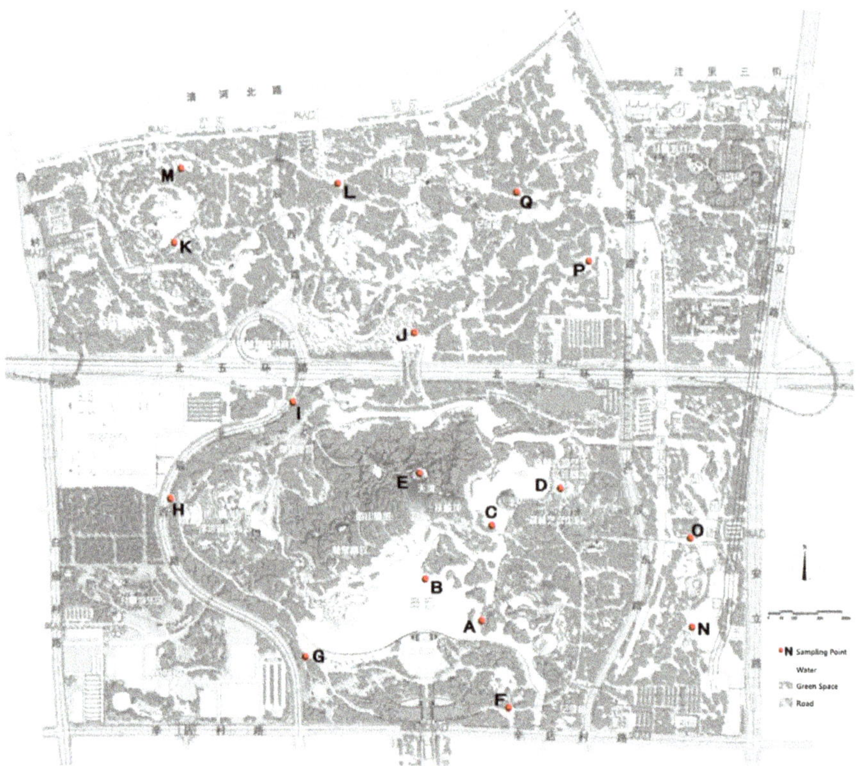

Figure 2. Seventeen experimental sample points in the green space of BOFP.

Table 1. Biological characteristics of plant communities at green space test sample points in BOFP (synchronous test data).

Sample Points	Community Structure	Community Type	Dominant Species (DS)	Other Types	Plant Height (m)	DBH (cm)	Crown Width (m)	Canopy Height (m)	Canopy Density (CD)
CK 1	-	-	-	-	-	-	-	-	-
CK 2	-	-	-	-	-	-	-	-	-
A	TG	DBP	Populus tomentosa	Pinus tabulaeformis, Sabina chinensis, Midget crabapple, Syringaoblata	10–12	25–30	2.0–2.5	5.0–6.0	0.65
B	-	-	Salix matsu-danacvpen-dula	Phragmites-communis, Purple loosestrife	4.5–5.5	20–25	4.0–4.5	2.0–2.5	0.25
C	TSG	DBP	Salix matsudana f.pendula	Lonicera maackii	5.5–6.0	20–25	3.5–4.0	3.0–3.5	0.75
D	TSG	CBP	Sabina chinensis; Sophora japonica	Salix matsudana	3.5–4.0/ 5.5–6.0	20–25	2.0–2.5/ 4.5–5.0	1.5–2.0/ 2.5–3.0	0.85
E	T	CP	Pinus tabulaeformis	Prunus armeniaa, Lonicera maackii, Vitex negundo	3.0–3.5	10–15	3.5–4.0	1.5–2.0	0.35
F	TG	DBP	Salix matsudana	Robinia pseudoacacia (young), Viburnum dilatatum	7.0–8.0	20–25	4.5–5.0	3.0–4.0	0.85
G	SG	S	Syringa oblata	Sophora japonica, Forsythia suspensa	2.5–3.0	–	2.0–2.5	1.5–2.0	0.75
H	SG	S	Caryopteris× clandonensis, 'Worcester Gold'	Euonymus japonicus	0.5–1.0	–	–	–	0.45
I	SG	S	Euonymus japonicus	Sophora japonica	0.5–1.0	–	–	–	0.45
J	SG	CP	Pinus tabulaeformis	Ulmus pumila, Fontanesia	4.5–5.0	10–15	2.5–3.0	2.0–2.5	0.55
K	SG	DBP	Prunus triloba	Pinus tabulaeformis, Sophora japonica	3.0–3.5	–	2.0–2.5	1.0–1.5	0.75
L	G	G	Lawn and ground cover plants	Prunus armeniaca	–	–	–	–	0.75
M	TS	CP	Pinus tabulaeformis	Sophora japonica, Syringaoblata	3.5–4.0	10–15	2.5–3.0	1.5–2.0	0.95
N	TSG	CBP	Populus tomentosa	Malus spectabilis, Weigela florida, Lespedeza	9.5–10.0	25–30	2.5–3.0	5.0–6.0	0.90
O	TSG	DBP	Sophora japonica	Syringa oblata, Forsythia suspensa	6.5–7.0	20–25	4.0–4.5	2.5–3.0	0.90
P	TG	DBP	Ginkgo biloba	-	4.5–5.0	15–25	2.5–3.0	2.0–2.5	0.75
Q	TG	DBP	Sophora japonica, Fraxi-nuschinensis	Syringa microphylla, Sorbaria sorbifolia, Prunuscerasifera	3.5–4.0	15–20	3.0–3.5/5–6	2–3/3–4	0.75

(2) Test methods

The test instrument was the 6 SWEMA Y-Boat-R multifunctional online environmental detection system (which has 30 channels that can simultaneously collect and store air temperature, radiation temperature, wind speed, and direction data and synchronously measure reference data such as air relative humidity, differential pressure, and heat flux). Three repetitions were set for each index data point. The test was conducted over 3 days between August 10 and 20, 2020. The meteorological conditions were sunny (cloud cover not more than 30%) with a calm wind (3–4 m/s). First, data on green plant community characteristics such as plant height of dominant species, DBH, crown width, canopy height, canopy coverage, and canopy density were obtained by field investigation within the quadrat. Quantitative plant community parameters such as the leaf area index were measured using the CI-110 plant canopy image analyser.

During the sample point measurement, the instrument automatically recorded and stored the measured data. In the data analysis, the arithmetic mean of the automatically recorded data obtained at intervals of 10 min from 08:00 to 18:00 (76 times in total), was taken as the daily mean. The average values automatically recorded at 5-min intervals between 08:50 and 09:10 (obtained at 08:50, 08:55, 09:00, 09:05, and 09:10), between 13:20 and 13:40, and between 17:20 and 17:40 were taken as the morning instantaneous value, noon instantaneous value, and afternoon instantaneous value, respectively. The average values of the comparative sample data were used as background comfort data for the urban area environment.

2.2.3. Data Analysis Methods

(1) PMV–PPD body comfort index

PMV is the estimated average thermal sensation index, which is a quantitative measure of people's thermal sensation in a specific outdoor thermal environment. It can be used to evaluate and judge whether a certain environmental state meets the requirements for human thermal comfort (the higher the absolute value of PMV, the greater people's discomfort, which is categorised as "thermal discomfort" or "cold discomfort", otherwise "comfort"). PPD is a quantitative index measuring people's predicted dissatisfaction with a specific thermal environment, i.e., the predicted dissatisfaction rate, which is expressed in percentage form (the higher the value, the greater the predicted dissatisfaction). The PMV–PPD index corresponds to the ASHRAE thermal sensation level 7 index. The range of values from cold to hot is −3 to 3, where 0 is the most comfortable state of the environment (as shown in Table 2). The mathematical relationship between PMV and PPD is shown in Figure 3.

Table 2. PMV thermal sensation scale (ASHRAE thermal sensation).

Thermal Sensation	Cold	Cool	Slightly Cool	Moderate	Slightly Warm	Warm	Hot
PMV Value	−3	−2	−1	0	1	2	3

The calculation formulas are given below.

$$PMV = \left(0.303e^{-0.036M} + 0.028\right) \times \left\{(M - W) - 3.0510^{-3} \times [5733 - 6.99\,(M - W) - P_a] - 0.42 \times [(M - W) - 58.15] - 1.7 \times 10^{-5}\,M\,(5867 - Pa) - 0.0014\,M\,(34t_a) - 3.96 \times 10^{-8}f_{cl} \times [(t_{cl} + 273)^4 - (\bar{t}_r + 273)^4] - f_{cl}h_c(t_{cl} - t_a)\right\} \tag{1}$$

$$PPD = 100 - 95 \times e^{-(0.03353 \times PMV^4 + 0.2179 \times PMV^2)} \tag{2}$$

$$\text{Formula}: t_{cl} = 35.7 - 0.028(M - W) - I_{cl}\left\{3.96 \times 10^{-8}f_{cl} \times \left[(t_{cl} + 273)^4 - (\bar{t}_r + 273)^4\right] + f_{cl}h_c(t_{cl} - t_a)\right\} \tag{3}$$

$$h_c = \begin{cases} 2.38|t_{cl} - t_a|^{0.25} & When\ 2.38|t_{cl} - t_a| > 12.1\sqrt{V_{ar}} \\ 12.1\sqrt{V_{ar}} & When\ 2.38|t_{cl} - t_a| < 12.1\sqrt{V_{ar}} \end{cases} \tag{4}$$

$$f_{cl} = \begin{cases} 1.00 + 1.290I_{cl} & When\ I_{cl} \leq 0.078\ \text{m}^2 \cdot \text{K/W} \\ 1.05 + 0.645I_{cl} & When\ I_{cl} > 0.078\ \text{m}^2 \cdot \text{K/W} \end{cases} \tag{5}$$

In Formula (1), *PMV* is the estimated average thermal sensation index; M is the metabolic rate of the human body when engaged in certain activities in the thermal environment (this study focused on aerobic exercise such as fast walking, jogging, and dance-based fitness), W/m^2; W is the heat consumed by external work (negligible for most activities), W/m^2; I_{cl} is the thermal resistance of clothing, $\text{m}^2 \cdot \text{K/W}$ (the background clothing in this study was set as trousers, shirt/T-shirt, shoes, and socks; coefficient = 0.110 $\text{m}^2 \cdot \text{K/W}$); f_{cl} is the ratio of body surface area when clothed to body surface area when exposed; t_a is the air temperature, °C; $\overline{t_r}$ is the average radiation temperature, °C; V_{ar} is the relative wind speed, m/s; P_a is the partial pressure of water vapour, P_a; h_c is the convective heat transfer coefficient, $W/(\text{m}^2 \cdot °C)$; and t_{cl} is the clothing surface temperature, °C.

h_c and t_{cl} can be obtained by formula iteration. PMV can be obtained from metabolic rate, clothing thermal resistance, air temperature, average radiation temperature, wind speed, and other parameters (std.samr.gov.cn).

In Formula (2), when the PMV value is 0, the PPD value is not also 0, because regardless of how comfortable the objective environment is, people's physical comfort is affected by multiple subjective and objective factors beyond the environment, such as physiological and psychological factors. It was assumed here that even when the objective environment was at its most comfortable, 5% of people would still be dissatisfied (i.e., PPD = 5%).

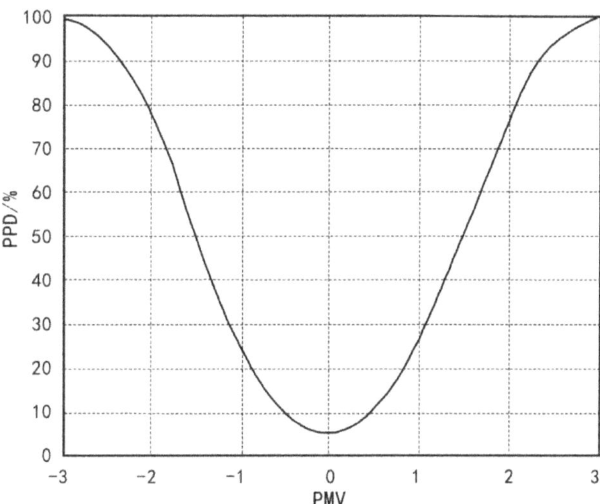

Figure 3. Mathematical function relationship between PMV and PPD.

3. Results

3.1. Differences in Body Comfort between Different Plant Community Structures in Green Space

Figure 4 (including Figure 4a–c) and Figure 5, respectively, show the instantaneous values and mean values of human thermal sensation in areas of green space with different green plant community structures. As shown in Figure 2, the level of morning instantaneous somatosensory comfort in the areas with TG, TS, SG, and G community structures was high. However, TG and TS show negative values in the figure, indicating "cool comfort", while the SG and G areas were associated with "warm comfort". As the SG and G community structure areas were directly exposed to solar radiation in the local plant community areas, their temperature rose quickly, while the TS and TG community structure areas did not receive much direct irradiation from sunlight due to their dense forest canopy, resulting in a slow rise in air temperature. The comparison sample also showed a rapid temperature rise, with its areas associated with "warm comfort". The results shown in Figure 4b,c are very similar, illustrating a gradual intensification of thermal sensation (from noon instantaneous to afternoon instantaneous), reaching the level of "thermal discomfort" before the end of the test and remaining at this level after the test.

Figure 5 presents two main sets of results. Somatosensory comfort in the TG and TS community structure areas was basically at the level of "thermal comfort", while that in the TSG, SG, and G community structure areas was at the level of "thermal discomfort". The reason for the former finding may lie in tree canopy shielding, which meant that direct solar radiation intensity and air relative humidity in the forest were controlled within a certain range. In addition, there were relatively few vegetation layers in the forest, which was conducive to the flow of air in the horizontal and vertical directions, making the temperature feel more comfortable. Although the TSG community structure had full canopy coverage, space in the forest was limited, which restricted airflow and thereby limited the dissipation of air humidity and heat. Although the SG and G community structure areas had good ventilation, they experienced strong direct radiation, making them feel hot.

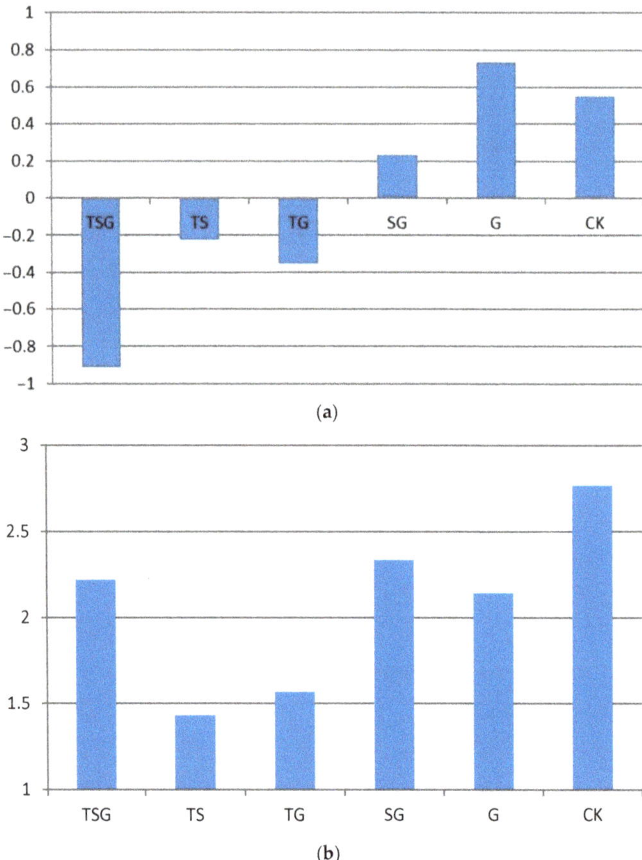

Figure 4. *Cont.*

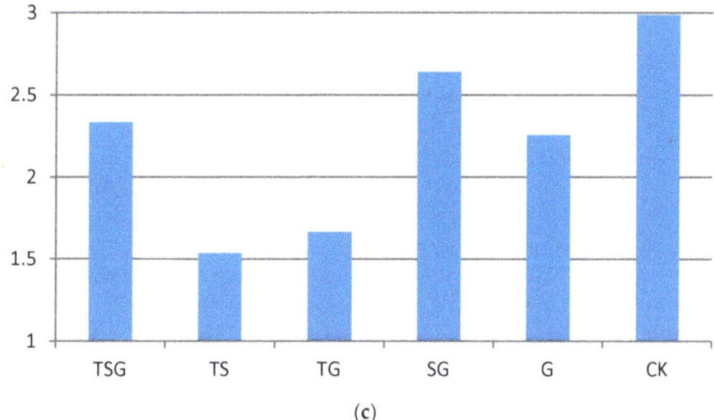

(c)

Figure 4. (**a**). Regional thermal sensation in green plant community structure areas (morning instantaneous values). (**b**). Regional body comfort in green plant community structure areas (noon instantaneous values). (**c**). Regional body comfort in green plant community structures (afternoon instantaneous values).

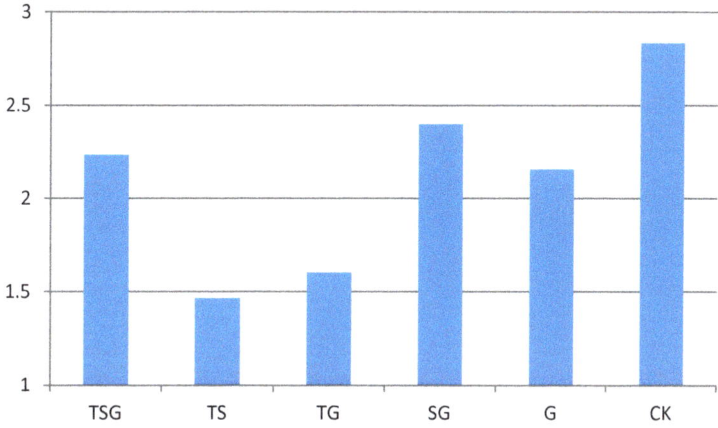

Figure 5. Regional body comfort in green plant community structure areas (mean values).

Figure 5 presents two main sets of results. Somatosensory comfort in the TG and TS community structure areas was basically at the level of "thermal comfort", while that in the TSG, SG, and G community structure areas was at the level of "thermal discomfort". The reason for the former finding may lie in tree canopy shielding, which meant that direct solar radiation intensity and air relative humidity in the forest were controlled within a certain range. In addition, there were relatively few vegetation layers in the forest, which was conducive to the flow of air in the horizontal and vertical directions, making the temperature feel more comfortable. Although the TSG community structure had full canopy coverage, space in the forest was limited, which restricted airflow and thereby limited the dissipation of air humidity and heat. Although the SG and G community structure areas had good ventilation, they experienced strong direct radiation, making them feel hot.

3.2. Differences in Somatosensory Comfort between Green Space with Different Plant Community Types

Figure 6 (including Figure 6a–c) and Figure 7, respectively, show the instantaneous values and mean values of perceived bodily heat across the five green plant community types. As shown in Figure 4a, all five community type areas felt comfortable in the morning.

Among them, the S and G community type areas were associated with "warm comfort", while the CP, DBP, and CBP community type areas were associated with "cool comfort". The reason for the "warm comfort" experienced in S and G may have been that the air in these areas was heated by direct solar radiation, while the DBP and CBP areas stayed cooler for longer due to their greater canopy closure. However, the CP community type area had small trees and thus no significant canopy coverage. Although direct solar radiation was strong in this area, according to the field observation, the regional environment was still experienced as cool. The reasons for this unexpected finding need to be further studied. The results in Figure 6b,c are similar, showing a continuous increase in thermal sensation from noon to afternoon. At noon (Figure 6b), some of the community type areas were still comfortable, but in the afternoon (Figure 6c), all of the community type areas were associated with "thermal discomfort". As shown in Figure 5, the somatosensory comfort level in the DBP community type area was "thermal comfort", while that in the other community type areas was "thermal discomfort". The reasons for this finding also need to be further studied.

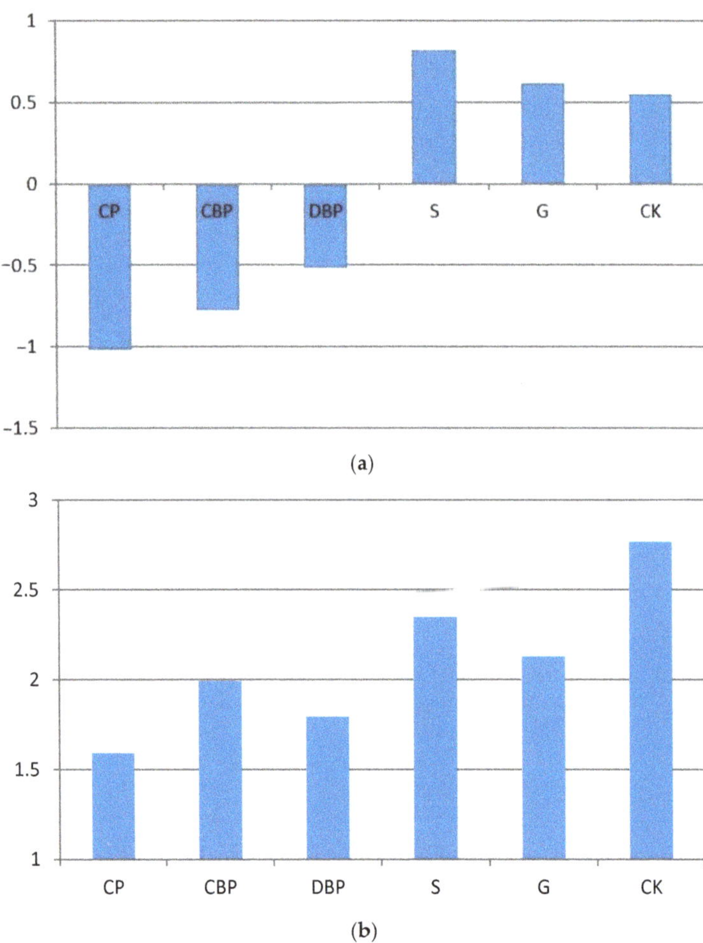

(a)

(b)

Figure 6. *Cont.*

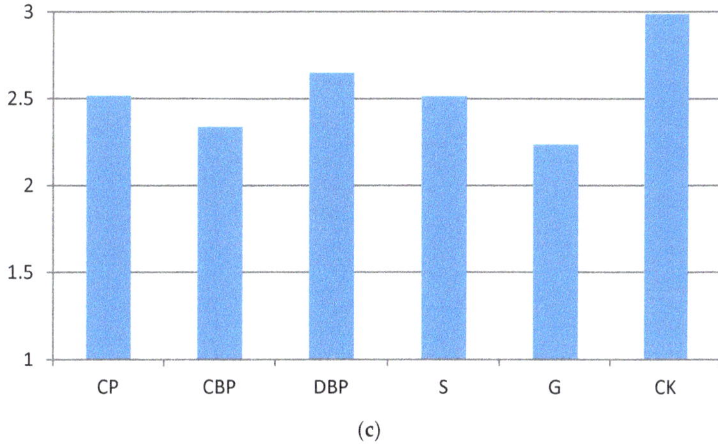

(c)

Figure 6. (**a**). Regional body comfort in green plant community type areas (morning instantaneous values). (**b**). Regional body comfort in green plant community type areas (noon instantaneous values). (**c**). Regional body comfort in green plant community type areas (afternoon instantaneous values).

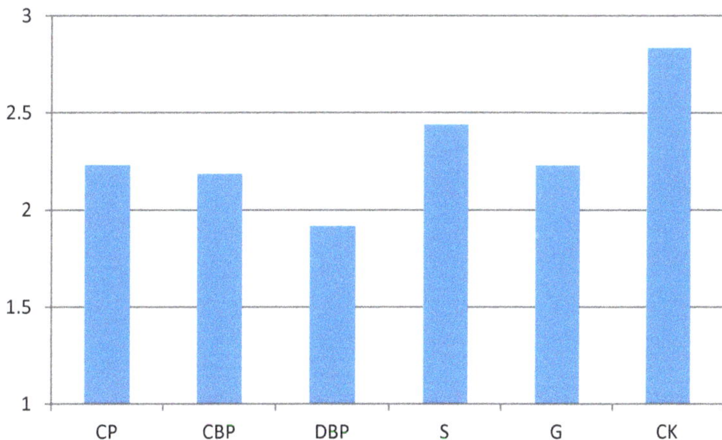

Figure 7. Regional body comfort in green plant community type areas (mean values).

3.3. Differences of Typical Landscape Environment and Body Comfort of Green Space

Figure 8 (including Figure 8a–c) and Figure 9, respectively, show the instantaneous values and mean values of thermal sensation in areas with different typical landscape environments. As shown in Figure 8a, all of the community types in the morning felt comfortable. The MPC, DPC, and WPC community structure areas were associated with "cool comfort", while the SPC and WS community structure areas were associated with "warm comfort". As the MPC community structure in this section is the same as the TSG community structure, the corresponding results are the same as those reported in Section 2.1. The results presented in Figure 8b,c are very similar, showing that thermal sensation in the typical landscape environment areas increased or remained roughly stable within the threshold range of "thermal discomfort" ($2 \leq |\mathrm{PMV}| \leq 3$). The results in Figure 9 show that thermal sensation in the three landscape environments represented by the MPC, DPC, and WPC community structures ranged from "thermal comfort" to "thermal discomfort". The WPC community area showed similar characteristics to the DPC community area, which may have been due to its open space and proximity to a water body, which is conducive to ventilation. The two landscape environments of SPC and WS were associated with significant "thermal discomfort", as they were subject to strong direct solar radiation.

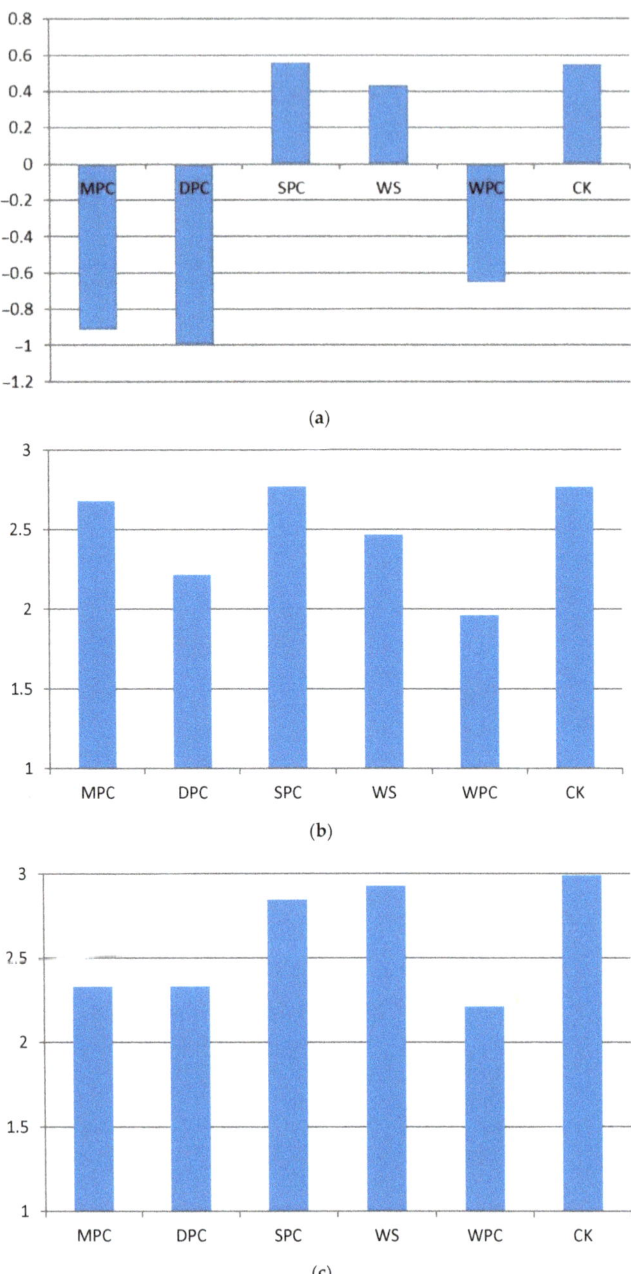

Figure 8. (**a**) Body comfort in typical landscape environment areas of green space (morning instantaneous values). (**b**) Body comfort in typical landscape environment areas of green space (noon instantaneous values). (**c**) Body comfort in typical landscape environment areas of green space (afternoon instantaneous values).

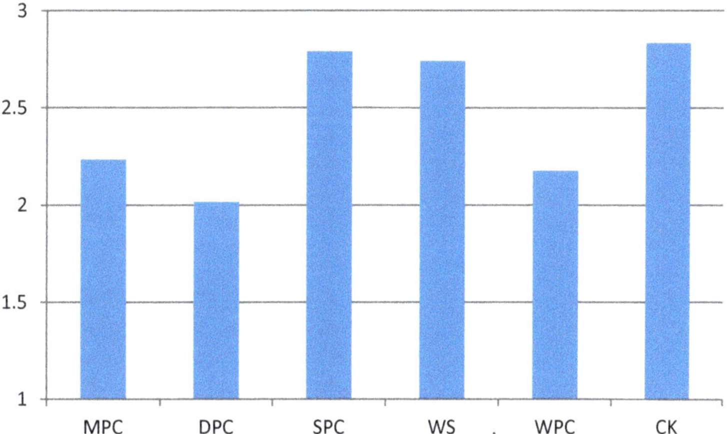

Figure 9. Body comfort in typical landscape environment areas of green space (mean values).

3.4. Fitting PMV–PPD Body Comfort Index to Observed Values across Landscape Types of Green Space

Figure 10 shows the results of fitting the PMV–PPD thermal comfort index to the observed values for the representative landscape types of green space. The background curve in the figure is the PMV–PPD mathematical function relationship (the complete curve is shown in Figure 3). Figure 10a fits the mean and instantaneous values of body comfort across landscape types to the PMV–PPD curve, and Figure 10b fits the mean values of body comfort across landscape types to the PMV–PPD curve. The distribution and degree of aggregation of the body comfort index values on the fitting curve directly reflect the differences in body comfort across green landscape types. As shown in Figure 10a, the instantaneous somatosensory comfort index values of the sample points with green space community structure as the influencing factor are closest to the ideal data points (PMV = 0, PPD = 5%), followed by the values for the waterfront plant community area. The morning instantaneous values for somatosensory comfort show a "comfortable" state ($|PMV| < 1$, PPD < 20%). As shown below in Figure 10a, most of the body comfort index values are located around the upper right part of the curve (PMV > 2, PPD > 80%), indicating "thermal discomfort". This may be related to the choice of August, a summer month, as the test period.

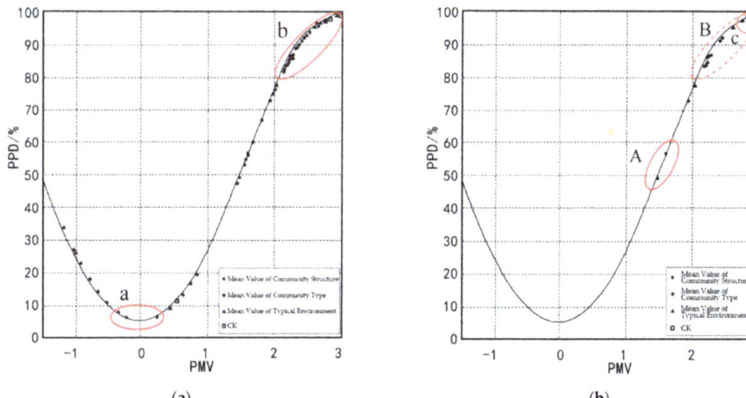

(a) (b)

Figure 10. (**a**). PMV–PPD instantaneous value/mean fitting. (**b**) PMV–PPD mean fitting. Index Value Fitting of PMV–PPD Body Comfort Across Landscape Types of Green Space. The red circle (a, b, c, A, B) in the Figure 10 shows the aggregation area of PMV-PPD mean value fitting points.

Figure 10b shows the fitting distribution of the mean values of the body comfort index during the test period. The figure shows that the body comfort index values of the sample points with green space community structure as the influencing factor are closest to the ideal data points (indicating thermal comfort), specifically the plant community area with arbour shrubs and grass structures (A in Figure 10b). The somatosensory comfort index values for the comparison sample (CK) and most of the landscape types are located around the upper right part of the curve (B and C in Figure 10b). This suggests that the overall environment was perceived as too hot (i.e., "thermal discomfort"), which may be related to the decision to collect data in August, when the temperature is high.

4. Discussion

4.1. Correlation between Plant Community Structure and Somatosensory Comfort in Green Space

According to previous studies, the three-dimensional size of a plant community is correlated with temperature, relative humidity, and the PMV index [15,21]. Compared with the other structural types, the TSG MPC community structure in this study had the largest volume of green space or greenery, helping to reduce the intensity of the UHI effect. However, as reported in Section 3.1, the level of somatosensory comfort in this environment was not high. This finding is not completely consistent with other studies [21]. In the process of landscape architecture planning and design, the TSG MPC community structure can be adopted in urban areas with a strong heat island effect. However, if a human activity site is located within or near a TSG MPC area, the canopy density of upper trees and coverage of shrubs in the forest should be controlled to maximise ventilation and light transmission and give full play to the advantages of this plant community structure.

Section 3.1 reports that the DPC community structure area with TS and TG had the highest level of somatosensory comfort compared with the other community structure areas, for three reasons:

(1) The arbour canopy acted as a "buffer zone", intercepting sunlight travelling into the forest and thereby controlling the temperature in the forest [20];
(2) Transpiration from forest canopy leaves controlled the relative humidity in the forest [16,18];
(3) The lack of inter-forest vegetation layers was conducive to horizontal and vertical airflow in the forest, helping to regulate somatosensory comfort. However, the quantitative relationship between canopy closure and somatosensory comfort needs to be further studied.

4.2. Correlation between Plant Community Types and Somatosensory Comfort in Green Space

Section 3.2 reports that the somatosensory comfort of some DBP community structure areas was greater than that of areas with other types of plant communities. This may be related to the high biochemical efficiency of DBP, but further research is needed. Beijing is located in the ecotone of two vegetation types, namely, CBP and DBP. DBP is the main local vegetation type. In the tree species composition of this vegetation type, native trees such as poplar, willow, elm, Sophora japonica, toon, and maple are prevalent. The somatosensory comfort provided by experimental sites with these kinds of plants should be considered. The P test site located in the north of BOFP was a TG DPC community structure area composed of Ginkgo biloba and grass. The physical comfort experienced at this sample point was poor, perhaps because Ginkgo biloba is not native to Beijing, so its growth potential is limited, and it lacks large and dense canopies.

4.3. Correlation between Typical Landscape Environment of Green Space and Physical Comfort

The analysis of the relationship between typical landscape environments and the somatosensory comfort provided by green space provided a multi-perspective verification of the analysis in Sections 3.1 and 3.2 The results were similar to those in Section 3.1, indicating that DPC community structure areas offered a high level of somatosensory comfort. Somatosensory comfort in WPC community areas was also high because both shade and ventilation were provided by upper trees [20,26,28,29]. Water bodies with a large surface area have always been a scarce landscape resource in Beijing, on account of the climatic conditions of northern China. The water body in the green space of BOFP is artificial. Such landscape features can only be constructed if the area of green space is large enough and investment is sufficient. Therefore, the conclusion that the WPC community areas provided high somatosensory comfort is actually of limited significance for practical application. However, this finding does show that ventilation is conducive to the improvement of somatosensory comfort and that the right ventilation conditions are derived from "wind guiding" or even "wind making". The scientific basis of this relationship needs to be further studied. Based on the numerical fitting in Section 3.4 combined with the analysis results in Sections 3.1–3.3, the improved thermal comfort provided by DPC community structure areas (i.e., TS, TG, and DBP) deserves careful attention in future research.

The research reported in this paper was carried out over 15 years. With the continuous growth and development of green space vegetation in BOFP, the three-dimensional quantity of greenery per unit of green space area will become larger and larger (until it reaches a stable value). The role of green space vegetation in improving the urban regional thermal microenvironment may become more and more significant. What are the intensity and scope of this UCI effect? The measurement methods used in this study revealed only the spatial differentiation of body comfort in discontinuous and point areas. How can continuous data be obtained? It may be effective to use remote sensing to gather data for interpretation and inversion while carrying out continuous monitoring. The results support the "evidence-based design" of landscape architecture, providing a scientific foundation for the planning and design of functional green spaces that efficiently improve microenvironmental conditions.

5. Conclusions

The above findings lead to suggestions for urban landscape environmental planning, design, and renewal practice. In areas in which the UHI effect is intense, TS and TG community structures with DBP as the dominant species should be adopted to significantly improve human thermal comfort. However, the focal microenvironmental factors show significant temporal and spatial differentiation due to the varying geographic, climatic, and other characteristics of the city in which BOFP is located, such as differences in landscape patterns, species composition, quantity of regional green space vegetation, and patterns of growth. Given the complexity of the research object—urban green space—the study was unable to comprehensively model the spatial composition of green space plant communities and may have ignored or failed to pay attention to important heterogeneous factors. Therefore, the findings require further testing to ensure their reliability and generalisability.

Author Contributions: Conceptualization, L.Z. and J.P.; methodology, J.P.; software, J.P.; validation, L.Z., H.X. and J.P.; formal analysis, J.P.; investigation, L.Z.; resources, J.P.; data curation, J.P.; writing—original draft preparation, L.Z.; writing—review and editing, L.Z.; visualization, J.P.; supervision, H.X.; project administration, J.P.; funding acquisition, L.Z., H.X. and J.P. All authors have read and agreed to the published version of the manuscript.

Funding: This research received the support from the funding of research enhancement project for young scholars(X21046 and X21044), Beijing University of Civil Engineering and Architecture and the national natural foundation of China (51641801).

Institutional Review Board Statement: Not applicable.

Informed Consent Statement: Not applicable.

Data Availability Statement: Not applicable.

Conflicts of Interest: The authors declare no conflict of interest.

References

1. Li, Y.; Zhang, J.; Gu, R. Research on the Relationship between Urban Greening and the Effect of Urban Heat Island. *Chin. Landsc. Archit.* **2004**, *20*, 72–75.
2. Chen, L.D.; Sun, R.H.; Liu, H.L. Eco-environmental effects of urban landscape pattern changes: Progresses, problems, and perspectives. *Acta Ecol. Sin.* **2013**, *33*, 1042–1050. [CrossRef]
3. Nieuwenhuijsen, M.J. New urban models for more sustainable, livable and healthier cities post covid19; reducing air pollution, noise and heat island effects and increasing green space and physical activity. *Environ. Int.* **2021**, *157*, 106850. [CrossRef] [PubMed]
4. Ying, X.; Fang, Z. Thermal environment effects of urban human settlements and influencing factors based on multi-source data: A case study of Changsha city. *Acta Geogr. Sin.* **2020**, *75*, 2443–2458.
5. Zhang, Y.; Yu, Q.; Li, M.; Huang, Y.; Yue, P.; Wang, J. Simulation of Land Surface Temperature in Haidian District Based on EnKF-3DVar Model. *Trans. Chin. Soc. Agric. Mach.* **2017**, *48*, 166–172.
6. Wang, X.; Zhu, Q.; Chen, S.; Liu, X.; Hu, Y. Remote Sensing Retrieval of Water and Heat Fluxes over Urban Green Space and Experimental Validation. *Areal Res. Dev.* **2010**, *29*, 63–66.
7. Chen, Q.; Cheng, Q.; Chen, Y.; Li, K.; Wang, D.; Cao, S. Analysis of the Influence of the Urban Building Sky View Factor on Land Surface Thermal Environment. *Sci. Surv. Mapp.* **2021**, *46*, 148–155.
8. Xie, J.; Cong, R.; Wang, Y.; Duan, M. Spatiotemporal Characteristics of Surface Temperature and Greening Role in Tongzhou District, Beijing. *Chin. Landsc. Archit.* **2021**, *37*, 41–45.
9. Bassett, R.; Janes-Bassett, V.; Phillipson, J.; Young, P.J.; Blair, G.S. Climate driven trends in London's urban heat island intensity reconstructed over 70 years using a generalized additive model. *Urban Clim.* **2021**, *40*, 100990. [CrossRef]
10. Li, Y.; Zhou, D.; Yan, Z. Spatiotemporal Variations in Atmospheric Urban Heat Island Effects and Their Driving Factors in 84 Major Chinese Cities. *Huanjingkexue* **2021**, *42*, 5037–5045.
11. Paranunzio, R.; Dwyer, E.; Fitton, J.M.; Alexander, P.J.; O'Dwyer, B. Assessing current and future heat risk in Dublin city, Ireland. *Urban Clim.* **2021**, *40*, 100983. [CrossRef]
12. Gao, S.; Sha, J.-M.; Shuai, C. Quantitative Study on the Relationship between Land Surface Temperature and Vegetation Coverin Xiamen City. *J. Fujian Norm. Univ.* **2019**, *35*, 14–21.
13. Liu, S.; Xie, M.; Wu, R.; Wang, Y.; Li, X. Influence of the choice of geographic unit on the response of urban thermal environment: Taking Beijing as an example. *Prog. Geogr.* **2021**, *40*, 1037–1047. [CrossRef]
14. Ye, W.T.; Chen, Y.H.; Lu, Y.H.; Wu, P. Spatio-temporal variation of land surface temperature and land cover responses in different seasons in Shengjin Lake wetland during 2000–2019 based on Google Earth Engine. *Remote Sens. Land Resour.* **2021**, *33*, 228–236.
15. Chen, B.; Xu, S.; Yang, D.; Wang, H. Thermal Environmental Effects of Park Landscape of Main Urban Region in Wuhan. *Remote Sens. Inf.* **2021**, *36*, 58–66.
16. Shi, L.; Zhao, M. Cool island effect of urban parks and impact factors in summer: A case study of Xi'an. *J. Arid. Land Resour. Environ.* **2021**, *34*, 154–161.
17. Fu, Y. Research on the Evolution Characteristics and Regulation Mechanism of Urban Thermal Environment and Comfort. Doctoral Dissertation, University of the Chinese Academy of Sciences, Beijing, China, 2020.
18. Zhan, H.; Xie, W.-J.; Sun, H.; Huang, H. Using ENVI-met to simulate 3D temperature distribution in vegetated scenes. *J. Beijing For. Univ.* **2014**, *36*, 64–74.
19. Chen, R. Microclimate Correlation Analysis of Landscape Elements in City Parks. *Landsc. Archit.* **2020**, *27*, 94–99.
20. Zhang, X.; Nie, Q.; Liu, J. Research on urban geothermal comfort improvement strategy based on ENVI-met. *Ecol. Sci.* **2021**, *40*, 144–155.
21. Geng, H.; Wei, X.; Zhang, M.; Li, Q. Influenceofvegetationandarchitecture on microclimate based on Envi-met: A case study of Nanjing Agricultural University. *J. Beijing For.* **2020**, *42*, 115–124.
22. Wang, K.; Xue, S. Correlation Analysis of Buildings, Green Layout and Human Comfort in Summer in Urban Residential Areas: Taking Zhengzhou as an Example in Cold Region. *Build. Sci.* **2021**, *37*, 53–60.
23. Singh, V.K.; Mughal, M.O.; Martilli, A.; Acero, J.A.; Ivanchev, J.; Norford, L.K. Numerical analysis of the impact of anthropogenic emissions on the urban environment of Singapore. *Sci. Total Environ.* **2022**, *806*, 150534. [CrossRef] [PubMed]
24. Lai, J.; Zhan, W.; Quan, J.; Liu, Z.; Li, L.; Huang, F.; Hong, F.; Liao, W. Reconciling Debates on the Controls on Surface Urban Heat Island Intensity: Effects of Scale and Sampling. *Geophys. Res. Lett.* **2021**, *48*, e2021GL094485. [CrossRef]
25. Chen, Y.; Song, S.; Hou, Y. Effect of Vertical Greening in Summer on Human Comfort in Nanjing. *Chin. Landsc. Archit.* **2020**, *36*, 64–69.
26. Zhao, H.Y.; Mao, B. Multi-Scale Optimization on Urban Wind Environment of Changchun City Based on Improved Ventilation and Thermal Comfort. *J. Hum. Settl. West China* **2020**, *35*, 24–32.
27. Pan, J.; Li, S. Study on Spatial Pattern of the Function of Thermal Comfort Improvement on Beijing City Parks. *Chin. Landsc. Archit.* **2015**, *37*, 91–95.

28. National Standard of the People's Republic of China. Ergonomics of the Thermal Environment: Analytical Determination and Interpretation of Thermal Comfort Using Calculation of the PMV and PPD Indices and Local Thermal Comfort Criteria. GB/T 18049–2017. Available online: https://openstd.samr.gov.cn/bzgk/gb/index (accessed on 27 January 2023).
29. Liu, K.; Ma, C.; Chen, W.; Liu, B. Exploration of Waterfront Landscape Planning and Design for Thermal Comfort in Microclimate. *Landsc. Archit.* **2020**, *27*, 104–109.

 *sustainability*

Article

Spatiotemporal Variations of Air Pollution during the COVID-19 Pandemic across Tehran, Iran: Commonalities with and Differences from Global Trends

Mohsen Maghrebi [1,*], Ali Danandeh Mehr [2,3,*], Seyed Mohsen Karrabi [4], Mojtaba Sadegh [5], Sadegh Partani [6], Behzad Ghiasi [1] and Vahid Nourani [3,7]

1   School of Environment, College of Engineering, University of Tehran, Tehran 1417853111, Iran
2   Civil Engineering Department, Antalya Bilim University, Antalya 07190, Turkey
3   Centre of Excellence in Hydroinformatics, Faculty of Civil Engineering, University of Tabriz, Tabriz 51666, Iran
4   Department of Civil Engineering, Faculty of Engineering, Ferdowsi University of Mashhad, Mashhad 9177948974, Iran
5   Civil Engineering Department, Boise State University, Boise, ID 83725, USA
6   Civil Engineering Department, University of Bojnord, Bojnord 9453155111, Iran
7   Faculty of Civil and Environmental Engineering, Near East University, Nicosia 99010, Cyprus
*   Correspondence: maghrebi.mohsen@ut.ac.ir (M.M.); ali.danandeh@antalya.edu.tr (A.D.M.)

Abstract: The COVID-19 pandemic has induced changes in global air quality, mostly short-term improvements, through worldwide lockdowns and restrictions on human mobility and industrial enterprises. In this study, we explored the air pollution status in Tehran metropolitan, the capital city of Iran, during the COVID-19 outbreak. To this end, ambient air quality data (CO, NO_2, O_3, PM_{10}, SO_2, and AQI) from 14 monitoring stations across the city, together with global COVID-19-related records, were utilized. The results showed that only the annual mean concentration of SO_2 increased during the COVID-19 pandemic, mainly due to burning fuel oil in power plants. The findings also demonstrated that the number of days with a good AQI has significantly decreased during the pandemic, despite the positive trend in the global AQI. Based on the spatial variation of the air quality data across the city, the results revealed that increasing pollution levels were more pronounced in low-income regions.

Keywords: COVID-19; air pollution; Tehran; AQI

Citation: Maghrebi, M.; Danandeh Mehr, A.; Karrabi, S.M.; Sadegh, M.; Partani, S.; Ghiasi, B.; Nourani, V. Spatiotemporal Variations of Air Pollution during the COVID-19 Pandemic across Tehran, Iran: Commonalities with and Differences from Global Trends. *Sustainability* **2022**, *14*, 16313. https://doi.org/10.3390/su142316313

Academic Editors: José Carlos Magalhães Pires and Álvaro Gómez-Losada

Received: 14 October 2022
Accepted: 4 December 2022
Published: 6 December 2022

Publisher's Note: MDPI stays neutral with regard to jurisdictional claims in published maps and institutional affiliations.

1. Introduction

Urban air pollution is known as a major human health challenge [1,2]. According to the World Health Organization (WHO), approximately seven million premature deaths occur annually across the globe due to air pollution [3]. While air quality has improved substantially in the US and many developed countries, unhealthy levels of air pollution remain a daunting challenge in many developing countries and are expected to worsen in some regions owing to a variety of natural and anthropogenic sources [4–7]. The lack of a comprehensive decision-making system, old public transportation fleets, inefficient planning and management in urban environments in the face of high population density, and inadequate financial means are the main barriers to achieving or maintaining clean air in developing countries [8–10]. Iran, rich in oil and gas resources, suffers greatly from the complex challenges caused by air pollution [4,11,12] and endures more than 49,000 air pollution-related deaths annually [13]. The air pollution status in densely populated areas, especially the capital city of Tehran, is at a critically concerning level. The main driving factors for the excessive air pollution in this metropolitan city are (i) special topography that allows the confinement of pollutants over the city: Tehran is engulfed from three sides by mountains i.e., Shemiran hills in the north, Damavand hills in the east, and Karaj hills in the west [14]; (ii) dry climatic and stagnant air conditions: there is no noticeable rainfall during half of a year, and winds usually do not have the necessary power to move pollution out of the urban area, with 70% of the winds having a speed of less than 3 m/s [15]; (iii) high population density: the population density in Tehran, 11,969 people per square kilometer, is more than 180 times higher than Iran's average [16]; (iv) excessive daily trip frequency: more than 20 million daily trips occur in the city and there are currently more than 4 million vehicles and 3 million motorcycles in use, twice of its ecological capacity [17]; and (v) old and inefficient vehicles: 71% of Tehran's pollution is due to mobile pollution sources. This undesirable air condition is the main cause of 4800 annual deaths and health costs exceeding 2 billion U.S. dollars per year [1,4].

The extreme outbreak of COVID-19 occurred in the presence of significant air quality challenges in Tehran. Lockdown policy and urban activity restrictions have the potential to reduce urban transportation and ultimately help improve air quality [18–22]. Different nations, however, with different economic and social conditions responded differently to the COVID-19 pandemic, which translated to various levels of change in the urban air quality status, with significant implications for environmental justice issues [23–28]. Therefore, despite the national stay-at-home orders during the COVID-19 outbreak, albeit infrequent, some residents of Tehran had to continue their daily activities to survive. To understand how different factors affected Tehran's air quality during the COVID-19 pandemic, this study investigates Tehran's air pollution status using daily records from 14 monitoring stations across this metropolitan area and compares it with global COVID-19-related records. The results are of paramount importance to show how different socioeconomic factors affect urban air quality.

2. Materials and Methods

Tehran, the capital of Iran, is located at $35°40'$ North and $51°19'$ East with an average altitude of 1190 m above sea level. Figure 1 demonstrates the 22 municipal regions of Tehran and the location of meteorological stations distributed across the city. This city has a length of approximately 27 km from north to south and 50 km from east to west. Figure 2 illustrates the main meteorological features of the city during the COVID-19 pandemic. This vibrant metropolitan area is home to more than 8.5 million people (10% of Iran's total population and 30% of Iran's urban population). Tehran's population increases by almost three million during the day as residents from surrounding communities move to Tehran for work and personal business [16]. This mega city alone accounts for more than 20% of total energy consumption in Iran [29]. One of the most important features of this city is its self-purification capacity due to its vast green space, a capacity that is overpowered by the city's power plants and the immense number of vehicles. Tehran has 2277 orchards with a total area of 5949 hectares (Figure S1). In addition, green space along Tehran's urban roads accumulates to 8253 hectares and the green belt around the city has an area of 42,855 hectares. Around Tehran, there are five fossil fuel power plants, including the Montazer Ghaem and Tarasht power plants in the west, and the Rey, Besat, and Parand power plants in the south of the city, with a capacity of 3000 megawatts (MW). The main power generation capacity is mainly located in the south of the city [30].

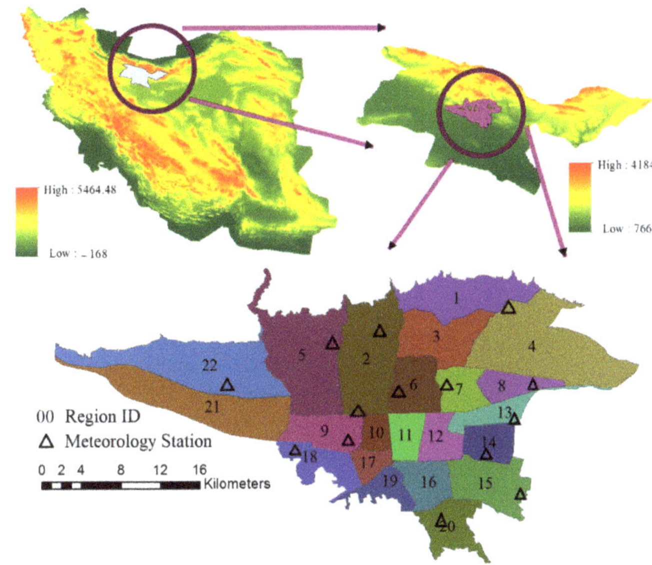

Figure 1. Study area and location of monitoring stations.

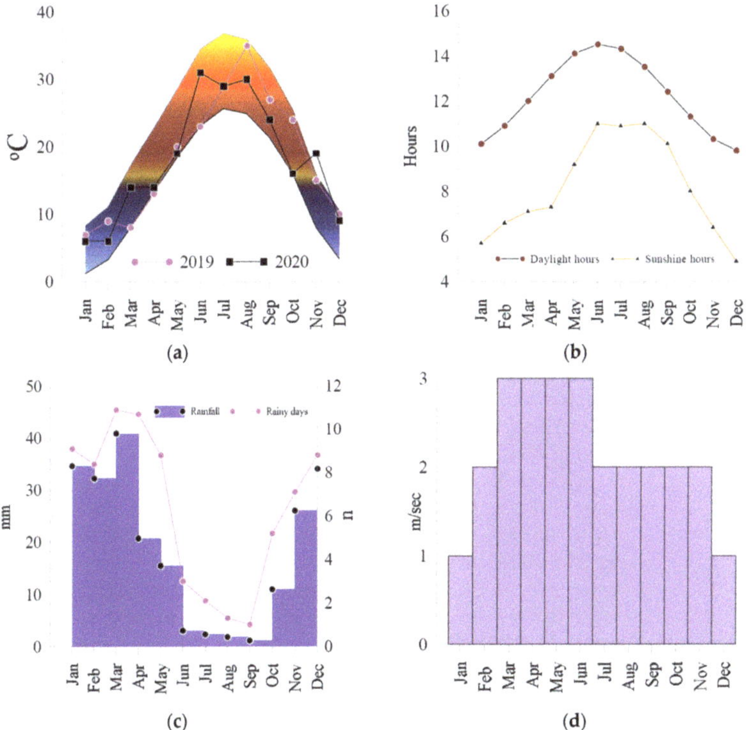

Figure 2. Main meteorological variables at Tehran synoptic station during the COVID-19 pandemic, (**a**) maximum and minimum mean temperature with mean temperature, (**b**) mean daylight and sunshine hours, (**c**) mean monthly rainfall and number of rainy days, and (**d**) mean wind speed.

Air Quality Indices

Ambient air quality data for critical pollutants used in this study include carbon monoxide (CO), nitrogen dioxide (NO_2), coarse particulate matter (PM_{10}), ozone (O_3), sulfur dioxide (SO_2), as well the as air quality index (AQI). CO is an odorless toxic pollutant in the air that is largely produced from fossil fuel combustion [31]. Long-term exposure to CO can reduce blood oxygen-carrying capacity and cause neurological and cardiovascular diseases in humans [32]. NO is a gaseous air pollutant that is formed when fossil fuels such as oil or coal are burned at high temperatures. In the atmosphere, it turns into NO_2 through a reaction with O_3. Public and private cars are the largest source of this pollutant, followed by power plants, heavy equipment, and other mobile engines and industrial boilers [33,34]. Long exposure to NO_2 can cause bronchitis, asthma attacks, and phlegm [35]. Tehran's NO_2 usual concentration is 85 ppb [1]. PM_{10} can cause serious environmental problems and certain types of cancer [36,37]. O_3 is produced by chemical reactions between nitrogen oxides and volatile organic compounds in the presence of sunlight and heat. Therefore, the possibility of increasing the O_3 levels to unhealthy levels on sunny and hot days is high. Ground-level O_3 is a grave environmental hazard [38]. People with asthma, children, older adults, and people with reduced intake of certain nutrients, such as vitamins C and E, are at greater risk from O_3 exposure. SO_2 is a colorless, bad-smelling, toxic gas that is emitted by the burning of fossils. Diesel vehicles and equipment (i.e., power plants) have long been a major source of SO_2 [39]. SO_2 can harm the human respiratory system and make breathing difficult [40]. At high concentrations, gaseous SO_2 can harm trees and plants by damaging foliage and decreasing growth. The AQI is a unitless measure of air quality that runs from 0 to 500. An AQI of 100 is usually associated with the regulatory pollutant level. The higher the AQI value, the greater the health threat from air pollution and the higher the environmental hazards [41]. To attain the AQI at each air quality monitoring station, first, the pollutant index (Ip) is calculated individually for each ambient pollutant using Equation (1) [42]. Then, the maximum acquired Ip is considered as the station's AQI.

$$Ip = \frac{I_{Hi} - I_{Lo}}{BP_{Hi} - BP_{Lo}}(C_p - BP_{Lo}) + I_{Lo} \tag{1}$$

where C_p is the truncated concentration of pollutant p ($\mu g/m^3$), BP_{Hi} is the concentration breakpoint that is greater than or equal to Cp, BP_{Lo} is the concentration breakpoint that is less than or equal to Cp, I_{Hi} is the AQI value corresponding to BP_{Hi}, and I_{Lo} is the AQI value corresponding to BP_{Lo}. It is required to mention that O_3 should be truncated to three decimal places, CO should be truncated to one decimal place, and SO_2, PM_{10}, and NO_2 should be truncated to an integer number. For details on pollutant-specific breakpoints/truncations and examples of the AQI calculation/classification procedure, the interested reader is referred to technical assistance on the AQI [42].

All the data were collected from 14 atmospheric monitoring stations in Tehran between 1 January 2019, and 30 December 2020, provided by the Air Quality Control Center of the Tehran municipality. COVID-19-related data was acquired from the "Our World in Data" website (https://ourworldindata.org/coronavirus/country/iran?country=~IRN, accessed on 21 April 2021). It should be noted that COVID-19 national cumulative data is published daily, and spatial classification of these data did not exist at the time of this study. To spatially interpret the data and estimate the values between measurements, the deterministic reverse distance weighting interpolation method was used. In this method, the effect of a known data point is inversely related to the distance from the location being estimated [9,42]. In this study, the most recent published values of the ambient air quality indices have been used as the baseline (usual) concentration. Finally, the well-known t-test was implemented to discover how significant the differences between the indices before and during COVID-19 are.

3. Results and Discussion

In Iran, the official onset date of COVID-19 was announced as February 19, 2020 (in the city of Qom near Tehran). On the same day, the Iranian government raised the COVID-19 alert to yellow [43]. The number of confirmed cases increased slightly from mid-March to early July followed by a slight decrease in August; since September, with the arrival of rapid diagnosis kits from South Korea, the number of confirmed cases escalated. To reduce the propagation of the illness, some of the restrictions imposed on daily life and social activities, such as the travel ban policy and prevention of intercity transportation, positively affected the air quality of big cities in Iran [44]. Referring to Tehran, the air pollution related to the travel ban policy was reduced significantly in 2020 [45]. A pioneering study on the effect of changes in traffic flow patterns across Iran has demonstrated that each one million intercity traffic was associated with a 3% increase in COVID-relevant mortality with a time lag of five weeks [46].

3.1. Carbon Monoxide (CO)

The concentration of CO in 2015 was considered here as a usual concentration in different districts of Tehran, which varied between 26.6 to 32.6 ppm [9]. The mean annual CO concentration in 2020 decreased to 25.2 ppm from 27.7 ppm in 2019, which is not significant (the significancy of variation is discussed later in Section 3.7), both of which are lower than the usual concentration. In 2020, the range of change in CO concentration was lower than its preceding at all stations (Figure S2). The highest inter-annual range of change in CO concentration in 2020 was 91 ppm (station 14), which was lower than that in 2019 at 157 ppm (station 13). The minimum and maximum CO concentrations in 2020 were also lower than in 2019. Minimum CO concentrations of 3 ppm and 4 ppm were measured at stations 5 and 6 in April 2020 and September 2019, respectively, whereas its maximum concentration levels were 99 ppm and 163 ppm in October 2020 and November 2019 at stations 14 and 13, respectively. The CO concentration only changed significantly in November and December (p-value < 0.05), and we cannot prove any significant trends in other months. Traditionally, the CO concentration is higher in winter and cold weather than in the rest of the year [47], a phenomenon that was observed in both years. The main reason for this is the confinement of pollution due to the special topography of Tehran, and temperature. Despite the 2020 average CO concentration levels across all stations being lower than that of 2019, there was significant heterogeneity among stations without a clear trend that holds for all (Figure 3). Only in November 2020 did Tehran witness a steady reduction in CO concentration at all stations, compared to the same month in 2019, which is attributed to the enforcement of strict traffic restrictions as the confirmed deaths due to COVID-19 soared. The southwestern part of the city had a higher concentration of CO from March to September 2020 compared to the previous year. A similar pattern was observed in the center and northwest of Tehran but during October and December. We highlight a decrease in CO concentration in the west and southwest of Tehran after the outbreak of COVID-19 compared to previous months.

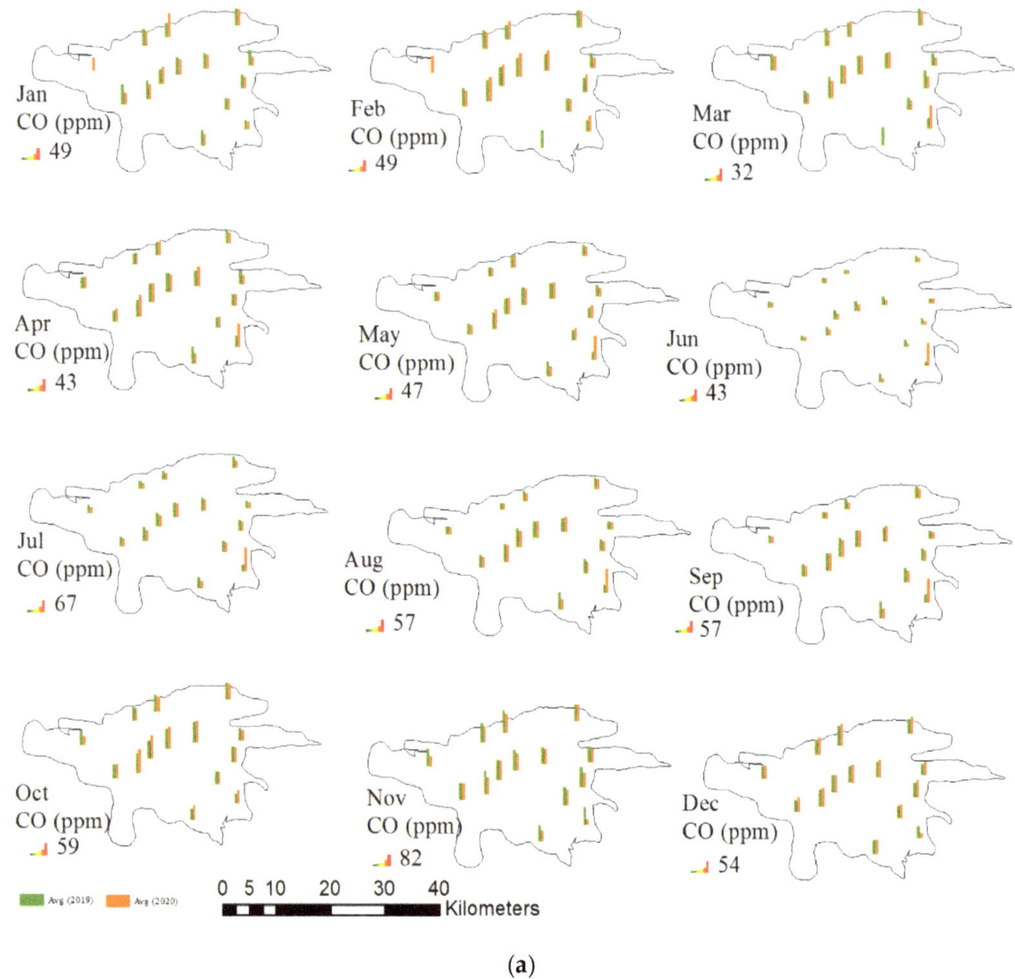

(**a**)

Figure 3. *Cont.*

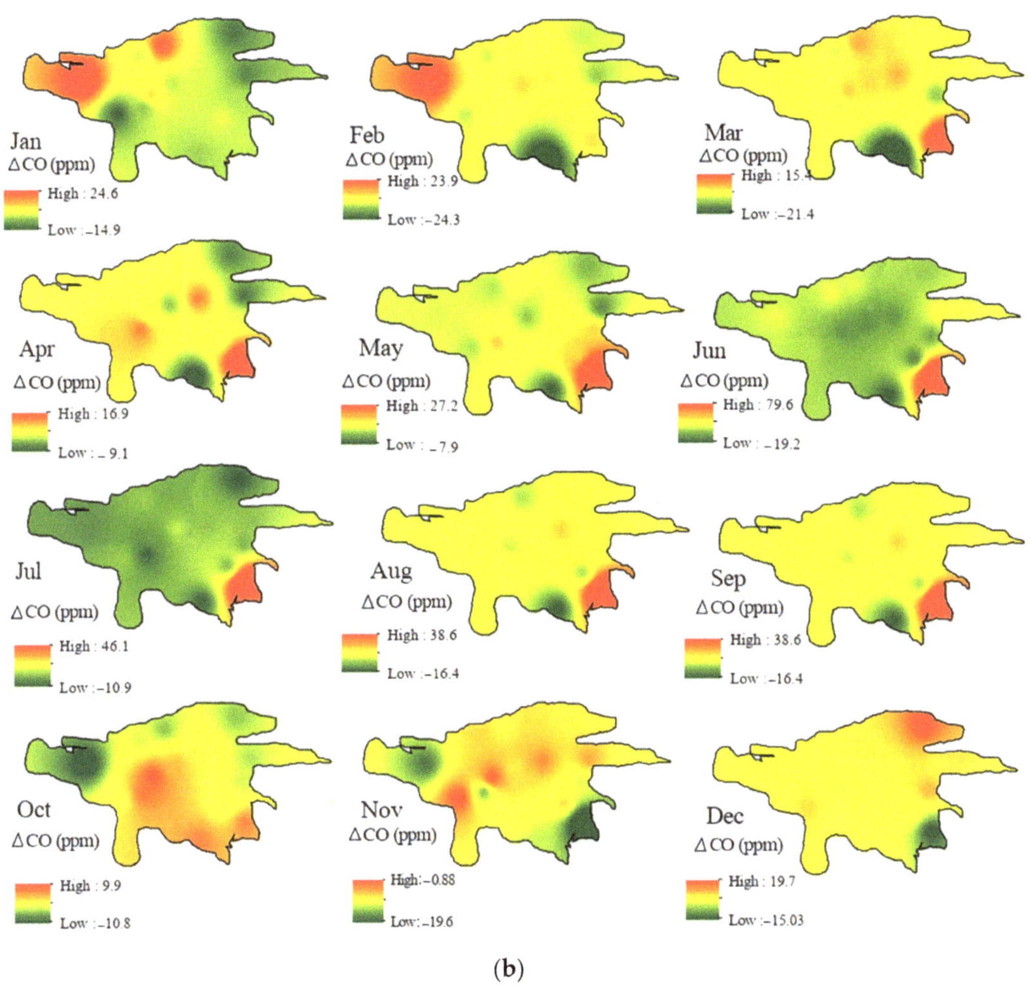

(b)

Figure 3. (**a**) Distribution map of maximum, minimum and mean of monthly CO concentration at each station and (**b**) spatial distribution of variations in the mean monthly CO concentration in 2020 compared to 2019.

3.2. Nitrogen Dixide (NO₂)

The mean annual NO_2 concentrations in 2020 and 2019 were 71.7 ppb and 71.8 ppb, respectively (Figure S3), which although being lower than the usual concentration level, are about 1.7 times higher than the standard threshold recommended by the World Health Organization (WHO). The mean monthly concentration decreased significantly only in March, April, and December (p-value < 0.05). The maximum NO_2 level in 2020 (2019) was 114 ppb (142 ppb), which was observed at station 8 (6) in December (November). The minimum NO_2 concentration in 2020 (2019) was 13 ppb (5 ppb), which was observed at station 12 (3) in August (March). The average NO_2 concentration was 85.9 ppb in 2020 as compared to 88.1 ppb in 2019 (Figure 4). While other studies show a sharp decrease in NO_2 concentrations globally due to the COVID-19 outbreak [48–51], a significant decrease in NO_2 was not observed in Tehran. Furthermore, changes in NO_2 concentrations follow a non–uniform pattern. Most increases in NO_2 concentrations from 2019 to 2020 were seen in the west, south, and southwest of Tehran, where the main out-of-town passenger centers and main city terminals are located. The highest monthly increase was observed in August and September when the highest intercity travels traditionally occur. The maximum increase in NO_2 levels occurring in these months could be a sign of a change in travel patterns and a higher rate of intercity public transport use—such as buses—in the shadow of the COVID-19 outbreak when the government prohibited the entrance of cars with out-of-province license plates into Tehran.

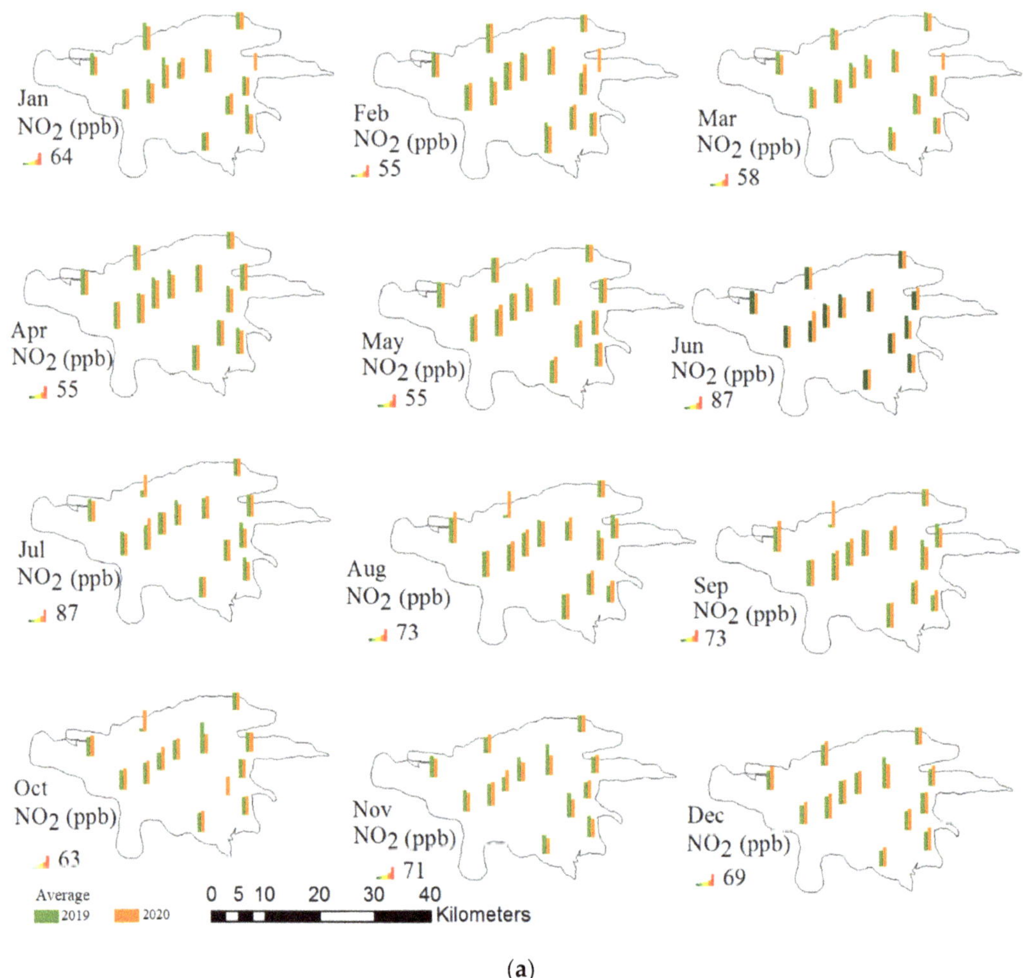

(a)

Figure 4. *Cont.*

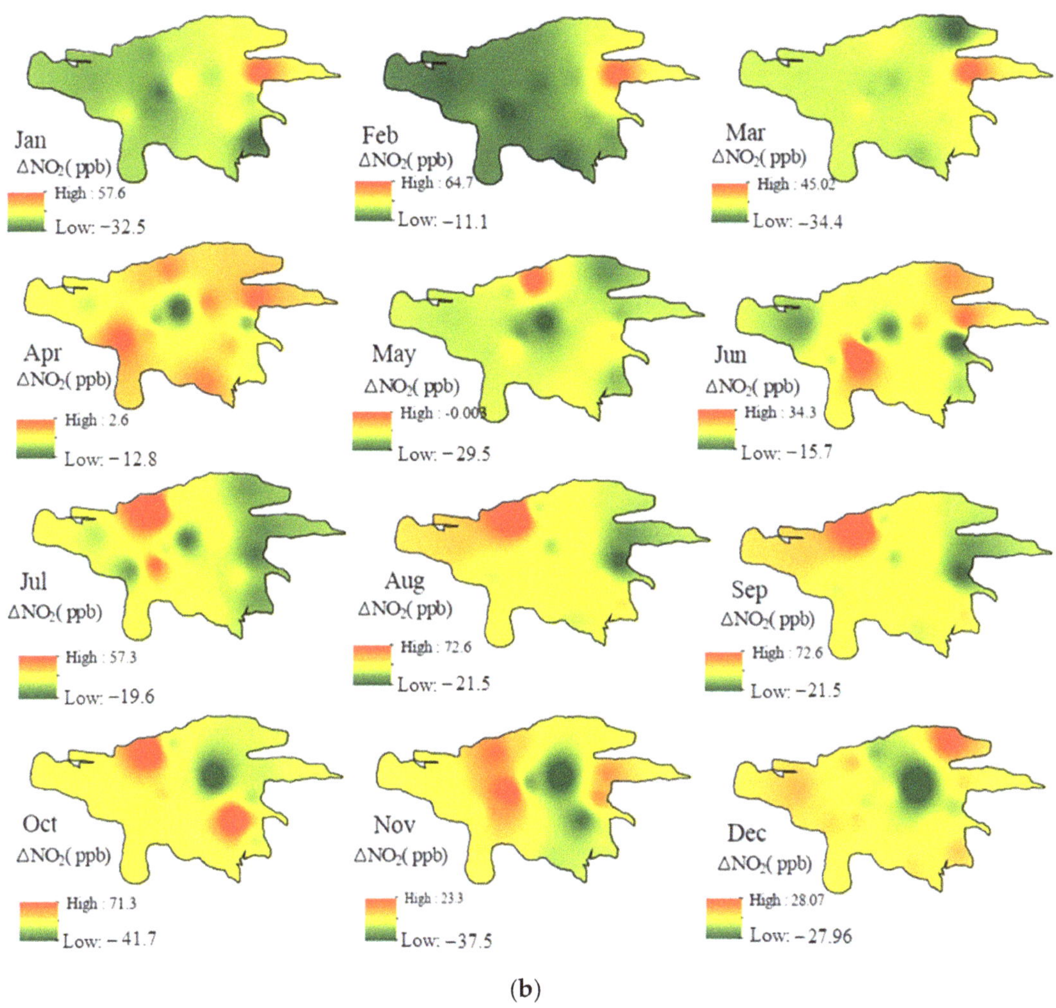

(b)

Figure 4. (**a**) Distribution map of mean monthly NO_2 concentration at each station and (**b**) spatial distribution of variations in the mean monthly NO_2 concentration in 2020 compared to 2019.

3.3. Coarse Particulate Matter (PM$_{10}$)

PM_{10} is one of the leading driving factors of air pollution-caused deaths in Tehran. The contribution of traffic flow emissions to particular matter concentrations and their negative impact on human health has been explored in many studies [52–54]. In Tehran, PM_{10}'s usual concentration is 90.6 $\mu g/m^3$ [1]. The mean annual PM_{10} concentration in 2020 decreased from 56.7 $\mu g/m^3$ to 55.8 $\mu g/m^3$ in 2019, which is not a significant reduction (p-value > 0.05) (Figure S4). This indicates that PM_{10} in Tehran was approximately three times higher than the standard threshold recommended by WHO (annual mean of 15 $\mu g/m^3$) [55]. In 2020 (2019), the highest concentration of PM_{10} was 139 $\mu g/m^3$ (204 $\mu g/m^3$) at station 7 (4) and in December (August). In addition, in 2020 (2019), the minimum concentration of PM_{10} was 5 $\mu g/m^3$ (8 $\mu g/m^3$) at station 2 (2), and it was observed in February (March). The results showed an increase in the mean PM_{10} concentration in six stations (Stations 5, 7, 8, 9, 11, and 13) in 2020 compared to 2019, especially in March, June, July, and November (p-value < 0.05). In general, the lowest PM_{10} concentrations were observed in February and March, whereas the highest levels were observed in September and October. This pattern may be a sign of stability in the atmospheric boundary condition and is consistent with the behavior presented by other researchers [56] (Figure 5).

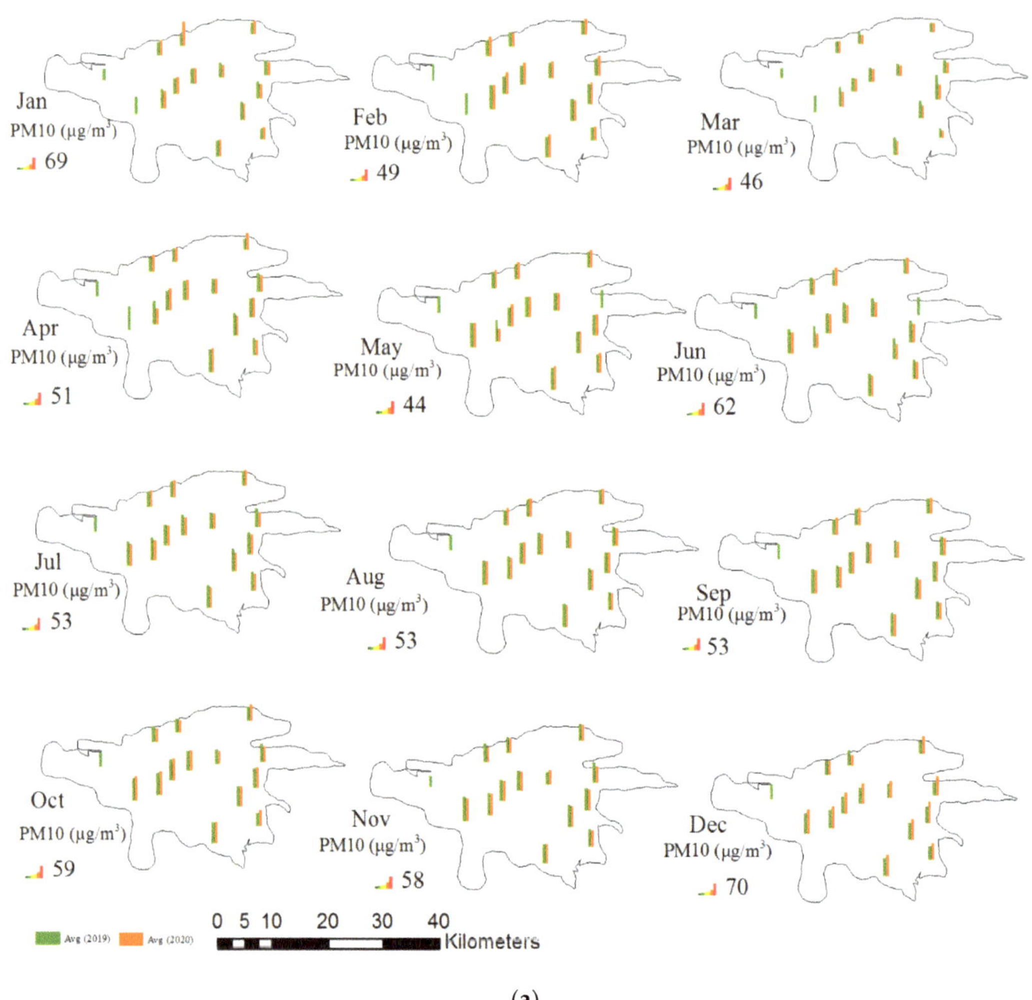

(**a**)

Figure 5. *Cont.*

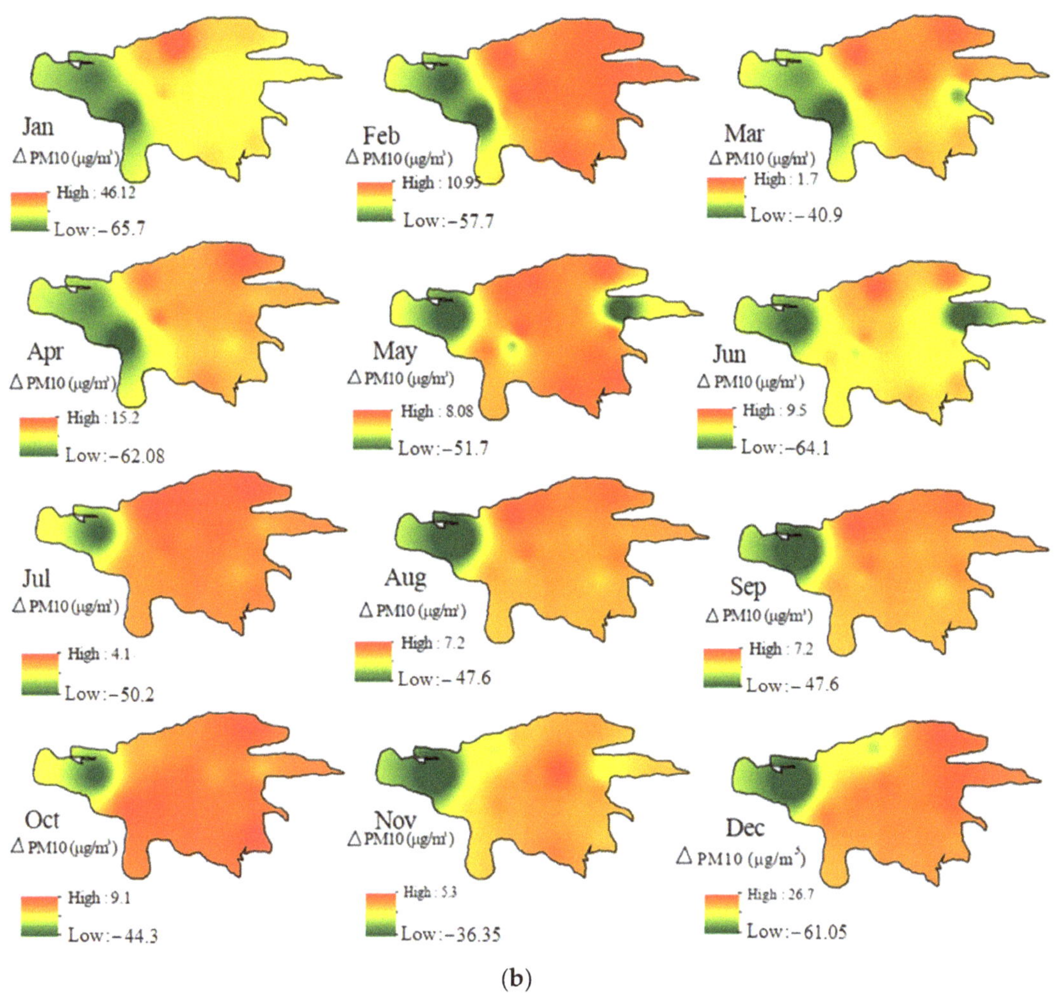

(b)

Figure 5. (**a**) Distribution map of the mean monthly PM$_{10}$ concentration at each station and (**b**) spatial distribution of variations in the mean monthly PM$_{10}$ concentration in 2020 compared to 2019.

3.4. Ozone (O$_3$)

Tehran's usual O$_3$ concentration is 68.8 ppb [1]. The mean O$_3$ concentration in 2020 compared to 2019 has increased from 42.4 ppb to 45.7 ppb, which is not a significant change (*p*-value > 0.05). The maximum O$_3$ concentration in 2020 (2019) was 235 ppb (204 ppb) and seen at station 2 (7) in July (June) (Figure 6). The minimum O$_3$ level is 2 ppb in both 2019 and 2020 and can be seen in several stations during November and December. Maximum and minimum O$_3$ concentrations are expected to occur in summer and winter due to temperature and solar radiation levels in these seasons (Figure S5). Therefore, it can be concluded that O$_3$ concentration changed significantly in June, August, September, and October (*p*-value < 0.05).

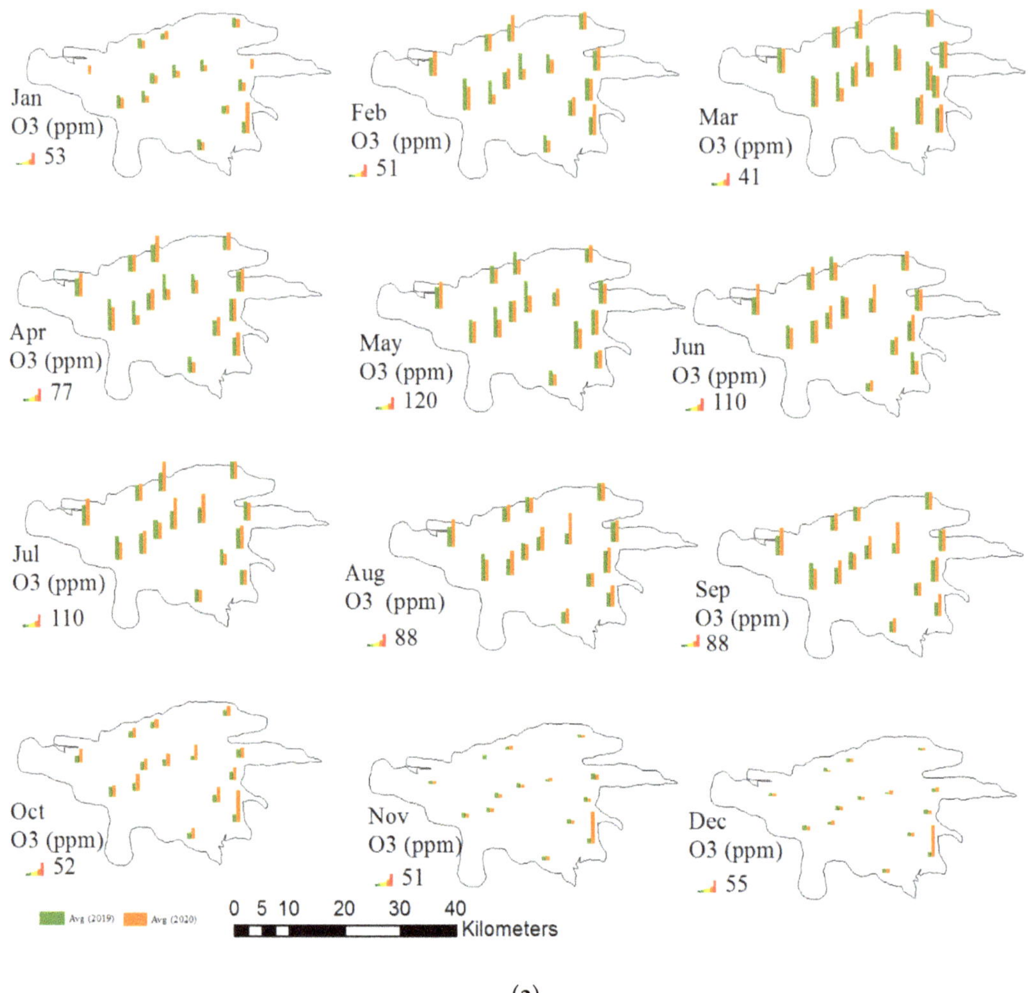

(**a**)

Figure 6. *Cont.*

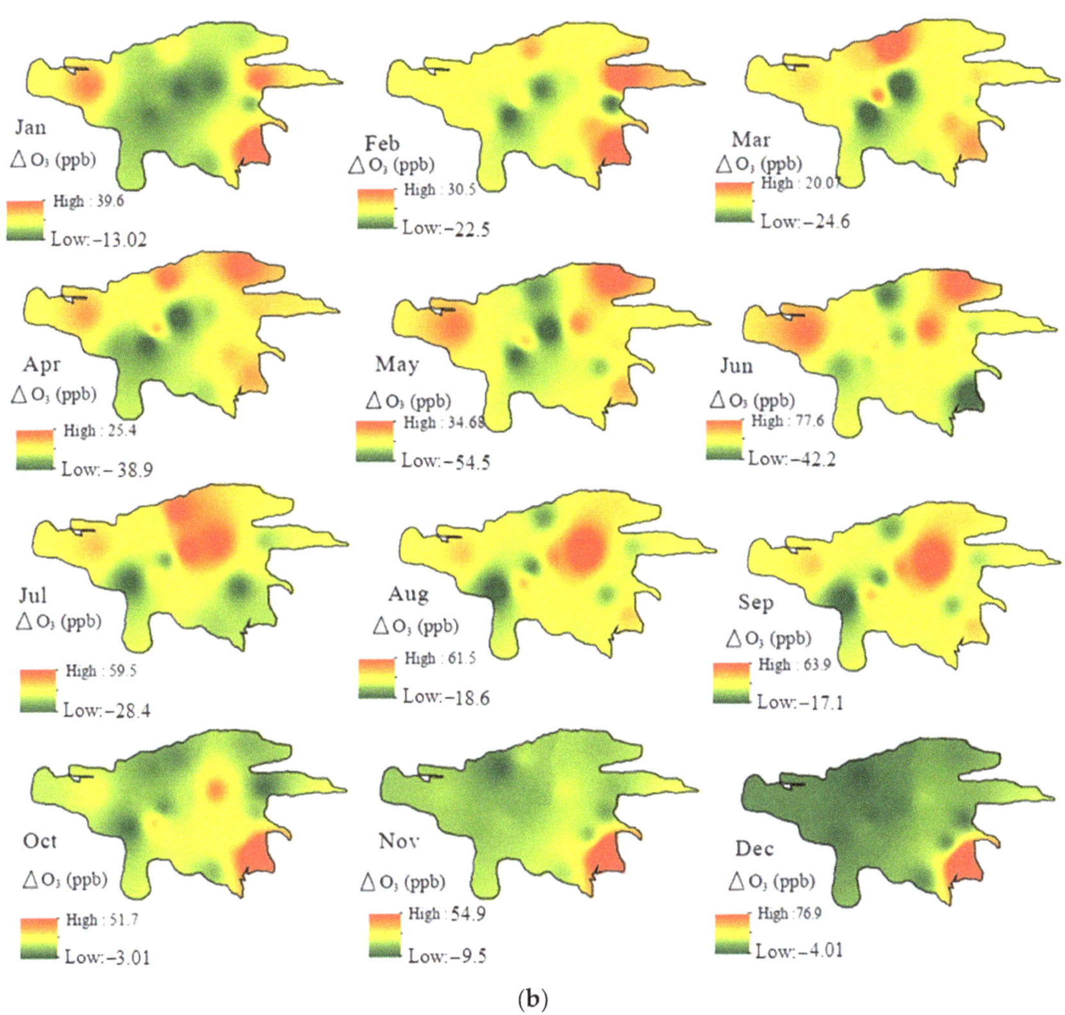

Figure 6. (**a**) Distribution map of mean monthly O_3 concentration at each station and (**b**) spatial distribution of variations in the mean monthly O_3 concentration in 2020 compared to 2019.

3.5. Sulfur Dioxide (SO₂)

Tehran's usual SO_2 concentration is 8.9 ppb, and its highest levels are usually seen in winter [1]. The mean annual SO_2 concentration increased from 7.4 ppb to 17.5 ppb from 2019 to 2020 (p-value < 0.05). The maximum SO_2 concentration in 2020 (2019) was 143 ppb (34 ppb) at station 7 (7) and in November (January) (Figure 7). The minimum SO_2 level ranges between 2 ppb to 3 ppb in 2019 and 2020 between January and March. In all stations, the maximum and mean monthly concentration increased in 2020 compared to 2019 (Figure 7). A major change in the source of fuel in Tehran's power plants is the main culprit of this pollution increase. In the cold winter of 2020, due to the need to maintain natural gas in home networks, the fuel of power plants was changed from natural gas to fuel oil. This decision, which made its way to the press a few months later, became a source of public controversy, and the media called it a wrong decision and a threat to public health. It is noteworthy that, in the northern region of the city, the SO_2 concentration decreased in all months after the COVID-19 outbreak. The wealthiest population of Tehran inhabits this region and it is not close to any power plants. In parts of the west and south of Tehran—where the power plants are located—we see an increase in SO_2 concentration (Figure S6). These regions house working-class and low-income communities, which suffered increasing air pollution during the pandemic when most of the global population benefited from cleaner air due to stay-at-home orders.

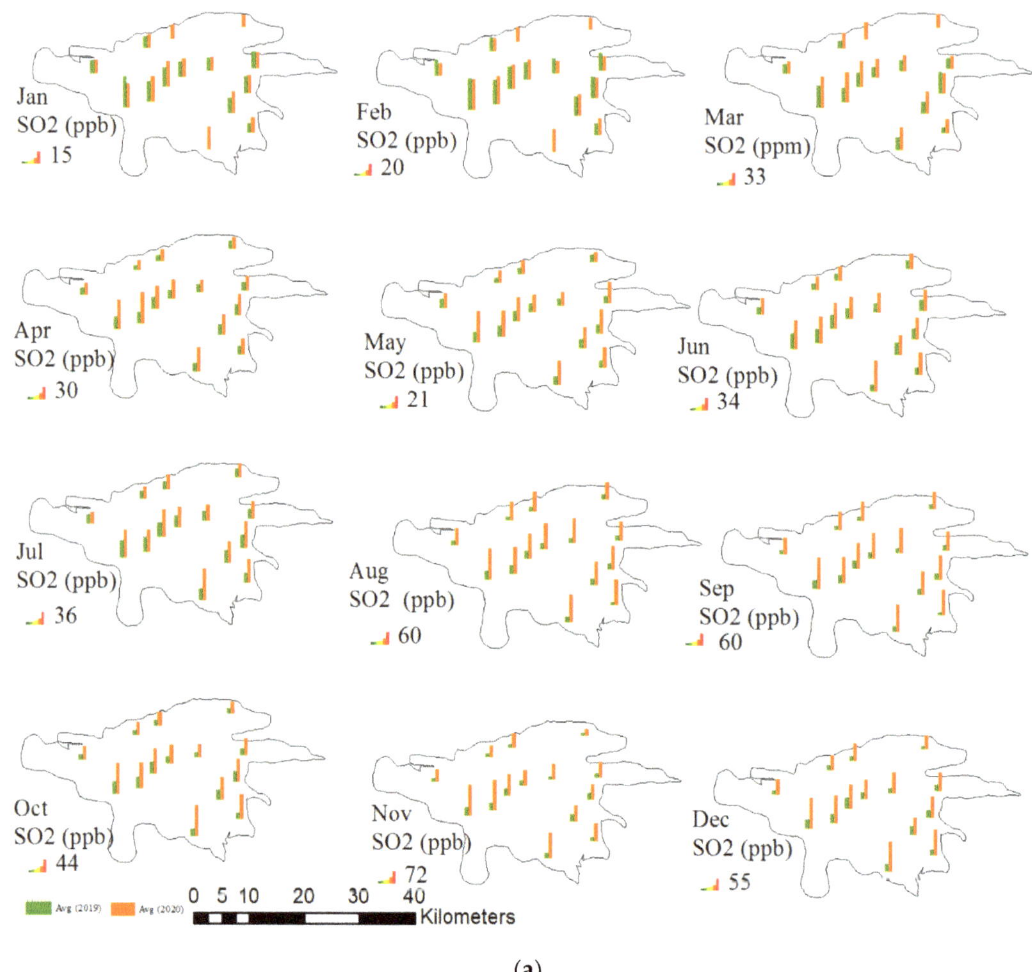

(a)

Figure 7. *Cont.*

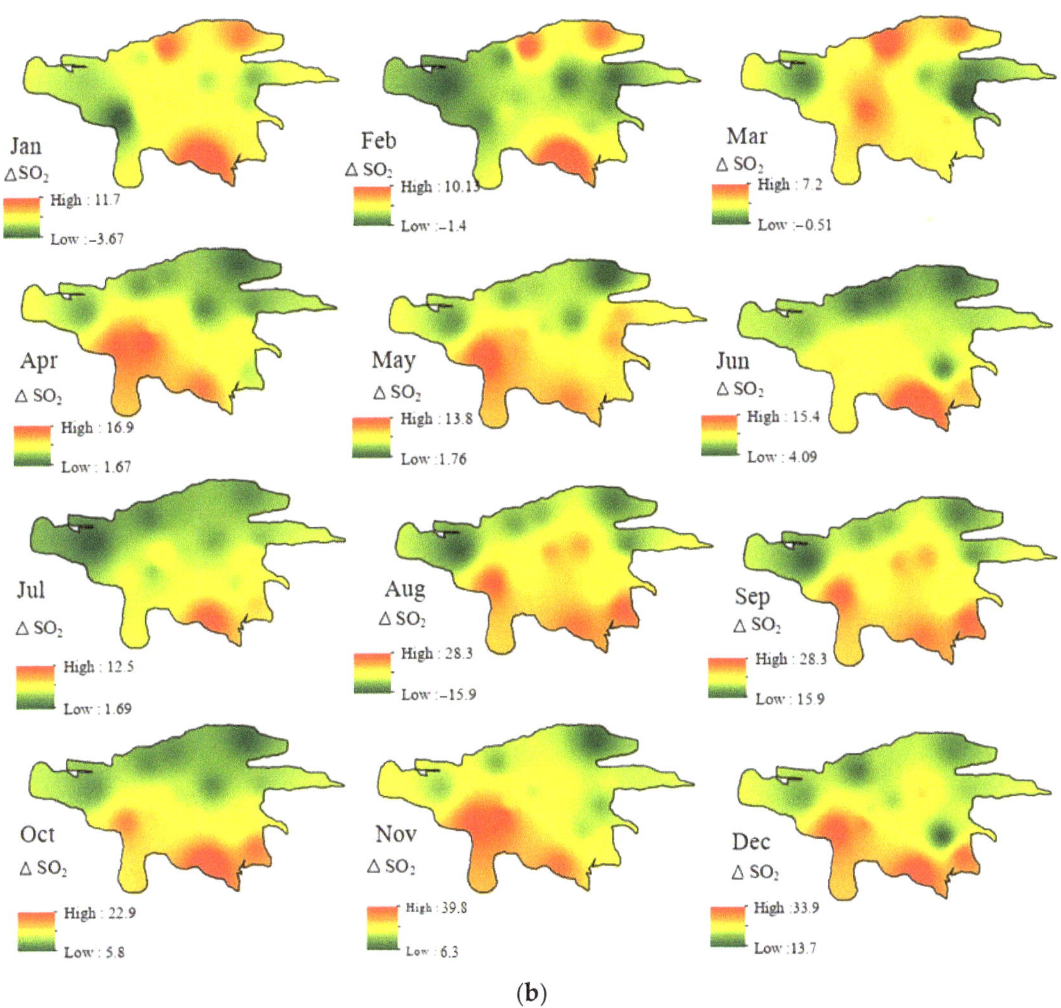

(b)

Figure 7. (**a**) Distribution map of mean monthly SO_2 concentration at each station and (**b**) spatial distribution of variations in the mean monthly SO_2 concentration in 2020 compared to 2019.

3.6. Air Quality Index—AQI

Figure 8 compares Tehran's AQI among 14 stations in 2019 and 2020. For example, station 1 witnessed 60 days of good AQI in both years. In 2020 (2019), the maximum AQI index was 235 (204), which occurred at station 2 (12) and was observed in May (June). In addition, its minimum was equal to 15 (6) and was observed at station 4 (11) and in February (January) (Figure S6). In all stations, the mean and maximum AQI significantly increased in 2020 compared to 2019 (p-value < 0.05) (except for stations 7, 9, 12, and 13, whose maximum decreased in 2020). This trend is contrary to the globally observed trends that showed improved air quality in 2020 [57]. In the rich northern region of Tehran, in all months the AQI decreased, and the air condition improved, a trend that is comparable to the rest of the globe. In the northeastern parts, however, the AQI increased and the air quality deteriorated. In almost all stations, we see an increase in the AQI in the intervals between stay-at-home orders. This can be attributed to the accumulation of transportation needs after periods of quarantine (Figure 9). While in 2019 the number of good days was

58, it was reduced to 30 days in 2020. On the other hand, the number of moderate days marginally increased from 202 in 2019 to 214 in 2020 (Figure 8).

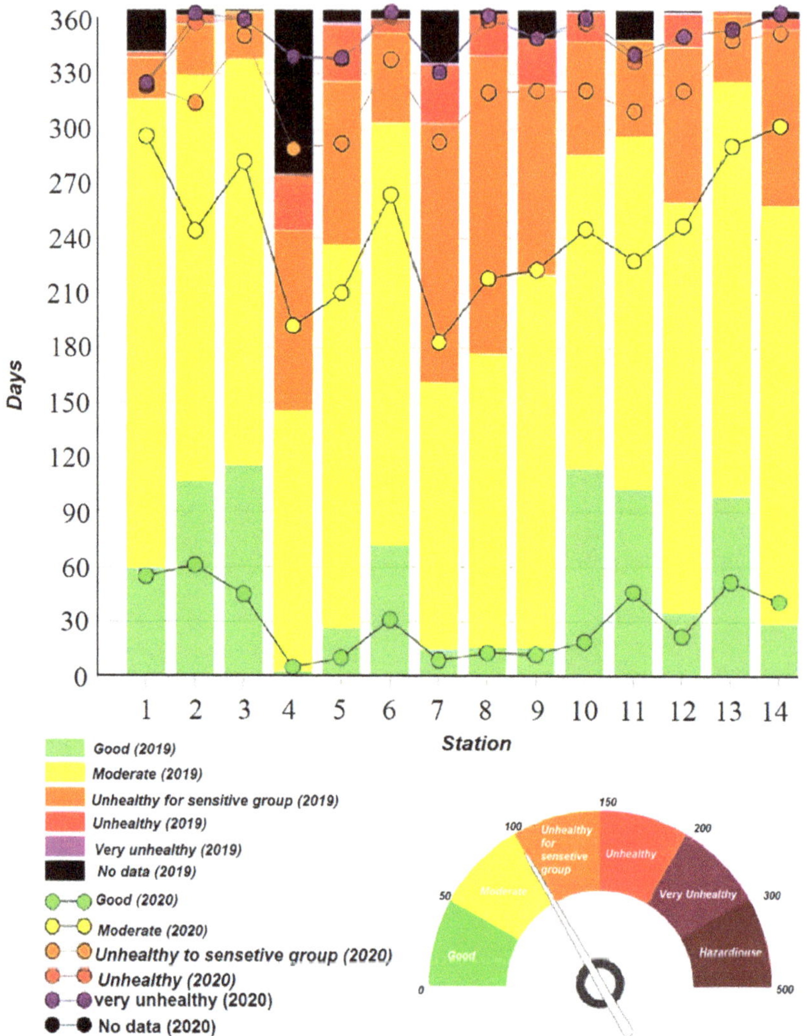

Figure 8. AQI index among investigated stations in 2019 and 2020.

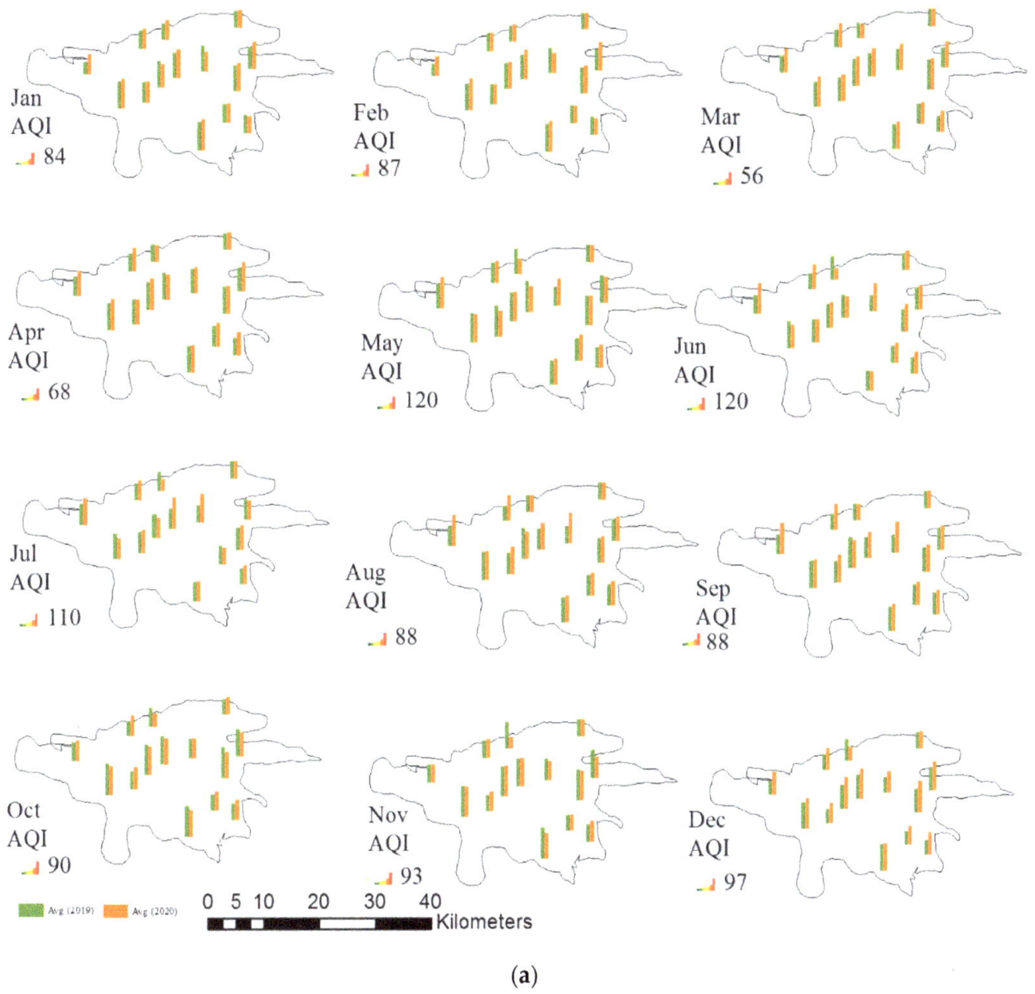

(**a**)

Figure 9. *Cont.*

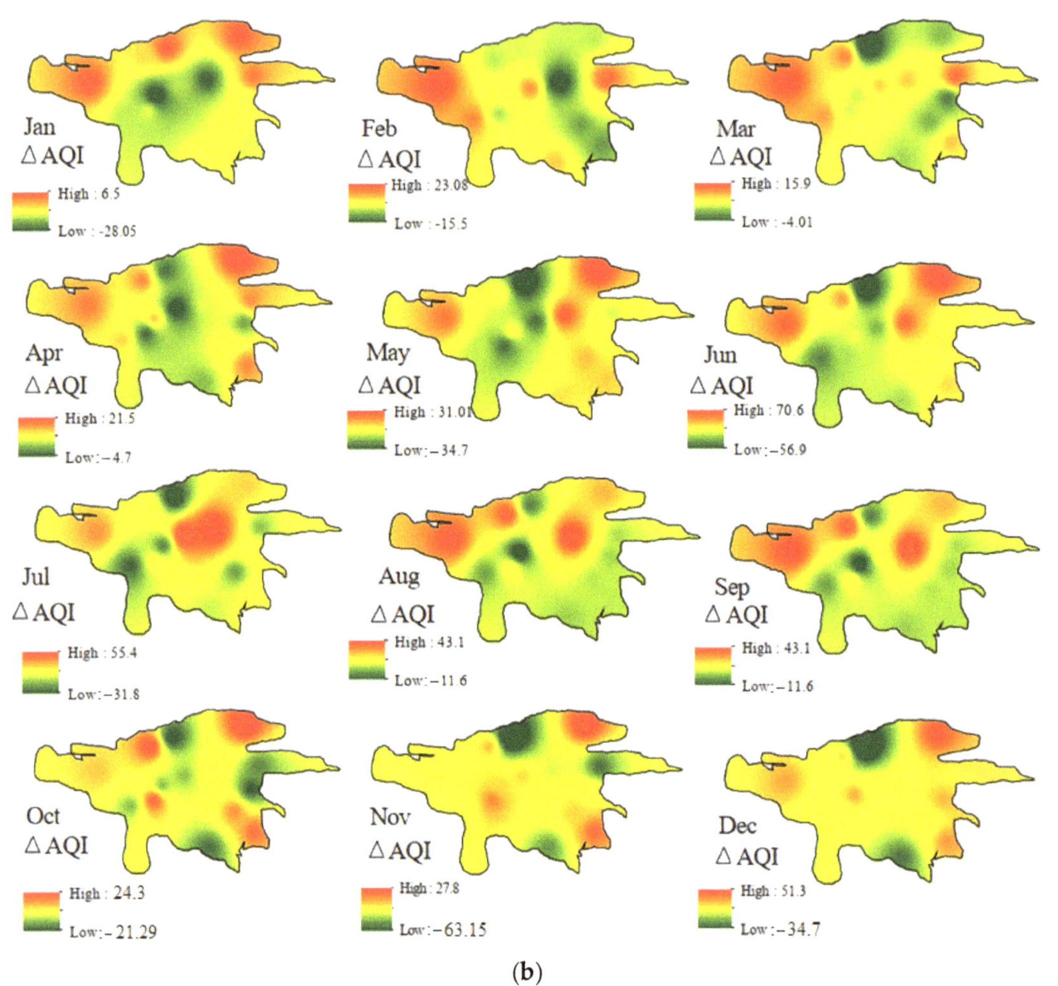

Figure 9. (**a**) Distribution map of the mean monthly AQI concentration at each station and (**b**) spatial distribution of variations in the mean monthly AQI concentration in 2020 compared to 2019.

3.7. Significance of Variations

As previously mentioned, a correlated pairs *t*-test was implemented to determine if the differences (results) were statistically significant. The null hypothesis is that the true difference in the mean value of each index before and during COVID-19 is zero. Table 1 summarizes the *p*-values achieved for the indices. Considering the 95% confidence interval (i.e., *p*-value ≤ 0.05), the null hypothesis is rejected in this study. Accordingly, significant change in SO_2 (AQI) was observed in all (almost all) months. Regarding CO and NO_2, significant variation in air quality was seen in two (December and November) and three (March, April, and December) months, respectively. The monthly significant changes in PM_{10} and O_3 indices were seen in four months during the pandemic.

Table 1. The attained *p*-value on air quality indices across Tehran.

Month \ Index	SO_2	CO	NO_2	PM_{10}	O_3	AQI
January	<0.05	0.12	0.12	0.41	0.71	<0.05
February	<0.05	0.62	0.43	0.54	0.98	<0.05
March	<0.05	0.23	<.05	<0.05	0.40	<0.05
April	<0.05	0.57	<.05	0.20	0.70	<0.05
May	<0.05	0.52	0.13	0.30	0.18	0.65
June	<0.05	0.81	0.63	<0.05	<0.05	<0.05
July	<0.05	0.53	0.87	<0.05	0.12	0.18
August	<0.05	0.82	0.17	0.22	<0.05	<0.05
September	<0.05	0.82	0.17	0.22	<0.05	<0.05
October	<0.05	0.51	0.19	0.54	<0.05	0.44
November	<0.05	<0.05	0.22	<0.05	0.64	0.44
December	<0.05	<0.05	<0.05	0.21	0.33	<0.05
Mean annual	<0.05	0.7	0.41	0.11	0.35	<0.05

Reductions in primary air pollutants during the COVID-19 outbreak were reported in several studies. Table 2 compares our findings with those attained in different regions.

Table 2. Summary of recent studies on COVID-19 effect on air pollution.

Study	Study Area	Findings
This article	Tehran	CO significantly reduced in December and November. NO_2 did not decrease sharply at the stations, but the daily mean concentration in March, April and December has significantly decreased. PM_{10} concentration increased in March, June, July, and November. O_3 has increased in June, August, September, and October. Mean annual and monthly SO_2 increased. Mean annual AQI has decreased.
[58]	New York	$PM_{2.5}$ and NO_2 has decreased. O_3 has increased.
[59]	Global	Primary air pollutants have reduced. Secondary PM and O_3 has increased in some cities.
[60]	India	AQI has improved and the tropospheric NO_2 and O_3 have reduced.
[61]	United Kingdom	NO_2 and $PM_{2.5}$ concentrations have reduced.
[62]	China	$PM_{2.5}$, PM_{10}, SO_2, NO_2, and CO have decreased
[63]	India	$PM_{2.5}$, NO_2, and AQI over Delhi, Mumbai, Hyderabad, Kolkata, and Chennai have declined.
[64]	Wuhan	AQI has decreased significantly. NO_2 has decreased, but O_3 has increased significantly.
[65]	Western Europe	NO_2 has decreased considerably. PM has reduced relatively. PM_{10}, $PM_{2.5}$ NO_2, and CO have reduced.
[66]	Delhi	The central and Eastern Delhi have experienced maximum improvement in air quality.
[67]	São Paulo	CO, NO, and NO_2 have decreased. O_3 has increased.

4. Conclusions

In this article, using daily data from 14 air quality monitoring stations across Tehran metropolitan city, air quality dynamics before and during the COVID-19 pandemic and its driving factors were examined. The results of this study compare monthly and mean annual pollution levels in 2019, 2020, and usual concentration for CO (20.17 ppm, 25.2 ppm, 26.6—32.1 ppm), NO_2 (71.8, 71.7, 85) ppb, PM_{10} (56.7, 55.8, 99.5) µg/m^3, O_3 (42.4, 45.7,68.8) ppb, SO_2 (7.4, 17.5, 8.9) ppb, respectively. According to the results, any significant reduction in annual concentration of CO, NO_2, PM_{10} and O_3 was not observed. However, the SO_2 concentrations increased significantly during the pandemic. These trends are attributed to the change in the main fuel source from natural gas to fuel oil in Tehran power plants. Political sanctions and undue economic pressure inhibited the maintenance and upgrade of Iran's infrastructure, which forced the decision makers to choose poor air quality (i.e., use of fuel oil in power plants) over cutting natural gas for urban consumers in the cold winter days. Spatial analysis shows that air pollution indices have drastic heterogeneity in Tehran, which are attributed to topography, population density and land use patterns. This heterogeneity leaves the poor with higher pollutant levels than the wealthy, with significant implications for environmental justice issues. Furthermore, changes in the pattern of inter-city travel, from personal travel to public bus travel due to the COVID-19 outbreak, have increased and concentrated pollution around suburban terminals. The AQI index analysis shows the days with a good quality index (AQI $\leq$ 50) in 2020 decreased when compared to 2019, whereas the number of moderate days (50 < AQI $\leq$ 100) increased. This trend is also in contrast to a global improvement in air quality, highlighting that economic restraints limited the effects of social activity reduction in Tehran. Finally, in the intervals between the lockdown periods, we see an increase in the AQI index. This could be due to the accumulation of social transportation needs after periods of quarantine.

Supplementary Materials: The following supporting information can be downloaded at: https://www.mdpi.com/article/10.3390/su142316313/s1, Figure S1: Green space and green area per capita across Tehran; Figure S2: Temporal distribution of CO content at each station, gray rectangular shows strict lockdown period. Figure S3: Temporal distribution of NO_2 content in all investigated station, gray rectangular shows strict lockdown period; Figure S4: Temporal distribution of PM_{10} content in all investigated station, gray rectangular shows strict lockdown period; Figure S5: Temporal distribution of O3 content in all investigated station, gray rectangular shows strict lockdown period; Figure S6: Temporal distribution of SO_2 content in all investigated station, gray rectangular shows strict lockdown period; Figure S7: Temporal distribution of the AQI indices in all investigated stations, the gray rectangles show strict lockdown periods.

Author Contributions: Conceptualization, M.M. and A.D.M.; methodology, M.M. and S.M.K.; software, M.M. and S.P.; validation, A.D.M., M.S. and V.N.; formal analysis, M.M. and B.G.; investigation, M.M. and A.D.M.; resources, B.G.; data curation, M.M. and S.P.; writing—original draft preparation, M.M., S.M.K. and A.D.M.; writing—review and editing, M.M., S.M.K. and A.D.M.; visualization, M.M. and A.D.M.; supervision, M.S. and V.N. All authors have read and agreed to the published version of the manuscript.

Funding: This research received no external funding.

Institutional Review Board Statement: Not applicable.

Informed Consent Statement: Not applicable.

Data Availability Statement: Data used in this study available from corresponding author upon a reasonable request.

Acknowledgments: The authors are thankful to three anonymous reviewers for their constructive comments.

Conflicts of Interest: The authors declare no conflict of interest.

References

1. Naddafi, K.; Hassanvand, M.S.; Yunesian, M.; Momeniha, F.; Nabizadeh, R.; Faridi, S.; Gholampour, A. Health impact assessment of air pollution in megacity of Tehran, Iran. *Iran. J. Environ. Health Sci. Eng.* **2012**, *9*, 28.
2. Tuygun, G.T.; Gundoğdu, S.; Elbir, T. Estimation of ground-level particulate matter concentrations based on synergistic use of MODIS, MERRA-2 and AERONET AODs over a coastal site in the Eastern Mediterranean. *Atmos. Environ.* **2021**, *261*, 118562. [CrossRef]
3. WHO (World Health Organization). *7 Million Premature Deaths Annually Linked to Air Pollution*; World Health Organization: Geneva, Switzerland, 2014.
4. Vafa-Arani, H.; Jahani, S.; Dashti, H.; Heydari, J.; Moazen, S. A system dynamic modeling for urban air pollution: A case study of Tehran, Iran. *Transp. Res. D-Transp. Environ.* **2014**, *31*, 21–36. [CrossRef]
5. Hosseini, V.; Shahbazi, H. Urban air pollution in Iran. *Iran. Stud.* **2016**, *49*, 1029–1046.

6. Kayalar, Ö.; Arı, A.; Konyalılar, N.; Doğan, Ö.; Can, F.; Şahin, Ü.A.; Gaga, E.O.; Kuzu, S.L.; Arı, P.E.; Odabası, M. Existence of SARS-CoV-2 RNA on ambient particulate matter samples: A nationwide study in Turkey. *Sci. Total Environ.* **2021**, *789*, 147976. [CrossRef]

7. Noori, R.; Hoshyaripour, G.; Ashrafi, K.; Araabi, B.N. Uncertainty analysis of developed ANN and ANFIS models in prediction of carbon monoxide daily concentration. *Atmos. Environ.* **2010**, *44*, 476–482.

8. Vasconcellos, E.A. *Urban Transport Environment and Equity: The Case for Developing Countries*, 1st ed.; Routledge: London, UK, 2014.

9. Habibi, R.; Alesheikh, A.A.; Mohammadinia, A.; Sharif, M. An assessment of spatial pattern characterization of air pollution: A case study of CO and PM2. 5 in Tehran, Iran. *ISPRS Int. J. Geoinf.* **2017**, *6*, 270. [CrossRef]

10. Cohen, B. Urbanization in developing countries: Current trends, future projections, and key challenges for sustainability. *Technol. Soc.* **2006**, *28*, 63–80. [CrossRef]

11. Atash, F. The deterioration of urban environments in developing countries: Mitigating the air pollution crisis in Tehran, Iran. *Cities* **2007**, *24*, 399–409. [CrossRef]

12. Akbarzadeh, A.; Vesali Naseh, M.; Nodefarahani, M. Carbon Monoxide Prediction in the Atmosphere of Tehran Using Developed Support Vector Machine. *Pollution* **2020**, *6*, 43–57.

13. Heger, M.; Sarraf, M. *Air pollution in Tehran: Health Costs, Sources, and Policies*; Environment and Natural Resources Global Practice Discussion Paper, No. 6; World Bank: Washington, DC, USA, 2018.

14. Bayat, R.; Ashrafi, K.; Motlagh, M.S.; Hassanvand, M.S.; Daroudi, R.; Fink, G.; Kunzli, N. Health impact and related cost of ambient air pollution in Tehran. *Environ. Res.* **2019**, *176*, 108547. [CrossRef] [PubMed]

15. Keyhani, A.; Ghasemi-Varnamkhasti, M.; Khanali, M.; Abbaszadeh, R. An assessment of wind energy potential as a power generation source in the capital of Iran, Tehran. *Energy* **2010**, *35*, 188–201. [CrossRef]

16. Madanipour, A. Urban planning and development in Tehran. *Cities* **2006**, *23*, 433–438. [CrossRef]

17. Shams, M.; Rahimi-Movaghar, V. Risky driving behaviors in Tehran, Iran. *Traffic Inj. Prev.* **2009**, *10*, 91–94. [CrossRef]

18. Contini, D.; Costabile, F. Does Air Pollution Influence COVID-19 Outbreaks? *Atmosphere* **2020**, *11*, 377. [CrossRef]

19. Liu, S.; Yang, X.; Duan, F.; Zhao, W. Changes in Air Quality and Drivers for the Heavy $PM_{2.5}$ Pollution on the North China Plain Pre- to Post-COVID-19. *Int. J. Environ. Res. Public Health* **2022**, *19*, 12904. [CrossRef]

20. Rumpler, R.; Venkataraman, S.; Göransson, P. An observation of the impact of COVID-19 recommendation measures monitored through urban noise levels in central Stockholm, Sweden. *Sustain. Cities Soc.* **2020**, *63*, 102469. [CrossRef]

21. Rahmani, A.M.; Mirmahaleh, S.Y.H. Coronavirus disease (COVID-19) prevention and treatment methods and effective parameters: A systematic literature review. *Sustain. Cities Soc.* **2021**, *64*, 102568.

22. Agarwal, N.; Meena, C.S.; Raj, B.P.; Saini, L.; Kumar, A.; Gopalakrishnan, N.; Kumar, A.; Balam, N.B.; Alam, T.; Kapoor, N.R. Indoor air quality improvement in COVID-19 pandemic. *Sustain. Cities Soc.* **2021**, *70*, 102942.

23. Velraj, R.; Haghighat, F. The contribution of dry indoor built environment on the spread of Coronavirus: Data from various Indian states. *Sustain. Cities Soc.* **2020**, *62*, 102371.

24. Kleinschroth, F.; Kowarik, I. COVID-19 crisis demonstrates the urgent need for urban greenspaces. *Front. Ecol. Environ.* **2020**, *18*, 318.

25. Casanova, L.M.; Jeon, S.; Rutala, W.A.; Weber, D.J.; Sobsey, M.D. Effects of air temperature and relative humidity on coronavirus survival on surfaces. *Appl. Environ.* **2010**, *76*, 2712–2717. [CrossRef]

26. Sun, C.; Zhai, Z. The efficacy of social distance and ventilation effectiveness in preventing COVID-19 transmission. *Sustain. Cities Soc.* **2020**, *62*, 102390.

27. Hashim, B.M.; Al-Naseri, S.K.; Al Maliki, A.; Sa'adi, Z.; Malik, A.; Yaseen, Z.M. On the investigation of COVID-19 lockdown influence on air pollution concentration: Regional investigation over eighteen provinces in Iraq. *Environ. Sci. Pollut. Res.* **2021**, *28*, 50344–50362.

28. Niu, H.; Zhang, C.; Hu, W.; Hu, T.; Wu, C.; Hu, S.; Silva, L.F.O.; Gao, N.; Bao, X.; Fan, J. Air Quality Changes during the COVID-19 Lockdown in an Industrial City in North China: Post-Pandemic Proposals for Air Quality Improvement. *Sustainability* **2022**, *14*, 11531. [CrossRef]

29. Nabavi-Pelesaraei, A.; Bayat, R.; Hosseinzadeh-Bandbafha, H.; Afrasyabi, H.; Chau, K.W. Modeling of energy consumption and environmental life cycle assessment for incineration and landfill systems of municipal solid waste management-A case study in Tehran Metropolis of Iran. *J. Clean Prod.* **2017**, *148*, 427–440.

30. Noroozian, A.; Bidi, M. An applicable method for gas turbine efficiency improvement. Case study: Montazar Ghaem power plant, Iran. *J. Nat. Gas. Sci. Eng.* **2016**, *28*, 95–105. [CrossRef]

31. Davies, H.W.; Vlaanderen, J.; Henderson, S.; Brauer, M. Correlation between co-exposures to noise and air pollution from traffic sources. *J. Occup. Environ. Med.* **2009**, *66*, 347–350. [CrossRef]

32. Chen, T.-M.; Kuschner, W.G.; Gokhale, J.; Shofer, S. Outdoor air pollution: Nitrogen dioxide, sulfur dioxide, and carbon monoxide health effects. *Am. J. Med. Sci.* **2007**, *333*, 249–256.

33. Kumar, R.; Joseph, A.E. Air pollution concentrations of PM 2.5, PM10 and NO_2 at ambient and kerbsite and their correlation in Metro City–Mumbai. *Environ. Monit. Assess.* **2006**, *119*, 191–199.

34. Rosofsky, A.; Levy, J.I.; Zanobetti, A.; Janulewicz, P.; Fabian, M.P. Temporal trends in air pollution exposure inequality in Massachusetts. *Environ. Res.* **2018**, *161*, 76–86. [PubMed]

35. Liu, J.C.; Peng, R.D. Health effect of mixtures of ozone, nitrogen dioxide, and fine particulates in 85 US counties. *Air Qual. Atmos. Health* **2018**, *11*, 311–324.
36. Al-Hemoud, A.; Al-Dousari, A.; Al-Shatti, A.; Al-Khayat, A.; Behbehani, W.; Malak, M. Health impact assessment associated with exposure to PM10 and dust storms in Kuwait. *Atmosphere* **2018**, *9*, 6.
37. Carugno, M.; Consonni, D.; Bertazzi, P.A.; Biggeri, A.; Baccini, M. Temporal trends of PM10 and its impact on mortality in Lombardy, Italy. *Environ. Pollut.* **2017**, *227*, 280–286. [PubMed]
38. Rovira, J.; Domingo, J.L.; Schuhmacher, M. Air quality, health impacts and burden of disease due to air pollution (PM_{10}, PM_{25}, NO_2 and O_3): Application of Air Q+ model to the Camp de Tarragona County (Catalonia, Spain). *Sci. Total Env.* **2020**, *703*, 135538.
39. Azimi, M.; Feng, F.; Yang, Y. Air pollution inequality and its sources in SO2 and NOx emissions among Chinese provinces from 2006 to 2015. *Sustainability* **2018**, *10*, 367.
40. Tucker, K.A. A Breath of Polluted Air: How Indiana's air pollution policies are impacting its citizens. *Ind. Health L. Rev.* **2020**, *17*, 339. [CrossRef]
41. Xue, J.; Xu, Y.; Zhao, L.; Wang, C.; Rasool, Z.; Ni, M.; Wang, Q.; Li, D. Air pollution option pricing model based on AQI. *Atmos. Pollut. Res.* **2019**, *10*, 665–674.
42. Air Quality Assessment Division of Office of Air Quality Planning and Standards. Technical Assistance Document for the Reporting of Daily Air Quality—The Air Quality Index (AQI), U.S. Environmental Protection Agency (USEPA), 2018, EPA 454/B-18-007. Available online: https://www.epa.gov/outdoor-air-quality-data/how-aqi-calculated (accessed on 10 April 2021).
43. Moradzadeh, R. The challenges and considerations of community-based preparedness at the onset of COVID-19 outbreak in Iran, 2020. *Epidemiol. Infect.* **2020**, *148*, e82.
44. Broomandi, P.; Karaca, F.; Nikfal, A.; Jahanbakhshi, A.; Tamjidi, M.; Kim, J.R. Impact of COVID-19 event on the air quality in Iran. *Aerosol Air Qual. Res.* **2020**, *20*, 1793–1804.
45. Aghashariatmadari, Z. The effects of COVID-19 pandemic on the air pollutants concentration during the lockdown in Tehran, Iran. *Urban Clim.* **2021**, *38*, 100882.
46. Nassiri, H.; Mohammadpour, S.I.; Dahaghin, M. How do the smart travel ban policy and intercity travel pattern affect COVID-19 trends? Lessons learned from Iran. *PLoS ONE* **2022**, *17*, e0276276. [CrossRef]
47. Khalesi, B.; Daneshvar, M.R.M. Comprehensive temporal analysis of temperature inversions across urban atmospheric boundary layer of Tehran within 2014–2018. *Model. Earth Syst. Environ.* **2020**, *6*, 967–982.
48. Berman, J.D.; Ebisu, K. Changes in US air pollution during the COVID-19 pandemic. *Sci. Total Env.* **2020**, *739*, 139864. [CrossRef]
49. Gautam, S. COVID-19: Air pollution remains low as people stay at home. *Air Qual. Atmos. Health* **2020**, *13*, 853–857. [CrossRef]
50. Fattorini, D.; Regoli, F. Role of the chronic air pollution levels in the COVID-19 outbreak risk in Italy. *Environ. Pollut.* **2020**, *264*, 114732. [CrossRef]
51. Wang, Q.; Li, S. Nonlinear impact of COVID-19 on pollutions–Evidence from Wuhan, New York, Milan, Madrid, Bandra, London, Tokyo and Mexico City. *Sustain. Cities Soc.* **2021**, *65*, 102629. [CrossRef]
52. Jandacka, D.; Durcanska, D. Seasonal Variation, Chemical Composition, and PMF-Derived Sources Identification of Traffic-Related PM1, PM2.5, and PM2.5–10 in the Air Quality Management Region of Žilina, Slovakia. *Int. J. Environ. Res. Public Health* **2021**, *18*, 10191.
53. Pant, P.; Harrison, R.M. Estimation of the contribution of road traffic emissions to particulate matter concentrations from field measurements: A review. *Atmos. Environ.* **2013**, *77*, 78–97. [CrossRef]
54. Wang, S.; Kaur, M.; Li, T.; Pan, F. Effect of Different Pollution Parameters and Chemical Components of PM2.5 on Health of Residents of Xinxiang City, China. *Int. J. Environ. Res. Public Health* **2021**, *18*, 6821.
55. WHO. *Air Quality Guidelines*; World Health Organization, Regional Office for Europe: Copenhagen, Denmark, 2000.
56. Givehchi, R.; Arhami, M.; Tajrishy, M. Contribution of the Middle Eastern dust source areas to PM10 levels in urban receptors: Case study of Tehran, Iran. *Atmos. Environ.* **2013**, *75*, 287–295. [CrossRef]
57. Venter, Z.S.; Aunan, K.; Chowdhury, S.; Lelieveld, J. COVID-19 lockdowns cause global air pollution declines. *Proc. Natl. Acad. Sci. USA* **2020**, *117*, 18984–18990. [CrossRef]
58. Ghiasi, B.; Alisoltani, T.; Jalali, F.; Tahsinpour, H. Effect of COVID-19 on transportation air pollution by moderation and mediation analysis in Queens, New York. *Air Qual. Atmos. Health* **2022**, *15*, 289–297.
59. Adam, M.G.; Tran, P.T.; Balasubramanian, R. Air quality changes in cities during the COVID-19 lockdown: A critical review. *Atmos. Res.* **2021**, *264*, 105823. [PubMed]
60. Naqvi, H.R.; Datta, M.; Mutreja, G.; Siddiqui, M.A.; Naqvi, D.F.; Naqvi, A.R. Improved air quality and associated mortalities in India under COVID-19 lockdown. *Environ. Pollut.* **2021**, *268*, 115691. [CrossRef] [PubMed]
61. Jephcote, C.; Hansell, A.L.; Adams, K.; Gulliver, J. Changes in air quality during COVID-19 'lockdown' in the United Kingdom. *Environ. Pollut.* **2021**, *272*, 116011. [PubMed]
62. Wang, M.; Liu, F.; Zheng, M. Air quality improvement from COVID-19 lockdown: Evidence from China. *Air Qual. Atmos. Health* **2021**, *14*, 591–604. [CrossRef]
63. Singh, R.P.; Chauhan, A. Impact of lockdown on air quality in India during COVID-19 pandemic. *Air Qual. Atmos. Health* **2020**, *13*, 921–928.
64. Lian, X.; Huang, J.; Huang, R.; Liu, C.; Wang, L.; Zhang, T. Impact of city lockdown on the air quality of COVID-19-hit of Wuhan city. *Sci. Total Environ.* **2020**, *742*, 140556.

65. Menut, L.; Bessagnet, B.; Siour, G.; Mailler, S.; Pennel, R.; Cholakian, A. Impact of lockdown measures to combat COVID-19 on air quality over western Europe. *Sci. Total Environ.* **2020**, *741*, 140426. [CrossRef]
66. Mahato, S.; Pal, S.; Ghosh, K.G. Effect of lockdown amid COVID-19 pandemic on air quality of the megacity Delhi, India. *Sci. Total Environ.* **2020**, *730*, 139086. [CrossRef]
67. Nakada, L.Y.K.; Urban, R.C. COVID-19 pandemic: Impacts on the air quality during the partial lockdown in São Paulo state, Brazil. *Sci. Total Environ.* **2020**, *730*, 139087.

sustainability

Article

Analysis of Atmospheric Pollutant Data Using Self-Organizing Maps

Emanoel L. R. Costa [1,†], Taiane Braga [2,†], Leonardo A. Dias [3,†], Édler L. de Albuquerque [4,*,†] and Marcelo A. C. Fernandes [1,5,*,†]

1. Laboratory of Machine Learning and Intelligent Instrumentation, Federal University of Rio Grande do Norte, Natal 59078-970, RN, Brazil
2. Federal Institute of Education, Science, and Technology of Bahia, Salvador 40301-015, BA, Brazil
3. Centre for Cyber Security and Privacy, School of Computer Science, University of Birmingham, Birmingham B15 2TT, UK
4. Department of Industrial Processes and Chemical Engineering, Federal Institute of Education, Science and Technology of Bahia, Salvador 40301-015, BA, Brazil
5. Department of Computer Engineering and Automation, Federal University of Rio Grande do Norte, Natal 59078-970, RN, Brazil
* Correspondence: edler@ifba.edu.br (É.L.d.A.); mfernandes@dca.ufrn.br (M.A.C.F.)
† These authors contributed equally to this work.

Citation: Costa, E.L.R.; Braga, T.; Dias, L.A.; Albuquerque, É.L.d.; Fernandes, M.A.C. Analysis of Atmospheric Pollutant Data Using Self-Organizing Maps. *Sustainability* **2022**, *14*, 10369. https://doi.org/10.3390/su141610369

Academic Editors: José Carlos Magalhães Pires and Álvaro Gómez-Losada

Received: 12 July 2022
Accepted: 17 August 2022
Published: 20 August 2022

Abstract: Atmospheric pollution is a critical issue in our society due to the continuous development of countries. Therefore, studies concerning atmospheric pollutants using multivariate statistical methods are widely available in the literature. Furthermore, machine learning has proved a good alternative, providing techniques capable of dealing with problems of great complexity, such as pollution. Therefore, this work used the Self-Organizing Map (SOM) algorithm to explore and analyze atmospheric pollutants data from four air quality monitoring stations in Salvador-Bahia. The maps generated by the SOM allow identifying patterns between the air quality pollutants (CO, NO, NO_2, SO_2, PM_{10} and O_3) and meteorological parameters (environment temperature, relative humidity, wind velocity and standard deviation of wind direction) and also observing the correlations among them. For example, the clusters obtained with the SOM pointed to characteristics of the monitoring stations' data samples, such as the quantity and distribution of pollution concentration. Therefore, by analyzing the correlations presented by the SOM, it was possible to estimate the effect of the pollutants and their possible emission sources.

Keywords: machine learning; atmospheric pollution; Self-Organizing Maps; Salvador-BA

1. Introduction

Air pollution is one of the crucial challenges of modern society. In recent years, pollution caused by industrial, vehicular, and toxic-chemical emission sources has increased significantly. This increase can be seen mainly in low- and middle-income countries, also called developing countries [1]. Despite the continuous pollution growth, awareness and pollution control programs are limited and receive little attention and financial resources from governments, international agencies, and philanthropic donors [1].

In addition, effectively managing regulations for controlling air pollution requires considerable knowledge about the costs and benefits. Currently, the primary efforts for measuring pollutants aim to avoid possible harm to people's health, such as respiratory or cardiovascular diseases that can result in hospitalizations and even death, usually affecting vulnerable groups of the population [2].

Complex mixtures of solid particles and gaseous pollutants contribute to air pollution. Among these are priority pollutants, commonly regulated by law and categorized as primary and secondary. The primary pollutants are substances that can be released directly into the atmosphere, while the secondary pollutants are substances derivated from the primary ones through photochemical reactions in the troposphere [3]. Regarding the gaseous pollutants, to be particulary mentioned are sulfur dioxide (SO_2), nitrogen dioxide (NO_2), carbon monoxide (CO), volatile organic compounds (VOCs), solid materials or liquids suspended in the atmosphere due to their small size (called particulate matter (PM)), and the ozone (O_3). The ozone is one of the major photochemical pollutants formed in the atmosphere by the reaction of nitrogen oxides (NO_x) and hydrocarbons such as VOCs in the presence of sunlight, similarly to particulate sulfate and nitrate aerosols created from SO_2 and NO_x [3].

The dispersion of atmospheric pollutants results from different elements such as temperature, relative humidity, atmospheric pressure, wind direction and speed, as well as topography [4]. Consequently, the complexity of analyzing and identifying pollutants and their primary sources in large-scale areas increases, which leads to the problem of positioning monitoring stations for data collection.

There are several emission sources of air pollutants, and a single source can emit several pollutants. For instance, the composition of fossil fuels used in motor vehicles can emit different pollutants during combustion and evaporation, or by the wear of tires and roads where vehicles run. Due to the increasing number of private vehicles, their emissions have become a dominant source of CO, CO_2, VOCs, NO_x and PM. Meanwhile, industrial processes normally include pollutants such as CO, PM, NO_x, and SO_2 [4–6].

Thus, monitoring the concentration of pollutants in the environment at specific points is essential. Identifying the main components enables understanding of the current condition of air pollution, variations, correlations, and possible emission sources, which leads to the development of public policies to raise awareness and reduce pollutants. Therefore, many researchers have proposed the analysis of environmental data mainly using multivariate statistical techniques [4,7,8].

Multivariate statistical methods such as correlation or cluster analysis [9–13], and principal component analysis [7,14,15] are commonly applied in various studies to identify the correlation among parameters that can influence air quality. Large databases that carry various information about air pollution require techniques to extract and identify characteristics inherent to the analyzed data.

In this context, machine learning has proved to be a great alternative to the traditional methods used [16,17]. A well-known algorithm that belongs to the group of unsupervised learning algorithms is self-organizing maps (SOM) [18]. The SOM supports data dimensionality reduction and clustering. In addition, the SOM does not need to make assumptions about the parameters' distribution, as it is capable of dealing with non-linear problems of great complexity and dimension and is effective in using noisy data [19].

The SOM algorithm is adopted in many applications to analyze data from atmospheric pollutants. For example, in Ref [20], the SOM is used to analyze data regarding air quality. In Ref. [21], the SOM is used to identify the level of pollution during foundry and land mining. The study carried out in [22] used the SOM to highlight the impact on air quality caused by the circulation of different air types, which alters the concentration of pollutants in the atmosphere. For this purpose, it is essential to identify suitable placements for positioning monitoring stations, as shown in [23]. Finally, the SOM has also been used to obtain particulate-matter characteristics in the atmosphere by evaluating its concentration in both internal and external exposure and connecting them to human activities. According to [24], the SOM can also function as a pollution identifier by defining limits to classify regions with low or high concentrations of a specific pollutant, such as ozone, enabling the evaluation of pollution zones.

Therefore, this work proposes an SOM implementation to study and analyze atmospheric pollutants to identify their patterns and characteristics. The main contributions are:

- A machine-learning-based approach for analyzing the air quality of Salvador monitoring stations, using the Government of Bahia State database—to the best of our knowledge, this work is the first to analyze this data using machine-learning algorithms.

- We discuss the common factors among meteorological parameters and pollutants and their clusters' impact on each monitoring station.

2. Methodology

2.1. Case Study

Salvador city (State of Bahia) has a territorial area of 693.453 km^2 and a population of 2,675,656 people. Located in the northeastern region of Brazil, it has an urban core and rugged topography formed by several columns and valleys, with a rainy tropical climate with no dry season and an average annual temperature of 25 °C.

The Government of Bahia State, through CETREL S. A., the company that operated the air monitoring stations from 2011 to 2016, provided the air quality database for this work. It contains the air quality data of a monitoring network constituted of eight stations. Nonetheless, we used data from four stations: Barros Reis (BR), Campo Grande (CG), Dique do Tororó (DT), and Itaigara (IT), due to their inherent characteristics. Figure 1 illustrates the stations' distribution in Salvador and highlights the four chosen. It is important to mention that this is the first air quality monitoring network ever installed in the city of Salvador. Therefore, this work portrays the first analysis of the pollutant and meteorological parameters in the database provided.

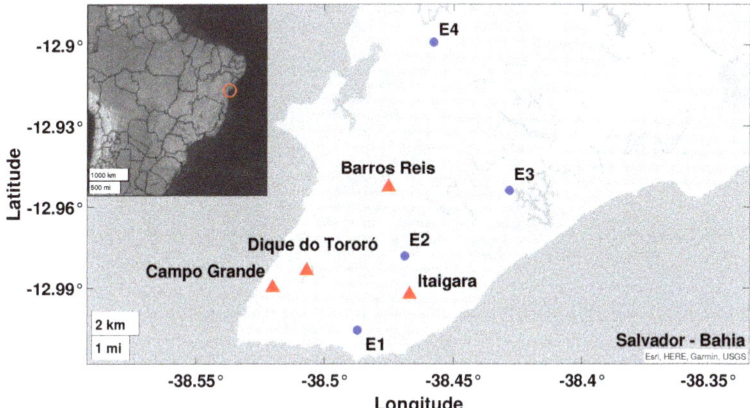

Figure 1. Location of the eigth air monitoring stations deployed in Salvador-BA.

2.2. Dataset

The dataset contains the hourly average of twelve features related to meteorological parameters and pollutants concentration. The meteorological parameters are wind speed (WS), ambient temperature (TEMP), relative air humidity (RH), the standard deviation of wind direction (STWD), rainfall, and wind direction. Meanwhile, the pollutants are SO_2, CO, O_3, particulate matter whose aerodynamic diameter is less than 10 μm (PM_{10}) and the oxides of nitrogen NO_2 and NO. We removed the rainfall and wind direction variables due to the small amount of data available; thus, only ten features were used in our analysis.

We performed a data preprocessing step by removing the null lines, the measurement errors (identified by a specific terminology), and the outliers to improve the quality of the analysis. The outliers were removed by investigating the data dispersion and symmetry and, subsequently, using the quartile separatrix measure [25] to divide the dataset into three quartiles: Q_1, Q_2 and Q_3. Finally, based on the interquartile range (AIQ) [25], outliers with value greater than $Q_3 + 3 \times AIQ$ and less than $Q_1 - 3 \times AIQ$, were removed from the database. We kept outliers with values greater than $Q_3 + 1.5 \times AIQ$ and less than $Q_1 - 1.5 \times AIQ$ to avoid a large reduction in the dataset. Table 1 presents the number of data samples for each monitoring station considered in our analysis and their period of operation.

Table 1. The operation period for each monitoring station provided by CETREL S. A., and the number of data samples available in the dataset before and after the preprocessing step.

Station	Operation Start Date	Operation End Date	Number of Rregistered Samples	Number of Samples after Preprocessing
Barros Reis (BR)	8 November 2013	31 December 2016	27,584	21,559
Campo Grande (CG)	2 July 2011	31 December 2016	48,234	24,559
Dique do Tororó (DT)	19 June 2011	31 December 2016	48,550	42,037
Itaigara (IT)	18 October 2013	30 April 2016	22,203	15,535

In the meantime, Tables 2–5 present the dataset for the Barros Reis (BR), Campo Grande (CG), Dique do Tororó (DT), and Itaigara (IT) stations, respectively. As can be observed, all pollutants and atmospheric data are shown after the preprocessing step for each station in a concentration of pollutants in parts per billion (ppb).

Table 2. Descriptive statistics of pollutants and atmospheric data from the Barros Reis station (P = 21,559 samples).

Parameters	Magniture	Maximum	Mean	Average	Standard Deviation	Variation Coefficient
SO_2	ppb	3.20	0.30	0.45	0.51	112.94%
CO	ppb	2180.00	570.00	601.60	335.70	55.80%
O_3	ppb	22.70	4.80	5.47	3.80	69.36%
PM_{10}	$\mu g/m^3$	129.80	37.30	40.10	19.88	49.58%
NO	ppb	206.40	44.40	52.47	38.01	72.50%
NO_2	ppb	49.20	13.30	14.15	7.61	53.84%
WS	m/s	10.80	2.20	2.62	1.75	67.00%
TEMP	°C	32.50	25.50	25.63	2.18	8.54%
RH	%	91.00	69.00	68.60	9.31	13.57%
STWD	°	73.30	31.30	31.61	11.61	36.73%

Table 3. Descriptive statistics of pollutants and atmospheric data from the Campo Grande station (P = 24,559 samples).

Parameters	Magniture	Maximum	Mean	Average	Standard Deviation	Variation Coefficient
SO_2	ppb	1.70	0.20	0.32	0.31	97.20%
CO	ppb	1830.00	360.00	396.60	292.70	73.81%
O_3	ppb	25.00	5.20	6.01	4.18	69.5%
PM_{10}	$\mu g/m^3$	77.30	19.30	21.10	12.53	59.38%
NO	ppb	139.00	25.10	28.03	23.37	83.38%
NO_2	ppb	44.00	13.30	13.37	6.42	48.00%
WS	m/s	5.10	1.20	1.42	0.93	66.01%
TEMP	°C	34.30	26.50	26.72	2.31	8.66%
RH	%	94.00	72.00	71.12	9.54	13.41%
STWD	°	79.60	53.20	52.01	13.31	25.57%

Table 4. Descriptive statistics of pollutants and atmospheric data from the Dique do Tororó station (P = 42,037 samples).

Parameters	Magniture	Maximum	Mean	Average	Standard Deviation	Variation Coefficient
SO_2	ppb	2.00	0.20	0.33	0.40	123.15%
CO	ppb	1000.00	220.00	239.40	163.90	68.44%
O_3	ppb	34.30	7.20	8.15	5.37	65.88%
PM_{10}	µg/m^3	75.60	20.00	22.13	12.42	56.12%
NO	ppb	73.60	12.40	13.77	11.54	83.78%
NO_2	ppb	31.30	8.20	8.67	5.02	57.92%
WS	m/s	6.90	1.50	1.63	1.01	61.72%
TEMP	°C	33.90	26.30	26.46	2.31	8.75%
RH	%	94.00	73.00	72.46	9.10	12.56%
STWD	°	78.80	33.00	38.45	15.34	39.90%

Table 5. Descriptive statistics of pollutants and atmospheric data from the Itaigara station (P = 15,535 samples).

Parameters	Magniture	Maximum	Mean	Average	Standard Deviation	Variation Coefficient
SO_2	ppb	1.60	0.10	0.2502	0.33	131.89%
CO	ppb	1210.00	190.00	226.48	207.26	91.51%
O_3	ppb	27.50	7.90	8.47	4.32	51.00%
PM_{10}	µg/m^3	67.40	13.60	16.16	10.98	67.94%
NO	ppb	70.70	11.40	15.50	13.45	86.77%
NO_2	ppb	31.10	7.30	8.21	5.15	62.72%
WS	m/s	10.20	2.70	2.76	1.58	57.24%
TEMP	°C	33.40	25.00	25.04	2.27	9.06%
RH	%	93.00	71.00	71.43	9.08	12.71%
STWD	°	51.30	22.80	24.32	8.13	33.42%

As can be observed, the BR station presents a higher concentration of SO_2, CO, and PM_{10}. The SO_2 has a maximum of 3.20 ppb and an average of 0.45 ppb due to the burning of fuels with sulfur. Meanwhile, the CO has a maximum of 2180 ppb and an average of 601.6 ppb, produced by burning organic fuels. The PM_{10} has an average of 40.10 ppb, almost double the value of other stations; it is a solid or liquid material that remains suspended in the atmosphere that can cause a significant impact on human health.

The CG station also has a high level of CO, with a maximum of 1830 ppb and an average of 396.6 ppb. Regarding the presence of nitrogen oxides (NO and NO_2), the CG and DT stations present higher average and maximum concentrations due to the combustion processes and atmospheric chemical reactions. Concerning the O_3, a secondary pollutant formed in the atmosphere indicating the presence of photochemical oxidants, it has its higher concentrations recorded at the DT and IT stations.

Therefore, the SO_2, CO, and NO pollutants present the most significant variations in concentration. These pollutants are mainly generated from the burning of fossil fuels. Hence, the station location and the intensity of the vehicle's traffic around its region can lead to different concentration records at certain times of the day. The datasets comprise 24 h of daily data collection.

All stations show similar measured values regarding the meteorological parameters, except for wind speed which has a high average at BR and IT stations, and the standard deviation of wind direction at CG. Note that the values were rescaled from 0 to 1 to improve the SOM results. In addition, this work performed the z-score normalization and logarithmic transformation, obtaining data with null mean and unit variance and reducing the data scale, respectively.

2.3. Self-Organizing Maps (SOM)

The Self-Organizing Map (SOM) is a neural network model widely applied to data dimensionality reduction and clustering [18,26]. The map consists of M neurons commonly arranged in a two-dimensional array representing the incoming data by shifting the neurons' position towards it. The maps' topology can be rectangular, hexagonal, or square, among others [18].

The N-dimensional input data sample can be characterized as

$$\mathbf{x} = [x_1, x_2, \dots, x_N].\tag{1}$$

Accordingly, each i-th neuron in the map is represented by a N-dimensional vector of weights expressed as

$$\mathbf{w}_i = [w_{i1}, w_{i2}, \dots, w_{iN}].\tag{2}$$

Therefore, the topology of a two-dimensional map with M neurons can be expressed as $(M_h \times M_v)$, where M_h is the number of neurons in the horizontal and M_v is the number of neurons vertically; thus, $M = M_h \times M_v$.

The SOM algorithm iteratively molds the neurons' map to the input data topological form, based on a similarity metric, according to the following steps [18]:

1. Randomly initialize the M neurons' weight vectors.
2. Calculate the distance of each p-th input data sample, $\mathbf{x}(p)$, to all M neurons.
3. Define the winning neuron, also known as best matching unit (BMU); it is the j-th nearest neuron to the input data defined based on a distance metric as follows:

$$j = \arg\min_i ||\mathbf{x}(p) - \mathbf{w}_i||, \ i = 1, 2, \dots, M.\tag{3}$$

4. Update the BMU neuron and its neighboring neurons' weights according to following

$$\mathbf{w}_i(t+1) = \mathbf{w}_i(t) + \eta(t)h_{i,j}(t)(\mathbf{x}(p) - \mathbf{w}_i)\tag{4}$$

where $\eta(t)$ is the learning rate (ranging from 0 to 1) and $h_{i,j}(t)$ represents the BMU neighborhood function at the t-th iteration. The neighborhood function is described as

$$h_{i,j}(t) = \exp\left(-\frac{d_{i,j}^2}{2\sigma^2(t)}\right)\tag{5}$$

where $d_{i,j}^2$ is the distance from the i-th neuron to the BMU (j-th neuron) and $\sigma^2(t)$ is the neighboring function size at the t-th iteration.

5. Repeat steps 2, 3 and 4 until the maximum number of iterations is reached, represented here by T.

The number of iterations must be enough to process the dataset samples several times; thus, $T = b \times P$, where b is the repetition number that every set of P samples is presented to the SOM. Moreover, increasing the iteration number (t) decreases the radius of the neighborhood function, $\sigma^2(t)$. Consequently, the number of neurons nearby the BMU to be updated is reduced, strengthening their connection and similarities. After training the network, each p-th entry $\mathbf{x}(p)$ is associated with a specific BMU in the output layer, and entries that share similar patterns will be associated with the same BMU or its neighbors, which can be understood as a grouping in the SOM.

We applied the SOM to each monitoring station shown in Table 1. Each p-th sample in the dataset has $N = 10$ dimensions, 6 regarding atmospheric pollutants (SO_2, CO, O_3, PM_{10}, NO, and NO_2) and 4 concerning meteorological parameters (WS, TEMP, RH, and STWD). Therefore, the SOM enables analyzing the influence and characteristics of these variables.

2.4. SOM Parameters

The map size is the first parameter to be defined. For this purpose, it is necessary to determine the number of neurons to be used during training; in addition, avoiding a large or small number of neurons is vital to prevent problems such as non-identification of characteristics and overfitting [27]. Commonly, the number of neurons can be determined using the following heuristic equation

$$M \approx 5\sqrt{P} \tag{6}$$

where P is the number of input data samples [27].

Subsequently, the map topology ($M_h \times M_v$) was defined according to quality measures commonly used for the SOM network, the quantization error (QE) and topographic error (TE) [28,29]. For each station, different values of M_h and M_v were tested, in which $M_h \times M_v = M$ (Equation (6)). Finally, to analyze the results, three different types of normalization were applied to the data: z-score, min–max, and logarithmic.

Hence, all tests were performed with $b = 500$, a hexagonal topology, and the training algorithm was applied in two steps. Firstly, the learning rate and neighborhood function were initialized as $\eta(0) = 0.5$ and $\sigma^2(0) = \frac{M_h}{2}$, respectively, and decreased over iterations. Secondly, these values were fixed as $\eta = 0.05$ and $\sigma^2 = 1$. Tables 6–9 present the quality measures obtained for each test.

Table 6. SOM quality measures for Barros Reis Station data (best values in bold).

($M_h \times M_v$)	M	z-Score		Min-Max		Logarithmic	
		QE	TE	QE	TE	QE	TE
(27×24)	648	1.4032	0.0649	0.2290	0.0636	0.7173	0.0606
(26×26)	676	1.3898	0.0687	0.2273	0.0661	0.7108	0.0616
(29×24)	696	1.3887	0.0668	0.2270	0.0668	0.7096	0.0607
(31×23)	713	1.3805	0.0631	0.2259	0.0607	0.7051	0.0593
(27×27)	729	1.3803	0.0612	0.2245	0.0653	0.7032	0.0629
(30×25)	750	1.3766	0.0660	0.2250	0.0667	0.7017	0.0616
(32×24)	768	1.3684	0.0649	**0.2232**	**0.0622**	0.6977	0.0601
(34×23)	782	1.3652	0.0701	0.2229	0.0644	0.6950	0.0644
(33×24)	792	1.3609	0.0655	0.2224	0.0673	0.6957	0.0587
(31×26)	806	1.3597	0.0658	0.2219	0.0663	0.6948	0.0622

Table 7. SOM quality measures for Campo Grande Station data (best values in bold).

($M_h \times M_v$)	M	z-Score		Min-Max		Logarithmic	
		QE	TE	QE	TE	QE	TE
(31×23)	713	1.4216	0.0666	0.2384	0.0626	0.7346	0.0625
(27×27)	729	1.4187	0.0660	0.2382	0.0667	0.7294	0.0584
(30×25)	750	1.4131	0.0650	0.2369	0.0664	0.7277	0.0626
(32×24)	768	1.4082	0.0648	0.2360	0.0640	0.7253	0.0630
(28×28)	784	1.4099	0.0619	0.2352	0.0685	0.7230	0.0610
(31×26)	806	1.3994	0.0645	0.2341	0.0670	0.7215	0.0589
(34×24)	816	1.3948	0.0642	**0.2340**	**0.0624**	0.7193	0.0592
(33×25)	825	1.3949	0.0636	0.2334	0.0636	0.7173	0.0593
(35×24)	840	1.3925	0.0651	0.2336	0.0669	0.7163	0.0630
(36×24)	864	1.3898	0.0619	0.2324	0.0643	0.7146	0.0594

Table 8. SOM quality measures for Dique do Tororó Station data (best values in bold).

$(M_h \times M_v)$	M	z-Score		Min-Max		Logarithmic	
		QE	TE	QE	TE	QE	TE
(38×25)	950	1.2812	0.0686	0.2175	0.0668	0.6834	0.0630
(37×26)	962	1.2773	0.0684	0.2172	0.0670	0.6814	0.0641
(38×26)	988	1.2742	0.0668	0.2163	0.0679	0.6802	0.0621
(36×28)	1008	1.2687	0.0733	0.2157	0.0676	0.6798	0.0619
(32×32)	1024	1.2667	0.0736	0.2152	0.0679	0.6777	0.0659
(40×26)	1053	1.2628	0.0702	**0.2146**	**0.0678**	0.6745	0.0644
(39×27)	1040	1.2639	0.0695	0.2152	0.0688	0.6750	0.0611
(38×28)	1064	1.2581	0.0721	0.2141	0.0680	0.6747	0.0660
(37×29)	1073	1.2600	0.0706	0.2136	0.0728	0.6718	0.0632
(40×27)	1080	1.2609	0.0657	0.2136	0.0717	0.6730	0.0633

Table 9. SOM quality measures for Itaigara Station data (best values in bold).

$(M_h \times M_v)$	M	z-Score		Min-Max		Logarithmic	
		QE	TE	QE	TE	QE	TE
(24×23)	552	1.4306	0.0584	0.2428	0.0591	0.7736	0.0510
(26×22)	572	1.4237	0.0603	0.2422	0.0566	0.7709	0.0485
(24×24)	576	1.4210	0.0618	0.2421	0.0557	0.7704	0.0503
(27×22)	594	1.4192	0.0548	0.2403	0.0589	0.7684	0.0547
(25×24)	600	1.4152	0.0593	0.2412	0.0565	0.7659	0.0477
(27×23)	621	1.4126	0.0574	0.2400	0.0585	0.7654	0.0444
(25×25)	625	1.4063	0.0573	**0.2399**	**0.0553**	0.7625	0.0458
(27×24)	648	1.4086	0.0572	0.2381	0.0561	0.7595	0.0472
(26×26)	676	1.3945	0.0640	0.2371	0.0556	0.7553	0.0525
(27×26)	702	1.3861	0.0578	0.2363	0.0559	0.7516	0.0538

Considering both QE and TE measures, the lowest values were obtained using min–max normalization. Thus, M_h and M_v were chosen according to the best result, being highlighted in each table.

3. Results

3.1. U-Matrix, Components Plane and Parameter Similarity

The SOM output can be represented by a unified distance matrix (U-matrix) and a component plane, both illustrated in Figure 2. The U-matrix provides a visualization of the relative distance between neurons in the map, which is evidenced through a color scale, and highlights the calculated distance between the adjacent neurons [18]. The closer the color approaches a dark blue in the U-matrix, the closer these neurons are, i.e., they have a more significant similarity. On the other hand, the closer the color approaches a dark red, the greater the distance between the neurons and their dissimilarity. In general, this form of representation allows us to consider that neurons with smaller distances form a cluster. In contrast, neurons with high distances can be considered as boundaries of a cluster.

The component plane shows the values of the weight vectors of each neuron through a color code, where the blue and red colors correspond to low and high values, respectively. This representation allows the recognition of parameter dependencies by comparing the patterns of each plane. The color gradient of a plane represents the parameters' value (component) for the analyzed samples. Each neuron is assigned a color according to the parameter value in that neuron; thus, it can be said that two or more parameters are related based on a comparison of their color gradients. A coherent gradient indicates a positive correlation, while an inverse gradient a negative correlation.

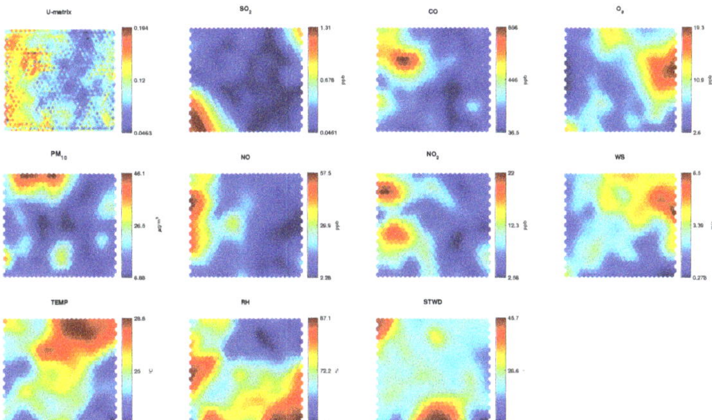

Figure 2. Unified distance matrix (U-matrix) and component planes of all analyzed variables (SO_2, CO, O_3, PM_{10}, NO, NO_2 , WS, TEMP, RH, and STWD) from the Itaigara station.

3.2. Itaigara Station

Analyzing the component planes in Figure 2, it is possible to note that the relative humidity (RH) and temperature (TEMP) planes display inverse gradients, indicating a negative correlation between these parameters—something already expected given their characteristics. For CO, NO, and NO_2 pollutants, their weight vectors present a dark red color on the left side of the components' plane, with a higher concentration of high values at the top left side; hence evidencing a certain similarity between them. These pollutants are generated by combustion, and incomplete burning of organic fuels, which are very common in cities with a large circulation of vehicles (the leading emitter) [5].

The O_3 pollutant can be formed by the reaction of nitrogen oxides with VOCs. However, it presents a different pattern than NO_2, which contributes to the formation of photochemical oxidants such as O_3. As can be seen in the O_3 component plane, its high-value region is concentrated on the right side, similar to the wind speed component plane. Therefore, it can be said that the O_3 presence at the Itaigara Station probably came from another region carried by the wind, as it has a low concentration near traffic routes and is generated by photochemical reactions.

The PM_{10} showed a different pattern than the other pollutants. Its main concentration region, with high weight vector values, is in the upper part of the plane. Since its emission sources are diverse, such as vehicles, biomass burning, industries, and dust resuspension, it is difficult to identify the major contributor pollutant. However, its formation can also be carried out in the atmosphere through VOCs, SO_2, and nitrogen oxides.

The most distinct pattern presented was by SO_2, with high values and concentration in the lower left part, it does not resemble any other component plane. This pollutant is released mainly by heavy vehicles burning diesel oil in urban areas.

An SOM arranges similar patterns in the same neighborhood region, clustering the network's output. Hence, an investigation into the clustering of samples provides important information about the data.

The U-matrix in Figure 2 illustrates how close or far the neurons are, showing their clusters. However, the cluster boundaries are not clearly represented, making it challenging to identify them. One of the methods for choosing the appropriate number of clusters is the so-called Davies–Bouldin index [30], an evaluation measure commonly used in SOM networks for validating clusters [31,32].

3.2.1. Sample Grouping with the SOM Algorithm

For the Davies–Bouldin index, the lowest value found indicates the best number of clusters for the analyzed problem. Thus, an experiment was conducted by varying the number of clusters from two to eight and observing the obtained values. The best result was achieved for a total of four clusters.

Aftwards, a hierarchical analysis was performed to define the neurons belonging to the four clusters. For this purpose, the Euclidean distance was used as the similarity metric and the Ward neuron linking criterion, illustrated by the dendrogram shown in Figure 3. A dendrogram threshold value is defined for that to which cluster each neuron belongs (horizontal line in Figure 3).

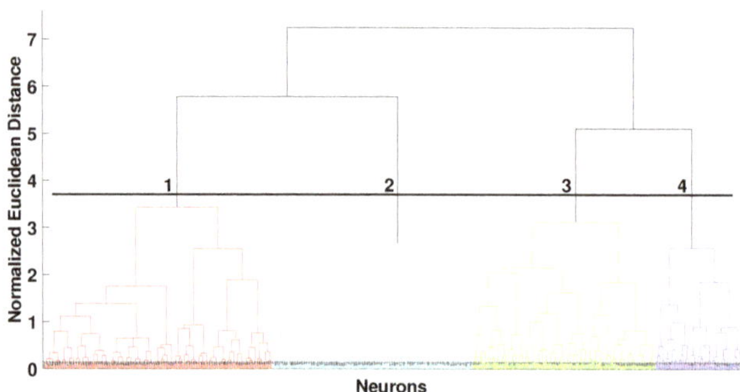

Figure 3. Hierarchical analysis of the neurons clusters using the Ward linkage method and Euclidean distance for the Itaigara station.

In addition, based on the hierarchical analysis, the SOM neurons were classified in four clusters, as shown in Figure 4. Therefore, the samples assigned to each cluster and its neurons present the characteristics of the distribution of pollutants and meteorological parameters. Table 10 shows the mean value of samples for each parameter and cluster.

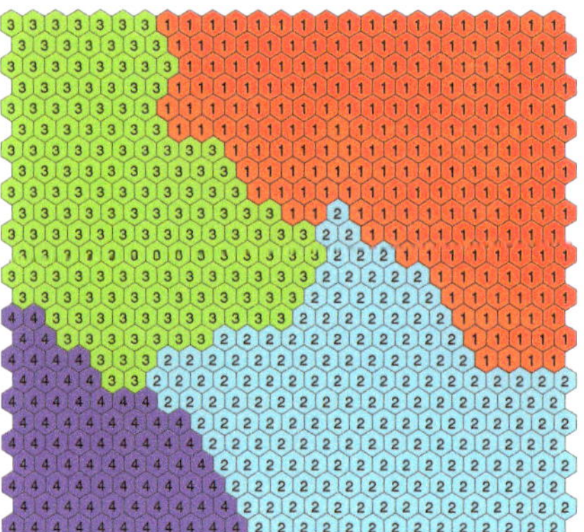

Figure 4. SOM neurons grouped into four clusters obtained by the hierarchical analysis of the Itaigara station.

Table 10. Parameters' average values for every cluster formed by the SOM network for the Itaigara station.

Parameters	Parameter Average Value per Cluster			
	1	2	3	4
SO_2 (ppb)	0.18	0.09	0.17	0.89
CO (ppb)	153.18	126.03	443.43	230.86
O_3 (ppb)	11.93	7.38	5.45	7.61
PM_{10} ($\mu g/m^3$)	17.73	12.92	17.97	15.83
NO (ppb)	9.20	9.15	29.64	19.28
NO_2 (ppb)	6.10	6.81	12.29	9.10
WS (m/s)	4.15	1.83	2.26	2.20
TEMP (°C)	26.44	24.10	24.80	23.97
RH (%)	64.30	76.97	73.15	74.38
STWD (°)	21.59	26.16	26.39	23.43
#Samples	5240	4469	3799	2027

According to Table 10, cluster 1 samples exhibit, in general, a low concentration of air pollutants, except for O_3 and PM_{10}, which have the highest average concentration. In addition, cluster 1 presents a wind speed and temperature considerably higher, and lower relative humidity. In total, about 34% of the data was assigned to cluster 1, thus sharing those characteristics.

Cluster 2, presented in Table 10, shows the lowest concentrations of SO_2, CO, PM_{10}, and NO pollutants, with intermediate values of O_3, and NO_2. It also presents the lowest average wind speed, intermediate temperature, and high relative humidity. In addition, cluster 2 is composed of 29% of the data, characterized by a low concentration of pollutants.

The highest concentrations of CO, PM_{10}, NO, and NO_2 are found in cluster 3, as can be observed in Table 10. In contrast, SO_2 and O_3 show low values (with O_3 having the lowest total average among all clusters). The wind speed, temperature, and relative humidity have intermediate values. A total of 24% of the data was assigned to cluster 3, characterized by high pollutant concentration values.

Finally, cluster 4 is mainly characterized by the high concentration of the SO_2 pollutant compared to the others. The other pollutants present intermediate concentration values, as well as wind speed, temperature, and relative humidity. In addition, cluster 4 has the lower amount of samples; a total of 2027 (13%) were assigned here.

3.2.2. Parameter Correlation

The component planes allow an initial and preliminary analysis of parameters through their visual gradients which, in a certain way, can turn out to be subjective and discretionary. Thus, to carry out a more objective and effective analysis of the results, a correlation analysis was applied between the component planes seen in Figure 2. Figure 5 shows the similarity between the planes (parameters) using the Ward criterion and the Pearson correlation coefficient, r.

As can be observed in Figure 5, two main branches are seen in the correlation analysis. The first branch, on the left of the figure, includes all the pollutants studied but O_3, whose origin is exclusively photochemical. Hence, O_3 is clustered with the wind speed and temperature.

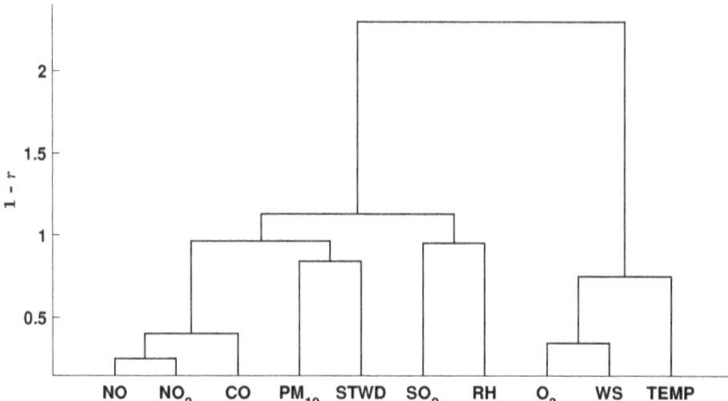

Figure 5. Parameter correlation using Ward criterion and distance $1 - r$, where r is Pearson coefficient, for the Itaigara station.

The NO, NO_2, and CO pollutants have a substantial similarity, probably due to a similar emission source such as vehicular, given the station allocation and the monitoring region. Those pollutants are correlated to PM_{10}, which also has a vehicular origin. In addition, the PM_{10} is connected to STWD, showing that intensive vertical turbulence (atmospheric instability), which is characterized by high STWD values, increases the PM_{10} concentration. Thus, it can be said that the wind movement is dragging out PM_{10} from other areas or causing the resuspension of particulate material at Itaigara station. In addition to vehicle influence, the particulate matter may also be dispersed by the existing vehicle movement, the wear of traffic lanes, and the vehicles' brake pads.

The similarity between RH and SO_2 shows the influence of RH on the formation or decomposition process of molecules during the heterogeneous procedure (liquid phase). In particular, the SO_2 can react with the air humidity and other oxidants in the atmosphere to form sulfuric acid H_2SO_4 and ammonium sulfate [33].

Meteorological parameters, such as wind speed, considerably influence the O_3 pollutant [24]. Given the similarity between O_3, the wind speed, and temperature (Figure 5), we consider that O_3 is not generated at the monitoring station site but instead transported by winds along with other pollutants such as VOCs. The temperature may also be responsible, since high temperatures result from the increase in the speed of chemical processes, generating ozone in the region.

3.3. Barros Reis Station

In the BR station component planes (Figure 6), the weight vectors for the PM_{10}, CO, NO, and NO_2 are displayed similarly across the map. The concentration of high values is on the upper left side, with average values in the nearby regions. The low values are located mainly in the lower right region of the map. All these pollutants can be formed from combustion processes, which shows the similarity obtained and, in particular, if they have a common source.

Unlike the pollutants discussed above, the O_3 component plane has its highest concentration at the bottom right of the map. O_3 is a secondary pollutant, i.e., its formation depends on atmosphere reactions from other pollutants, such as NO_2. Still, its plane does not resemble the planes of primary pollutants. Similarly, PM_{10} is also a secondary pollutant but is formed by SO_2, and no similarity is seen in their plane. However, PM_{10} can also be obtained from VOCs and nitrogen oxides, showing a relationship between their planes.

The SO_2 plane displays a unique pattern, with its highest values concentrated in the upper right region of the map, showing no similarity with the other pollutants. The component planes referring to meteorological parameters showed different distributions, with a negative correlation between TEMP and RH. At the same time, the high WS values are concentrated in the upper central region, and STWD with values dispersed throughout the map.

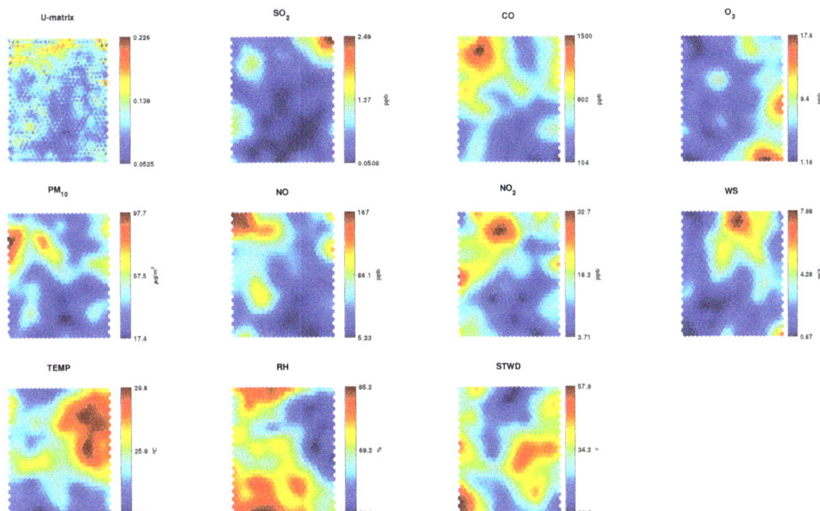

Figure 6. Unified distance matrix (U-matrix) and component planes of all analyzed variables (SO_2, CO, O_3, PM_{10}, NO, NO_2, WS, TEMP, RH and STWD) from the Barros Reis station.

3.3.1. Sample Grouping with the SOM Algorithm

Figure 6 presents the clusters through the U-matrix, representing the neurons with their distance to adjacent neurons. The cluster number was defined with the Davies–Bouldin index by varying it from two to eight, reaching the best result for three clusters.

Subsequently, a hierarchical analysis was performed to define the neurons belonging to the three clusters. Thereupon, the Ward criterion and the Euclidean distance were used as similarity metrics. Figure 7 displays the dendrogram obtained with the threshold value used for segregation. Meanwhile, Figure 8 shows how the clusters were arranged on the map.

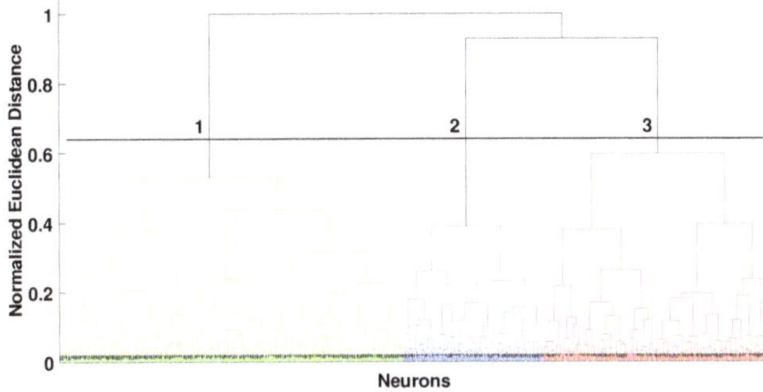

Figure 7. Hierarchical analysis of the neurons clusters using the Ward linkage method and Euclidean distance for the Barros Reis station.

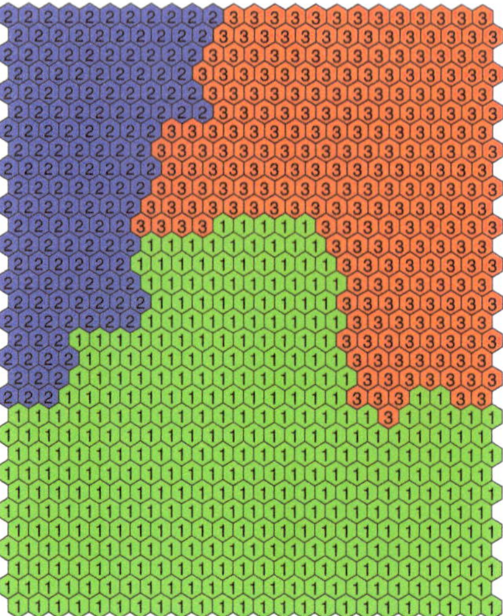

Figure 8. SOM neurons grouped into three clusters obtained by the hierarchical analysis of the Barros Reis station.

The samples are linked to a particular neuron belonging to one of the three clusters, allowing the analysis of the sample's distribution regarding the clusters.

Table 11 shows the average values of every parameter according to the cluster. As can be seen, cluster 1 represents the samples with the lowest pollutant concentration, except for O_3 which has a median value among the others. Meteorological parameters such as wind speed, temperature, and relative humidity also have low values. In total, the cluster has 10,599 samples with these characteristics, corresponding to 49.16% of the station data.

Table 11. Parameters average values for every cluster formed by the SOM network for the Barros Reis station.

Parameters	Parameter Average Value per Cluster		
	1	2	3
SO_2 (ppb)	0.28	0.64	0.59
CO (ppb)	442.11	973.00	621.71
O_3 (ppb)	5.42	3.03	7.07
PM_{10} ($\mu g/m^3$)	33.31	54.00	42.13
NO (ppb)	37.54	91.22	51.88
NO_2 (ppb)	10.74	20.23	15.68
WS (m/s)	1.93	1.93	4.12
TEMP (°C)	24.66	24.74	27.68
RH (%)	72.64	72.19	60.05
STWD (°)	32.73	31.04	30.19
#Samples	10,599	4183	6777

In the meantime, cluster 2 exhibits the highest concentration of pollutants, displaying a considerable difference from the values of other clusters except for O_3, which has the lowest average value obtained. Similar to cluster 1, the wind speed, temperature, and relative humidity also have low values. Cluster 2 has 4183 samples, equivalent to 19.40% of the data.

Finally, the samples assigned to cluster 3 have an intermediate value of pollutants concentration, with average values between the clusters 1 and 2 range, except for O_3 which has the highest average concentration recorded. In addition, cluster 3 has 31.44% of the station data with the highest wind speed and the lowest relative humidity.

3.3.2. Parameter Correlation

The component planes, shown in Figure 6, present the correlation between parameters. Meanwhile, Figure 9 presents the parameters similarity obtained using the Ward linking method and the Pearson correlation coefficient.

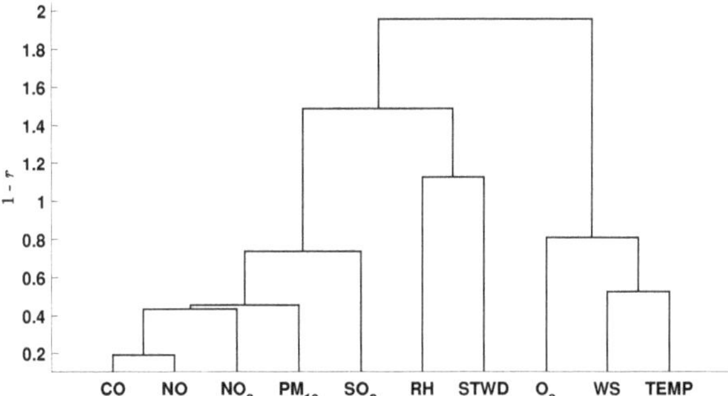

Figure 9. Parameter correlation using Ward criterion and distance $1 - r$, where r is Pearson coefficient, for the Barros Reis station.

As shown in Figure 9, there is a substantial similarity between CO and NO pollutants. Given the BR station characteristics (located in between two avenues), it can be said that motor vehicles are the primary emission source of those pollutants. Likewise, the NO_2 and PM_{10} pollutants are also emitted by combustion in vehicles; in addition, they can be formed secondarily by photochemical processes. Regarding SO_2, it can be said that the primary emission source is the burning process of fuels, such as diesel and gasoline, from heavy vehicles such as trucks, buses, microbuses, and light vehicles.

Unlike other pollutants, the O_3 showed a clear relationship with meteorological parameters such as wind speed and temperature, similar to the Itaigara station. Nonetheless, this relationship with meteorological parameters is not strong as in other stations.

The STWD indicates the local atmospheric stability. Its inverse relationship with RH can be related to the regions' water molecules' dissipation. Hence, the data regarding pressure and heat could improve the analysis precision by demonstrating the influence of the wind direction. The RH and STWD present a negative relationship with the other pollutants, consequently leading to the non-contribution or reduction in the present concentrations.

3.4. Campo Grande Station

Figure 10 illustrates the component planes for the CG station. Concerning the planes of nitrogen oxide, a significant similarity between NO_2 and CO can be observed, with high values concentrated in the central part of the map. The NO plane is also similar to the CO and NO_2, but the high values are concentrated in the region to the right, while median values are concentrated in the map center. The emission source of these pollutants is fuel combustion, especially from vehicles.

The SO_2 has high values concentrated in the lower right region of the map. The PM_{10}, on the other hand, did not show significant pattern similarities with other planes, having a higher concentration in the upper part of the map and moderate concentration in the lower part, equivalent to small regions of the SO_2 and NO_2 planes. Likewise, the O_3 pollutant also shows no similarity with other component planes. Despite its formation, resulting from the reaction between NO_2 and VOCs, its concentration of high values is located at the map edges, having similarities with the concentration regions of high values of meteorological parameters, such as WS, TEMP, RH, and STWD.

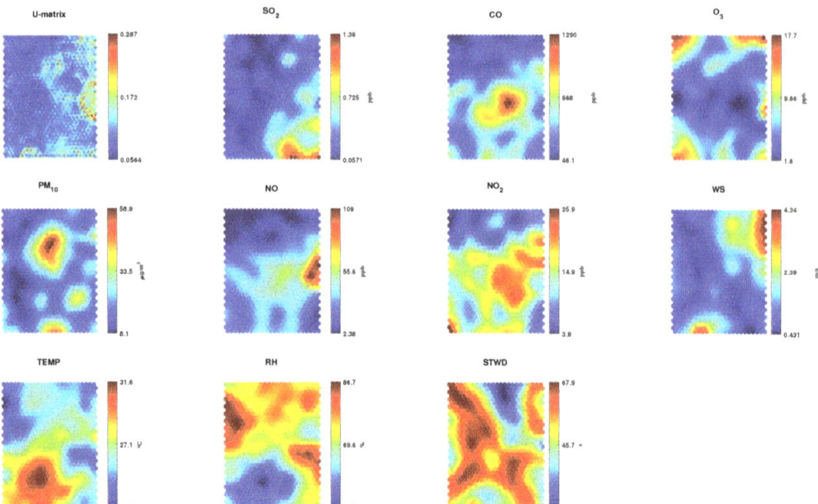

Figure 10. Unified distance matrix (U-matrix) and component planes of all analyzed variables (SO$_2$, CO, O$_3$, PM$_{10}$, NO, NO$_2$, WS, TEMP, RH and STWD) from the Campo Grande station.

3.4.1. Sample Grouping with the SOM Algorithm

To identify the CG station clusters through the U-matrix, illustrated in Figure 10, the Davies–Bouldin was used and the cluster number varied from two to eight. The best result was obtained for five clusters. Aftward, the neurons belonging to each cluster were obtained according to a hierarchical analysis defined based on the *Ward* method and Euclidean distance. Figure 11 shows the resulting dendrogram and the segregation threshold. Meanwhile, Figure 12 displays the neurons distribution regarding the clusters.

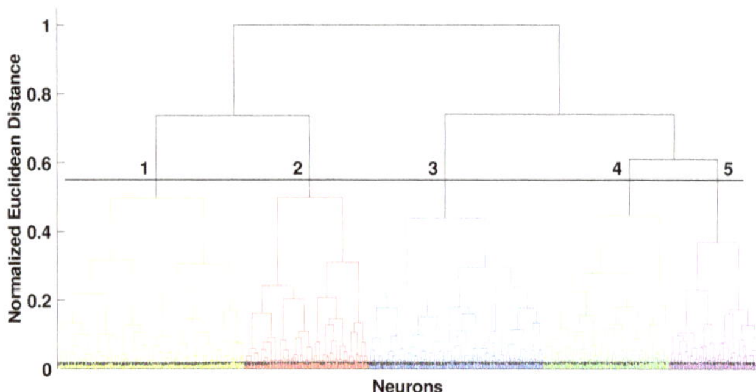

Figure 11. Hierarchical analysis of the neurons clusters using the Ward linkage method and Euclidean distance for the Campo Grande station.

Figure 12. SOM neurons grouped into five clusters obtained by the hierarchical analysis of the Campo Grande station.

Each CG station dataset sample was integrated into the cluster with the neuron it most resembles. Thus, an analysis was performed regarding the samples' distribution by cluster based on the average values of parameters, as shown in Table 12.

Table 12. Parameters average values for every cluster formed by the SOM network for the Campo Grande station.

Parameters	Parameter Average Value per Cluster				
	1	2	3	4	5
SO_2 (ppb)	0.14	0.26	0.23	0.38	0.82
CO (ppb)	250.98	269.62	456.78	655.67	404.37
O_3 (ppb)	6.12	7.07	6.59	4.15	5.72
PM_{10} ($\mu g/m^3$)	20.01	23.71	17.64	24.45	22.21
NO (ppb)	15.07	18.97	30.47	56.49	24.42
NO_2 (ppb)	10.34	11.12	14.33	19.15	13.12
WS (m/s)	0.89	2.70	1.40	1.36	0.94
TEMP (°C)	25.21	25.73	29.22	26.26	26.90
RH (%)	77.31	75.14	61.52	73.56	68.58
STWD (°)	57.82	42.83	53.05	50.50	52.28
#Samples	6640	4166	6223	4229	3301

According to Table 12, cluster 1 has the lowest average values of concentration for the SO_2, CO, NO, and NO_2 pollutants, while the PM_{10} and O_3 show intermediate values. Moreover, the wind speed and temperature are the lowest of all. Cluster 1 consists of 6640 data samples, equivalent to 27.04% of the dataset.

Cluster 2 also presents low average values of the concentrations of the pollutants, with values slightly higher than those obtained in cluster 1, except for the O_3 pollutant, which has a higher concentration average. Similar behavior can be seen for the meteorological parameters except for the wind speed, which shows the highest average among all clusters. In total, 16.96% of the data was assigned to cluster 2.

The samples assigned to cluster 3 present intermediate values for all pollutants concentration and meteorological parameters, where the temperature has the highest average and the relative humidity the lowest. This cluster has 25.34% of the data.

The highest concentrations of CO, PM_{10}, NO, and NO_2 are found in cluster 4, with an intermediate concentration of SO_2 and the lowest concentration of O_3. Meanwhile, all meteorological parameters showed intermediate values compared to other clusters. Cluster 4 has a total of 17.22% of the data.

Meantime, cluster 5 stands out with the highest average concentration of the SO_2 pollutant. The other pollutants, as well as the meteorological parameters, present intermediate average values. In total, 7.44% of the data was assigned to cluster 5.

3.4.2. Parameter Correlation

The hierarchical representation for the CG station was obtained with the *Ward* method and the *Pearson* correlation coefficient. Figure 13 presents the parameters' correlation obtained.

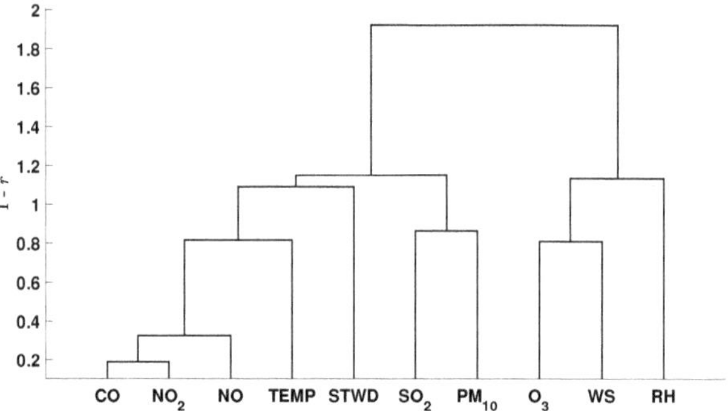

Figure 13. Parameter correlation using *Ward* criterion and distance $1 - r$, where r is *Pearson* coefficient, for the Campo Grande station.

First of all, the similarity between CO, NO, and NO_2 pollutants can be seen. These pollutants are emitted in urban areas mainly by motor vehicles, and their similarity validates the idea of a potential common emission source. The temperature is also similar to those three pollutants as it contributes to chemical processes that form them—for example, the NO_2 results from the sunlight action on NO. Thus, the temperature can impact the amount of those pollutants present in every season.

The PM_{10} is a primary and secondary pollutant, and it is correlated to SO_2. Thus, its atmospheric formation can be linked to gases turning into particles due to chemical reactions in the air, such as sulfur dioxide. The SO_2 is generated from the burning of fuels with sulfur in its composition, such as diesel oil or industrial fuel oil, and it appears to be related to the PM_{10} due to motor vehicle emissions, among other processes.

The photochemical oxidant, O_3, has a certain correlation with the wind speed, but with a much lower similarity than that presented by the Itaigara station. In addition, there is no apparent relationship with the temperature. The RH has a negative relationship with O_3 and wind speed, which may be a consequence of solar radiation; low RH concentrations are related to a high solar incidence and, therefore, a greater disposition to the O_3 formation.

3.5. Dique do Tororó Station

The SOM network component planes for the DT station are shown in Figure 14. The pollutants that are mainly emitted by combustion processes, such as CO, NO, NO_2, and PM_{10} showed similar distribution patterns of values, with the highest concentration from the left side to the upper left side of the map. In contrast, the PM_{10} has higher values at the bottom of the map, similar to the temperature and wind speed.

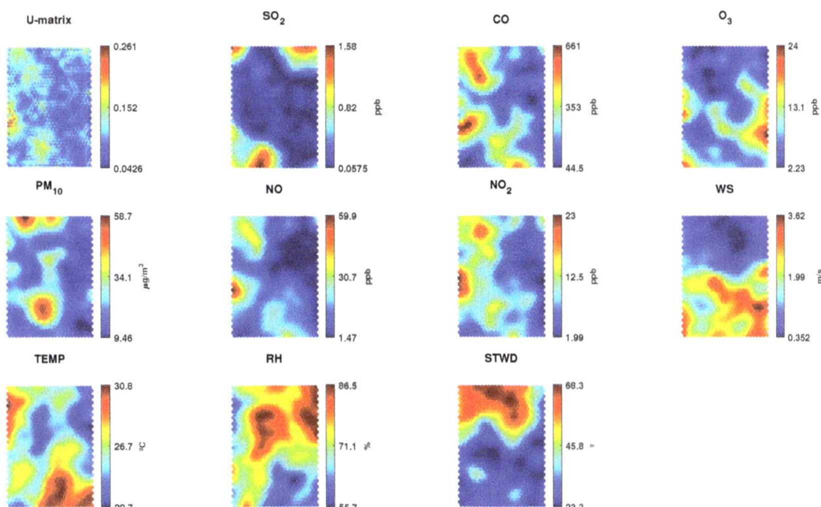

Figure 14. Unified distance matrix (U-matrix) and component planes of all analyzed variables (SO$_2$, CO, O$_3$, PM$_{10}$, NO, NO$_2$, WS, TEMP, RH and STWD) from the Dique do Tororó station.

As in the other stations, the SO$_2$ showed a different pattern from the other pollutants, with regions of high values concentration at the edges of the map. However, one of the high-concentration edges slightly coincides with those of the CO, NO, and NO$_2$. Lastly, the O$_3$ displays high values at the lower right region of the map, with a similar distribution to the wind speed plane. The other planes, such as relative humidity and STWD (which can influence the concentration of pollutants), showed patterns with well-defined regions at the top of the map.

3.5.1. Sample Grouping with the SOM Algorithm

The map neurons, represented by their respective distances to adjacent neurons in the U-matrix (Figure 14), were used to visualize and determine the clusters. For this purpose, the Davies–Bouldin index was used, and the number of clusters varied from two to eight, resulting in the best amount with three clusters. Again, hierarchical analysis was carried out using the Ward criterion and Euclidean distance. Figure 15 illustrates the dendrogram, and Figure 16 the segregation borders of the map.

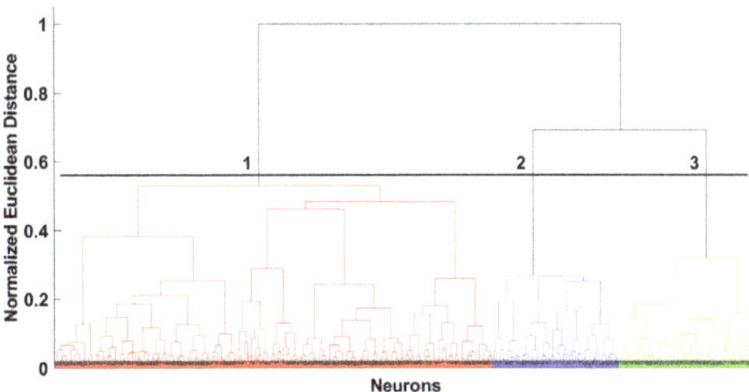

Figure 15. Hierarchical analysis of the neurons clusters using the Ward linkage method and Euclidean distance for the Dique do Tororó station.

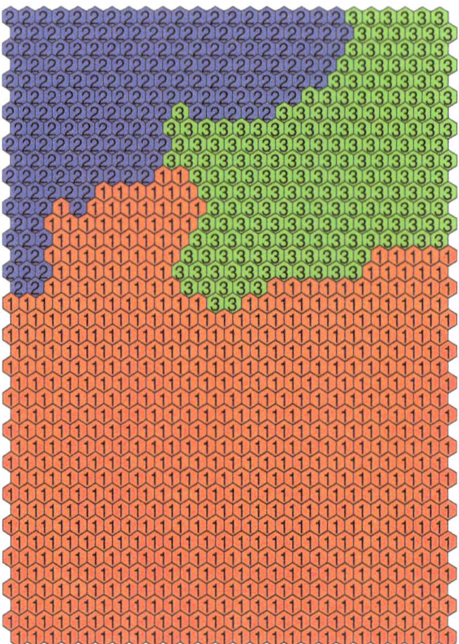

Figure 16. SOM neurons grouped into three clusters obtained by the hierarchical analysis of the Dique do Tororó station.

Each cluster was assigned a certain number of samples according to their characteristics. Table 13 presents the concentration averages of each pollutant according to the cluster.

Table 13. Parameters average values for every cluster formed by the SOM network for the Dique do Tororó station.

Parameters	Parameter Average Value per Cluster		
	1	2	3
SO_2 (ppb)	0.30	0.36	0.34
CO (ppb)	245.23	342.03	130.06
O_3 (ppb)	9.68	5.18	6.05
PM_{10} ($\mu g/m^3$)	21.23	29.66	18.23
NO (ppb)	15.14	20.06	3.91
NO_2 (ppb)	8.71	12.13	5.45
WS (m/s)	2.23	0.67	0.66
TEMP (°C)	26.90	27.22	24.40
RH (%)	69.65	71.88	81.69
STWD (°)	29.18	60.49	47.53
#Samples	26,101	7511	8425

As can be seen in Table 13, cluster 1 represents the samples with the highest mean value of O_3 and intermediate values of the other pollutants (SO_2, CO, NO, NO_2, and PM_{10}). The highest concentration value is the wind speed, while relative humidity and STWD are the lowest. Cluster 1 has 26,101 samples sharing its characteristics, equivalent to 62.09% of the station data.

The pollutants in cluster 2 had the highest average concentration, except for O_3 which showed the lowest concentration among all clusters. The wind speed presents low values, and the temperature parameter is the highest. In total, 17.87% of data constitutes this cluster.

Finally, the pollutants in cluster 3, that is, CO, NO, NO_2, and PM_{10}, had the lowest average concentrations, with SO_2 and O_3 showing intermediate values. The temperature and wind speed parameters have the lowest values found and the relative humidity the highest. Cluster 3 represents 20.04% of the station data with 8425 samples.

3.5.2. Parameter Correlation

The DT station component planes, shown in Figure 14, presents the parameters correlation. Meanwhile, Figure 17 illustrates the parameter similarity obtained through the *Ward* criterion and the *Pearson* correlation coefficient.

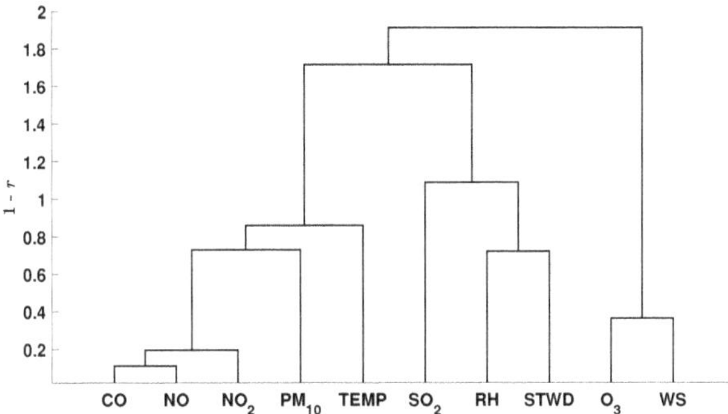

Figure 17. Parameter correlation using *Ward* criterion and distance $1 - r$, where r is *Pearson* coefficient, for the Dique do Tororó station.

As can be seen in Figure 17, the CO, NO, and NO_2 pollutants have the most significant similarity, a characteristic also observed for other stations. All stations are located in urban centers with a large flow of vehicles, leading to the possibility of a common emission source of these pollutants, mainly coming from the local vehicular fleet. The PM_{10} also showed a certain similarity with those pollutants, indicating a possible emission from fuel burning. The temperature parameter at the DT is also related to the mentioned pollutants, different from other stations where it is associated with O_3. In addition, the temperature can contribute to NO_2 formation and PM_{10} in secondary processes.

Like Barros Reis station, the RH and the STWD at the DT station are somewhat similar but with a positive coefficient. The RH and STWD can be influenced by atmospheric parameters such as pressure and heat and, consequently, the wind conditions and water particles.

The SO_2, different from the Itaigara station, is not correlated to either the PM_{10} or RH, as it is probably being generated by an independent source and not reacting to other pollutants.

Given that O_3 is a secondary pollutant, it was only correlated with wind speed, with no apparent similarity with temperature or nitrogen oxides. Therefore, its concentration at the DT station may be transported by the wind accompanied by other pollutants.

4. Discussion

The SOM implementation presented in the previous sections identifies the correlation among different air quality parameters for many monitoring stations. The SOM component planes provide a visual representation of the similarities between pollutants and meteorological parameters, simplifying their analysis and highlighting peculiarities.

Usually, the CO, NO, and NO_2 pollutants were related, showing higher similarities. On the other hand, the meteorological parameters differed from PM_{10} and SO_2. The RH and STWD parameters at Barros Reis station showed a negative correlation, unlike at Dique de Tororo station, where a positive correlation was presented. At Itaigara station, the influence of atmospheric stability was identified through the relationship between STWD and PM_{10}. Meanwhile, Campo grande station shows some degree of similarity between PM_{10} and SO_2. These relations are essential to identify the influence of meteorology on air-pollutants concentrations and information employed to create strategies for mitigating air-pollution critical episodes.

Unlike other pollutants, the O_3 presents a more significant link with meteorological parameters such as WS, as seen at Itaigara, Dique do Tororó and Campo grande stations. Thus, we can infer that the wind is mainly responsible for the transport of O_3. In addition, the correlation of the TEMP, WS, and O_3 parameters at Barros Reis and Itaigara stations indicates an increase in O_3 resulting from chemical processes, probably due to the influence of solar radiation.

The data of Dique do Tororo and Barros Reis stations were grouped only into three clusters, with their cluster 1 emphasizing a large number of samples with higher concentrations of O_3. In contrast, the other clusters present a sample distribution with intermediate to high concentrations for the CO, NO, NO_2, PM_{10}, and SO_2 pollutants. Meanwhile, the Itaigara station has four clusters, with one mainly characterized by the SO_2 pollutant; the remaining clusters are defined by higher concentrations of CO, NO, NO_2, PM_{10}, and O_3. Similar to Itaigara, the Campo Grande station has one cluster (out of five) where SO_2 is predominant, while the other clusters display low and high concentrations.

Commonly, studies about atmospheric pollutants rely on methods such as principal component analysis (PCA) and hierarchical analysis to define clusters based on similarity. For example, the studies carried out by [8,34] describe the clusters' characteristics according to the percentage of their main components' variance, thus, indicating which variables have more significance for their definition. Meanwhile, by applying a hierarchical classification on the SOM neurons, we can obtain the variables' concentration value and influence on defining each cluster.

In the meantime, in [13,35], the number of clusters is fixed for all monitoring stations, and the k-nearest neighbors provide a relationship between the defined clusters of each station. However, the SOM also allows an individual characteristic analysis of each pollutant, like in [35].

Thereby, the SOM enables finding similarities and estimating the link between parameters more deeply. As described in this work, the SOM can obtain data patterns and cluster characteristics and demonstrate the parameters' influence, which is not trivial in other techniques. Additionally, it can also deal with the non-linearity complexity of air pollution data [36], simplifying the analysis process and increasing its precision; this shows the advantage of using a machine-learning-based approach compared to traditional methods.

5. Conclusions

We implemented an SOM to analyze the air-quality data of four stations in the monitoring network of Salvador, Brazil. A detailed discussion regarding pollutants and their correlation with meteorological parameters is provided, assisting in estimating possible common emission sources and the influence of meteorological parameters. The latter permits the establishment of relations between meteorology and pollutants concentration, which is vital for developing, for example, alert systems to identify critical episodes of air pollution or for assisting in developing strategies to improve air quality.

The SOM outputs enabled the identification of data particularities concerning the parameters analyzed. For example, the data samples' concentration of Dique do Tororo and Barros Reis stations showed a cluster with a high concentration of O_3. In contrast, the other clusters presented well-defined contributions of remaining pollutants. The Itaigara and Campo Grande stations presented a more detailed definition regarding the clusters of (1) CO, NO, NO_2; (2) MP_{10}; (3) O_3; and (4) SO_2. Thus, the SOM also allows an analysis of the particularities of each cluster.

The results showed that the SOM could identify characteristics, describe similarities, recognize patterns, and define clusters of air-pollution problems. Unlike traditional methods, the SOM proved to be a good tool for studying atmospheric pollutants, providing several aspects that can contribute to and improve discussions in this area. To the best of our knowledge, this is the first study to analyze Salvador's air-quality monitoring database. Therefore, the tool developed and the results presented and discussed here can assist further studies and aid in the development of public policies for pollution management.

Author Contributions: All the authors have contributed in various degrees to ensure the quality of this work (e.g., E.L.R.C., T.B., L.A.D., É.L.d.A. and M.A.C.F. conceived the idea and experiments; E.L.R.C., T.B., L.A.D., É.L.d.A. and M.A.C.F. designed and performed the experiments; E.L.R.C., T.B., L.A.D., É.L.d.A. and M.A.C.F. analyzed the data; E.L.R.C., T.B., L.A.D., É.L.d.A. and M.A.C.F. wrote the paper. É.L.d.A. and M.A.C.F. coordinated the project). All authors have read and agreed to the published version of the manuscript.

Funding: This study was financed in part by the Coordenação de Aperfeiçoamento de Pessoal de Nível Superior (CAPES)—Finance Code 001.

Institutional Review Board Statement: Not applicable.

Informed Consent Statement: Not applicable.

Data Availability Statement: Not applicable.

Acknowledgments: The authors wish to acknowledge the financial support of the Coordenação de Aperfeiçoamento de Pessoal de Nível Superior (CAPES) for their financial support. The authors want to thank the "CETREL S. A. Company and Bahia State Government" for the availability of the monitoring data in Salvador.

Conflicts of Interest: The authors declare no conflict of interest.

References

1. Landrigan, P.J.; Fuller, R.; Acosta, N.J.R.; Adeyi, O.; Arnold, R.; Basu, N.N.; Baldé, A.B.; Bertollini, R.; Bose-O'Reilly, S.; Boufford, J.I.; et al. The Lancet Commission on pollution and health. *Lancet* **2017**, *391*, 462–512. [CrossRef]
2. Zivin, J.G.; Neidell, M. Air pollution's hidden impacts. *Science* **2018**, *359*, 39–40. [CrossRef] [PubMed]
3. Turner, M.C.; Andersen, Z.J.; Baccarelli, A.; Diver, W.R.; Gapstur, S.M.; Pope, C.A., III; Prada, D.; Samet, J.; Thurston, G.; Cohen, A. Outdoor air pollution and cancer: An overview of the current evidence and public health recommendations. *CA A Cancer J. Clin.* **2020**, *70*, 460–479. [CrossRef] [PubMed]
4. Zhang, J.; Zhang, L.; Du, M.; Zhang, W.; Huang, X.; Zhang, Y.; Yang, Y.; Zhang, J.; Deng, S.; Shen, F.; et al. Indentifying the major air pollutants base on factor and cluster analysis, a case study in 74 Chinese cities. *Atmos. Environ.* **2016**, *144*, 37–46. [CrossRef]
5. Zhang, K.; Batterman, S. Air pollution and health risks due to vehicle traffic. *Sci. Total Environ.* **2013**, *450–451*, 307–316. [CrossRef]
6. Bai, L.; Wang, J..; Ma, X.; Lu, H. Air Pollution Forecasts: An Overview. *Int. J. Environ. Res. Public Health* **2018**, *15*, 780. [CrossRef]
7. Núñez-Alonso, D.; Pérez-Arribas, L.V.; Manzoor, S.; Cáceres, J.O. Statistical Tools for Air Pollution Assessment: Multivariate and Spatial Analysis Studies in the Madrid Region. *J. Anal. Methods Chem.* **2018**, *2019*, 9753927. [CrossRef]
8. Tian, D.; Fan, J.; Jin, H.; Mao, H.; Geng, D.; Hou, S.; Zhang, P.; Zhang, Y. Characteristic and Spatiotemporal Variation of Air Pollution in Northern China Based on Correlation Analysis and Clustering Analysis of Five Air Pollutants. *J. Geophys. Res. Atmos.* **2020**, *125*, e2019JD031931. [CrossRef]
9. Manimaran, P.; Narayana, A.C. Multifractal detrended cross-correlation analysis on air pollutants of University of Hyderabad Campus, India. *Phys. A Stat. Mech. Its Appl.* **2018**, *502*, 228–235. [CrossRef]
10. Bai, Y.; Jin, X.; Wang, X.; Wang, X.; Xu, J. Dynamic Correlation Analysis Method of Air Pollutants in Spatio-Temporal Analysis. *Int. J. Environ. Res. Public Health* **2020**, *17*, 360. [CrossRef]
11. Zhao, S.; Yu, Y.; Yin, D.; He, J.; Liu, N.; Qu, J.; Xiao, J. Annual and diurnal variations of gaseous and particulate pollutants in 31 provincial capital cities based on in situ air quality monitoring data from China National Environmental Monitoring Center. *Environ. Int.* **2016**, *86*, 92–106. [CrossRef] [PubMed]
12. Yin, D.; Zhao, S.; Qu, J. Spatial and seasonal variations of gaseous and particulate matter pollutants in 31 provincial capital cities, China. *Air Qual. Atmos. Health* **2016**, *10*, 359–370. [CrossRef]
13. Li, C.; Wang, Z.; Li, B.; Peng, Z.; Fu, Q. Investigating the relationship between air pollution variation and urban form. *Build. Environ.* **2019**, *147*, 559–568. [CrossRef]
14. Periš, N.; Buljac, M.M.B.; Buzuk, M.; Brinić, S.; Plazibat, I. Characterization of the Air Quality in Split, Croatia Focusing Upon Fine and Coarse Particulate Matter Analysis. *Anal. Lett.* **2015**, *48*, 553–565. [CrossRef]
15. Wang, C.; Zhao, L.; Sun, W.; Xue, J.; Xie, Y. Identifying redundant monitoring stations in an air quality monitoring network. *Atmos. Environ.* **2018**, *190*, 256–268. [CrossRef]
16. Ran, Z.Y.; Hu, B.G. Parameter Identifiability in Statistical Machine Learning: A Review. *Neural Comput.* **2017**, *29*, 1151–1203. [CrossRef]
17. Capizzi, G.; Sciuto, G.L.; Monforte, P.; Napoli, C. Cascade Feed Forward Neural Network-based Model for Air Pollutants Evaluation of Single Monitoring Stations in Urban Areas. *Neural Comput.* **2015**, *61*, 327–332. [CrossRef]
18. Kohonen, T. *Self-Organizing Maps*, 3rd ed.; Springer: Berlin/Heidelberg, Germany, 2001.
19. Asan, U.; Ercan, S. An Introduction to Self-Organizing Maps. In *Computational Intelligence Systems in Industrial Engineering: With Recent Theory and Applications*; Atlantis Press: Paris, France, 2012; pp. 295–315. [CrossRef]

20. Pearce, J.L.; Waller, L.A.; Chang, H.H.; Klein, M.; Mulholland, J.A.; Sarnat, J.A.; Sarnat, S.E.; Strickland, M.J.; Tolbert, P.E. Using self-organizing maps to develop ambient air quality classifications: A time series example. *Environ. Health* **2014**, *11*, 56. [CrossRef]
21. Zhong, B.; Wang, L.; Liang, T.; Xing, B. Pollution level and inhalation exposure of ambient aerosol fluoride as affected by polymetallic rare earth mining and smelting in Baotou, north China. *Atmos. Environ.* **2017**, *167*, 40–48. [CrossRef]
22. Jiang, N.; Scorgie, Y.; Hart, M.; Riley, M.L.; Crawford, J.; Beggs, P.J.; Edwards, G.C., Chang, L.; Salter, D.; Virgilio, G.D. Visualising the relationships between synoptic circulation type and air quality in Sydney, a subtropical coastal-basin environment. *Int. J. Climatol.* **2017**, *37*, 1211–1228. [CrossRef]
23. Moosavi, V.; Aschwanden, G.; Velasco, E. Finding candidate locations for aerosol pollution monitoring at street level using a data-driven methodology. *Atmos. Meas. Tech.* **2015**, *8*, 3563–3575. [CrossRef]
24. Li, D.; Liao, Y. Pollution zone identification research during ozone pollution processes. *Environ. Monit. Assess.* **2020**, *192*, 591. [CrossRef] [PubMed]
25. Fávero, L.P.L.; Belfiore, P.P. *Manual de Análise de Dados: Estatística e Modelagem Multivariada com Excel, SPSS e Stata*, 1st ed.; Elsevier: Rio de Janeiro, Brazil, 2017.
26. Kohonen, T.; Oja, E.; Simula, O.; Visa, A.; Kangas, J. Engineering applications of the self-organizing map. *Proc. IEEE* **1996**, *84*, 1358–1384. [CrossRef]
27. Vesanto, J.; Alhoniemi, E. Clustering of the self-organizing map. *IEEE Trans. Neural Netw.* **2000**, *11*, 586–600. [CrossRef] [PubMed]
28. Pölzlbauer, G. Survey and Comparison of Quality Measures for Self-Organizing Maps. In *Proceedings of the Fifth Workshop on Data Analysis (WDA'04)*; Elfa Academic Press: Vysoké Tatry, Slovakia, 2004; pp. 67–82.
29. Kiviluoto, K. Topology preservation in self-organizing maps. In Proceedings of the Proceedings of International Conference on Neural Networks (ICNN'96), Washington, DC, USA, 3–6 June, 1996; Volume 1, pp. 294–299. [CrossRef]
30. Davies, D.L.; Bouldin, D.W. A Cluster Separation Measure. *IEEE Trans. Pattern Anal. Mach. Intell.* **1979**, *1*, 224–227. [CrossRef]
31. Li, T.; Sun, G.; Yang, C.; Liang, K.; Ma, S..; Huang, L. Using self-organizing map for coastal water quality classification: Towards a better understanding of patterns and processes. *Sci. Total. Environ.* **2018**, *628–629*, 1446–1459. [CrossRef]
32. Li, Y.; Wright, A.; Liu, H.; Wang, J.; Wang, G.; Wu, Y.; Dai, L. Land use pattern, irrigation, and fertilization effects of rice-wheat rotation on water quality of ponds by using self-organizing map in agricultural watersheds. *Agric. Ecosyst. Environ.* **2019**, *272*, 155–164. [CrossRef]
33. Turalıoğlu, F.S.; Nuhoğlu, A.; Bayraktar, H. Impacts of some meteorological parameters on SO2 and TSP concentrations in Erzurum, Turkey. *Chemosphere* **2005**, *59*, 1633–1642. [CrossRef]
34. Dominick, D.; Juahir, H.; Latif, M.T.; Zain, S.M.; Aris, A.Z. Spatial assessment of air quality patterns in Malaysia using multivariate analysis. *Atmos. Environ.* **2012**, *60*, 172–181. [CrossRef]
35. Iizuka, A.; Shirato, S.; Mizukoshi, A.; Noguchi, M.; Yamasaki, A.; Yanagisawa, Y. A Cluster Analysis of Constant Ambient Air Monitoring Data from the Kanto Region of Japan. *Int. J. Environ. Res. Public Health* **2014**, *11*, 6844. [CrossRef]
36. Yeganeh, B.; Motlagh, M.; Rashidi, Y.; Kamalan, H. Prediction of CO concentrations based on a hybrid Partial Least Square and Support Vector Machine model. *Atmos. Environ.* **2012**, *55*, 357–365.[CrossRef]

 *sustainability*

Article

Regional Differences, Distribution Dynamics, and Convergence of Air Quality in Urban Agglomerations in China

Yuting Xue [1] and Kai Liu [1,2,*]

[1] College of Geography and Environment, Shandong Normal University, Jinan 250358, China; 18220679378@163.com
[2] Collaborative Innovation Center of Human-Nature and Green Development in Universities of Shandong, Shandong Normal University, Jinan 250358, China
* Correspondence: liukaisdnu@163.com

Abstract: The urban agglomeration (UA), with a high concentration of population and economy, represents an area with grievous air pollution. It is vital to examine the regional differences, distribution dynamics, and air quality convergence in UAs for sustainable development. In this study, we measured the air quality of ten UAs in China through the Air Quality Index (AQI). We analyzed regional differences, distribution dynamics, and convergence using Dagum's decomposition of the Gini coefficient, kernel density estimation, and the convergence model. We found that: the AQI of China's UAs shows a downward trend, and the index is higher in northern UAs than in southern UAs; the differences in air quality within UAs are not significant, but there is a gap between them; the overall difference in air quality tends to decrease, and regional differences in air quality are the primary contributor to the overall difference; the overall distribution and the distribution of each UA move rightward; the distribution pattern, ductility, and polarization characteristics are different, indicating that the air quality has improved and is differentiated between UAs; except for the Guanzhong Plain, the overall UA and each UA have obvious σ convergence characteristics, and each UA presents prominent absolute β convergence, conditional β convergence, and club convergence.

Keywords: urban agglomeration (UA); air quality; regional difference; distribution dynamics; convergence

Citation: Xue, Y.; Liu, K. Regional Differences, Distribution Dynamics, and Convergence of Air Quality in Urban Agglomerations in China. *Sustainability* **2022**, *14*, 7330. https://doi.org/10.3390/su14127330

Academic Editors: José Carlos Magalhães Pires and Álvaro Gómez-Losada

Received: 4 May 2022
Accepted: 14 June 2022
Published: 15 June 2022

1. Introduction

The average annual growth rate of China's GDP was as high as 14.3% from 1978 to 2021 [1]. However, with this rapid economic growth [2], the coal-dominated energy structure and extensive economic development have led to excessive consumption of resources and aggravated environmental and ecological damage [3,4]. Air pollution, one of the most prominent environmental problems at present [5], has exerted a series of negative impacts on human health [6], social harmony and stability [7], and sustainable economic growth [8]. In response to this problem, the Chinese government has enacted a series of laws and regulations to oversee the conduct of all walks of life, protect and improve the atmospheric environment, and advance the construction of an ecological culture. At the same time, extensive academic efforts have been made to figure out how to improve air quality.

Research on air quality has been conducted with regard to two aspects: the spatiotemporal distribution and the influencing factors of air quality. Firstly, relevant studies reveal obvious regional differences in air quality [9–12]. China's air quality index (AQI) shows a spatial distribution pattern of "high in the north and low in the south" [13]. Urban air pollution is "severe in the east and minor in the west, severe in the north and minor in the south" [14]. In other words, the air pollution is severe in the east of the Heihe-Tengchong Line and the north of the Yangtze River and minor in the south of the Yangtze River and west of the Heihe-Tengchong Line [15]. Regional differences in air quality have also been confirmed in Catalonia in Spain, Detroit, and northern South America [16–18]. Air quality also has obvious temporal differences. Specifically, China's air pollution is high in autumn and winter and low in spring and summer [19–21], and there is a U-shaped trend in the monthly average concentration, while the concentration of pollutants in South Africa peaks between June and August [22]. There are also differences in the degree of air pollution in the two different time periods under normal and dusty weather [23]. Secondly, scholars have explored the impact of socioeconomic factors and natural conditions on air quality. The socioeconomic factors influencing air quality include population [24,25], the economy [26–29], energy consumption [30–32], environmental policy [33], income inequality [34], land use [35,36], transportation infrastructure [37,38], etc., with even the large-scale sale of industrial land [39], increase in urbanization rate [40], and foreign direct investment [41] also increasing air pollution. Among the natural factors, meteorological elements such as temperature, precipitation, relative humidity, maximum wind speed, air pressure, and sunshine hours have an important association with air quality [42–45], and geographical environment elements such as topography [46,47], climate [48,49], and vegetation [50–54] also have an important impact on air quality. In some studies [13–15], cluster analysis, geographical concentration index, and spatial autocorrelation analysis are used to measure regional differences in air quality.

The above research results are productive, but little attention has been paid to regional differences in air quality from the perspective of distribution dynamics, and research on the convergence of air quality is extremely scant. Moreover, the urban agglomeration (UA), as the hardest-hit area of air pollution, warrants more attention. Therefore, in the present study, we measured the air quality of ten UAs in China through the AQI and analyzed regional differences, distribution dynamics, and convergence using Dagum's decomposition of the Gini coefficient, kernel density estimation (KDE), and the convergence model. These methods allow us to discover the sources of regional differences in air quality in UAs and analyze the distribution and convergence of air quality dynamically.

Compared with the existing literature, the contribution of this paper is as follows: firstly, the AQI can objectively and precisely reflect the air quality of UAs, which provides a theoretical basis for air pollution control in UAs; secondly, the regional differences and distribution dynamics of air quality in UAs enrich the research in this field; thirdly, the convergence of air quality in UAs is helpful in revealing the development trend of air quality gaps in UAs.

This paper is organized as follows. The next section of this paper describes the methods and data, including Dagum's decomposition of the Gini coefficient, KDE, the convergence model, and the data source. The following section presents the findings, including the AQI of UAs, regional differences and their decomposition, distribution dynamics, and convergence. The last section elaborates on the conclusions of this paper and proposes policy implications.

2. Methods and Data

2.1. Methods

2.1.1. Dagum's Decomposition of the Gini Coefficient

Dagum's decomposition of the Gini coefficient [55] was used to discuss the regional differences in air quality in various UAs in China, which was calculated as follows:

$$G = \frac{\sum_{j=1}^{k} \sum_{h=1}^{k} \sum_{i=1}^{n_j} \sum_{r=1}^{n_h} |y_{ji} - y_{hr}|}{2\mu n^2} \tag{1}$$

where k represents the total number of UAs under investigation; j and h are the subscripts of UAs; n stands for the number of cities under investigation; i and r refer to the subscripts of the cities; $n_j(n_h)$ is the number of cities in the $j(h)$th UA; $y_{ji}(y_{hr})$ represents the AQI of city $i(r)$ in the $j(h)$th UA; and μ denotes the mean value of the AQI of all cities under investigation. Dagum's decomposition of the Gini coefficient can be decomposed into intra-UA difference contribution (G_w), inter-UA net difference contribution (G_{nb}), and inter-UA intensity of transvariation (G_t). The specific calculations were based on Equations (2), (4), and (5).

$$G_w = \sum_{j=1}^{k} G_{jj} p_j s_j \tag{2}$$

$$G_{jj} = \frac{1}{2\overline{y_i}} \sum_{i=1}^{c_j} \sum_{r=1}^{c_j} |y_{ji} - y_{jr}| / c_j^2 \tag{3}$$

$$G_{nb} = \sum_{j=2}^{k} \sum_{h=1}^{j-1} G_{jh} (p_j s_h + p_h s_j) D_{jh} \tag{4}$$

$$G_t = \sum_{j=2}^{k} \sum_{h=1}^{j-1} G_{jh} (p_j s_h + p_h s_j) \left(1 - D_{jh}\right) \tag{5}$$

$$G_{jh} = \sum_{i=1}^{c_j} \sum_{r=1}^{c_h} |y_{ji} - y_{jr}| / c_j c_h (\overline{y_i} + \overline{y_h}) \tag{6}$$

where $p_j = c_j / c$, $s_j = c_j \overline{y_j} / c\overline{y}$, and $\sum p_j = \sum s_j = \sum_{j=1}^{k} \sum_{h=1}^{k} p_j s_h = 1$; D_{jh} is the relative impact of air quality between UAs j and h, calculated by Equation (7); d_{jh} represents the air quality difference between UAs, calculated by Equation (8), indicating the mathematical expectation for the sum of all $y_{ji} - y_{hr} > 0$ samples in UAs j and h; p_{jh} stands for the hyper-variable first-order matrix, calculated by Equation (9), which represents the mathematical expectation for the sum of all $y_{hr} - y_{ji} > 0$ samples in UAs j and h; and $F_j(F_h)$ refers to the cumulative density distribution function of UA $j(h)$.

$$D_{jh} = \left(d_{jh} - p_{jh}\right) / \left(d_{jh} + p_{jh}\right) \tag{7}$$

$$d_{ij} = \int_0^{\infty} dF_j(y) \int_0^{y} (y - x) dF_h(x) \tag{8}$$

$$p_{jh} = \int_0^{\infty} dF_h(y) \int_0^{y} (y - x) dF_j(x) \tag{9}$$

2.1.2. Kernel Density Estimation

KDE can describe the distribution pattern of random variables with continuous density curves, thereby reflecting variables' distribution location, pattern, and ductility characteristics [56,57]. In this paper, kernel density estimation was used to analyze the distribution characteristics of air quality in China's UAs. The density function of random variable X was set to be $f(x)$, and the density function of point x was estimated by Equation (10), where N, X_i and $K(x)$ are the number of observations, independent and identically distributed observations, bandwidth, and kernel function. The kernel function used in this paper was calculated by Equation (11).

$$f(x) = (1/Nh) \sum_{i=1}^{N} K[(X_i - x)/h] \tag{10}$$

$$K(x) = \left(1/\sqrt{2\pi}\right) exp\left(-x^2/2\right) \tag{11}$$

2.1.3. Convergence Model

The evolution trend of air quality differences in UAs was tested by σ convergence, β convergence, and club convergence. In Equation (12), j and i represent the UA and the cities included therein; n_j is the number of cities included in UA j; $\overline{EE_{ij}}$ is the average air quality of UA j; and $EE_{i,t+1}$ and $EE_{i,t}$ refer to the air quality of UA i at time $t+1$ and t, respectively.

$$\sigma = \frac{\sqrt{\Sigma_i^{n_j}\left(EE_{ij} - \overline{EE_{ij}}\right)/n_j}}{\overline{EE_{ij}}} \tag{12}$$

$$ln\left(\frac{EE_{i,t+1}}{EE_{i,t}}\right) = \alpha + \beta ln(EE_{i,t}) + \mu_i + \eta_t + \varepsilon_{it} \tag{13}$$

where β is the convergence coefficient. If β is positive, the air quality of the UA is in a divergent trend; otherwise, when it is negative, the air quality of the UA has β convergence. μ_i, η_t, and ε_{it} represent the space effect, time effect, and interference term, respectively.

$$ln\left(\frac{EE_{i,t+1}}{EE_{i,t}}\right) = \alpha + \beta ln(EE_{i,t}) + \rho \sum_{j=1}^{n} w_{ij} ln\left(\frac{EE_{j,t+1}}{EE_{j,t}}\right) + \theta \sum_{j=1}^{n} w_{ij} ln\left(EE_{j,t}\right) + \mu_i + \eta_t + \varepsilon_{it} \tag{14}$$

Common spatial econometric models include the spatial autoregressive model (SAR), spatial error model (SEM), and spatial Durbin model (SDM). The absolute β-convergence model was transformed using the SDM, as shown in Equation (14).

$$ln\left(\frac{EE_{i,t+1}}{EE_{i,t}}\right) = \alpha + \beta ln(EE_{i,t}) + \rho \sum_{j=1}^{n} w_{ij} ln\left(\frac{EE_{j,t+1}}{EE_{j,t}}\right) + \mu_i + \eta_t + \varepsilon_{it} \tag{15}$$

where W_{ij} is the element of the ith row and the jth column of the spatial weight matrix W. The spatial weight matrix used in this paper is mainly based on the adjacency matrix. Generally, different degrees of spatial dependence exist between UAs, which must be considered when constructing the model. If the test shows that there is indeed a significant spatial correlation between UAs, the SAR model (Equation (15)) or the SEM (Equation (16)) should be selected.

$$ln\left(\frac{EE_{i,t+1}}{EE_{i,t}}\right) = \alpha + \beta ln(EE_{i,t}) + \mu_i + \eta_t + \varepsilon_{it} \qquad \varepsilon_{it} = \lambda \sum_{j=1}^{n} w_{ij}\varepsilon_{jt} + \sigma_{it} \tag{16}$$

Further, the conditional β-convergence model was constructed using the SDM as follows:

$$ln\left(\frac{EE_{i,t+1}}{EE_{i,t}}\right) = \alpha + \beta ln(EE_{i,t}) + \rho \sum_{j=1}^{n} w_{ij} ln\left(\frac{EE_{j,t+1}}{EE_{j,t}}\right) + \delta ln X_{i,t+1}$$
$$ + \theta \sum_{j=1}^{n} w_{ij} ln\left(EE_{j,t}\right) + \gamma \sum_{j=1}^{n} w_{ij} ln X_{j,t} + \mu_i + \eta_t + \varepsilon_{it} \tag{17}$$

where $X_{i,t+1}$ is a set of $k \times 1$-dimensional control variables; δ is a $1 \times k$-dimensional coefficient vector; and θ and γ are $1 \times (k+1)$-dimensional coefficient vectors. In this paper, the influencing factors of the air quality in China's urban agglomerations need to be selected when conducting the conditional convergence test, and accurately selecting these factors is of great significance for improving urban air quality. Referring to the previous studies [58–62], the conditional β convergence control variables we selected in this paper include population, technological progress, government financial resources, economic development, and industrial structure. The disorderly growth of population deteriorates resource consumption of UAs and aggravates environmental pollution, whereas an orderly population policy helps to improve air quality by intensive utilization of resources and environmental protection; technological progress helps to expand the scope of urban space activities, improve people's living standards, bring together innovative elements, intensify the use of resources, improve the efficiency of urban economic operation, and reduce waste emissions, thereby improving air quality; government financial resources are of great significance to promote the development of people's livelihood and the regular operation of social production activities; the improvement of economic development has a particular effect on enhancing people's awareness of environmental protection, satisfying the growing demand for a better life, and improving the quality of the environment; and the industrial structure plays a vital role in adjusting the structure of energy consumption, and its upgrade helps reduce the emission of pollutants, which is beneficial to the improvement of air quality.

In club convergence, the first layer was national-level UAs, including Beijing-Tianjin-Hebei, Yangtze River Delta, Pearl River Delta, the middle reaches of the Yangtze River, and Chengdu-Chongqing UAs; and the second layer was the regional UAs, including the central and southern Liaoning, Shandong Peninsula, the Central Plains, the Guanzhong Plain, and Western Taiwan Strait.

2.2. Data Source

We took ten UAs in China from 2014 to 2021 as the research object, and their spatial distribution is shown in Figure 1. The selection of these ten UAs was based on Xiao Jincheng and Yuan Zhu's study [63] because they are economically well developed. These ten UAs include a total of 157 prefecture-level cities. They are divided into national-level and regional UAs. National-level UAs include Beijing-Tianjin-Hebei (1), Yangtze River Delta (2), Pearl River Delta (3), the middle reaches of the Yangtze River (4), and Chengdu-Chongqing UAs (5), while regional UAs consist of the central and southern Liaoning (6), Shandong Peninsula (7), the Central Plains (8), the Guanzhong Plain (9), and Western Taiwan Strait (10).

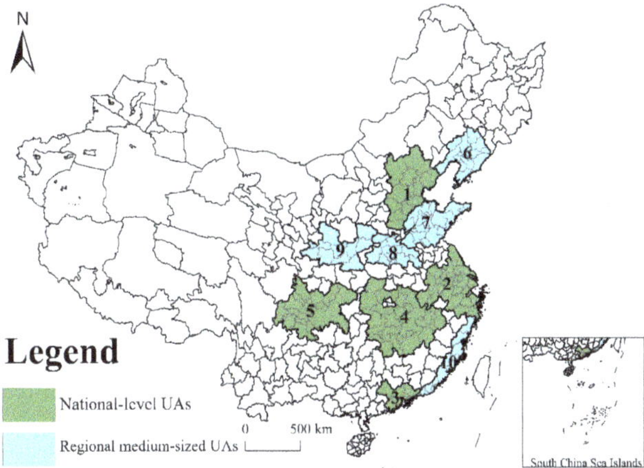

Figure 1. Spatial distribution of the ten UAs in China.

The AQI measures the air quality of various UAs in China, and its data is derived from https://www.aqistudy.cn/historydata, accessed on 3 May 2022. The average monthly AQI of the cities was available on this website, based on which the annual data were obtained. The AQI can comprehensively reflect and quantitatively evaluate the air quality of various UAs. This value is between 0 and 500. The higher the index, the more serious the air pollution, and the more threatening to human health [64]. The AQI monitors pollutants such as SO_2, NO_2, PM_{10}, $PM_{2.5}$, CO, and O_3. All pollutants were synthesized according to the Ambient Air Quality Standards (https://www.mee.gov.cn/ywgz/fgbz/bz/bzwb/dqhjbh/dqhjzlbz/201203/t20120302_224165.shtml, accessed on 2 June 2022) and Technical Regulation on Ambient Air Quality Index (on trial) (https://www.mee.gov.cn/ywgz/fgbz/bz/bzwb/jcffbz/201203/t20120302_224166.shtml, accessed on 2 June 2022). Interestingly, PM_{10} and O_3 accounted for a high proportion in the AQI of all UAs.

The data selected for the control variables in this paper were all from the China Economic and Social Big Data Research Platform (https://data.cnki.net, accessed on 3 May

2022). The missing values were filled by imputation. The descriptive statistics of the data studied in this paper are shown in Table 1.

Table 1. Descriptive statistics of sample data.

Variable	Designation	Definition/Unit	Mean	SD	Max	Min	Observation
X1	Population	Resident population/10,000	560.3802	434.8196	3212.43	68.93	1256
X2	Technological progress	Science and technology expenditure/CNY 100 million	23.0596	57.56474	554.98	0.0786	1256
X3	Government financial resources	Financial revenue/CNY 100 million	411.1661	796.1389	7771.8	20.06	1256
X4	Economic development	GDP per capita/CNY	65,493.46	33,805.19	19,1942	15,852	1256
X5	Industrial structure	The share of secondary industry in GDP/%	44.45925	8.29326	81.13335	15.83376	1256

3. Results

3.1. AQI

Table 2 shows the annual and mean values of AQI for the ten UAs in China from 2014 to 2021. The AQI of each UA in 2021 was smaller than that of 2014, and the index showed a fluctuating downward trend from 2014 to 2021, indicating that the air quality improved to a certain extent. Among the UAs, Beijing-Tianjin-Hebei had the most significant annual decline in AQI, from 128 in 2014 to 85 in 2021, with an average annual rate of 5.58%. The Guanzhong Plain had the smallest decline, with an average annual decline of 1.71%. From the spatial differences in the AQI of UAs, the AQI of UAs in the north was significantly higher than those in the south.

Table 2. The AQI of UAs.

	2014	2015	2016	2017	2018	2019	2020	2021	Mean
Beijing-Tianjin-Hebei	128	111	108	109	93	97	88	85	102
Yangtze River Delta	89	86	82	85	74	78	69	68	79
Pearl River Delta	70	61	63	69	62	67	55	60	63
The middle reaches of the Yangtze River	89	82	81	80	70	77	65	66	76
Chengdu-Chongqing	90	84	84	84	69	69	66	66	77
Central and southern Liaoning	89	89	86	83	70	77	73	68	79
Shandong Peninsula	118	114	105	102	90	100	89	89	101
Central Plains	116	112	110	98	94	108	95	93	103
Guanzhong Plain	93	89	101	102	79	91	81	83	90
Western Taiwan Strait	63	58	57	61	57	57	54	55	58

3.2. Regional Differences and Decomposition of Air Quality

3.2.1. Differences in Air Quality within UAs

Dagum's decomposition of the Gini coefficient was used to calculate the air quality differences within the ten UAs, and the results are shown in Table 3. From 2014 to 2021, the Gini coefficients of the nine UAs other than the Guanzhong Plain showed a fluctuating downward trend. Among them, Beijing-Tianjin-Hebei witnessed the most significant decline, from 0.1387 in 2014 to 0.0627 in 2021, with an average annual rate of 10.73%, followed by Western Taiwan Strait, with an average annual decline of 7.81%. The Gini coefficient of air quality in Guanzhong Plain displayed a fluctuating upward trend, increasing from 0.0716 in 2014 to 0.1043 in 2021, with an average annual increase of 5.51%. From the mean value of the Gini coefficient, the air quality of Guanzhong Plain, Beijing-Tianjin-Hebei, and the middle reaches of the Yangtze River were relatively high, with values of 0.1308, 0.0986, and 0.0751, respectively. The phenomenon of air quality imbalance within the UA was prominent, attributed to the prominent position of cities such as Xi'an, Beijing, and Wuhan in the development of each UA. The Gini coefficients of Central Plains and Western Taiwan Strait were relatively low, at 0.0360 and 0.0460, respectively, with a slight difference within each UA. In general, the Gini coefficient of air quality in the ten UAs was low, and the air quality gap within the UA was insignificant.

Table 3. Dagum's decomposition of Gini coefficient of air quality within the UA.

Year	Beijing-Tianjin-Hebei	Yangtze River Delta	Pearl River Delta	Middle Reaches of the Yangtze River	Chengdu-Chongqing	Central and Southern Liaoning	Shandong Peninsula	Central Plains	Guanzhong Plain	Western Taiwan Strait
2014	0.1387	0.0638	0.0809	0.1019	0.0554	0.0495	0.0811	0.0423	0.0716	0.0674
2015	0.1193	0.0542	0.0549	0.1010	0.0596	0.0518	0.0944	0.0604	0.0585	0.0654
2016	0.1044	0.0504	0.0530	0.0707	0.0531	0.0549	0.0859	0.0387	0.1082	0.0544
2017	0.1042	0.0500	0.0643	0.0559	0.0596	0.0495	0.0729	0.0454	0.1354	0.0467
2018	0.0938	0.0597	0.0549	0.0636	0.0449	0.0516	0.0783	0.0181	0.1380	0.0365
2019	0.0891	0.0584	0.0564	0.0774	0.0453	0.0412	0.0581	0.0241	0.1140	0.0287
2020	0.0770	0.0458	0.0595	0.0634	0.0502	0.0440	0.0622	0.0253	0.1007	0.0307
2021	0.0627	0.0551	0.0531	0.0672	0.0513	0.0330	0.0630	0.0341	0.1043	0.0381
Mean	0.0986	0.0547	0.0596	0.0751	0.0524	0.0469	0.0745	0.0360	0.1038	0.0460

3.2.2. Differences in Air Quality between UAs

Dagum's decomposition of the Gini coefficient of air quality among the ten UAs is shown in Figure 2. Among the regional differences in air quality, Beijing-Tianjin-Hebei and Pearl River Delta fluctuated the most in air quality, followed by that between Beijing-Tianjin-Hebei and Western Taiwan Strait. The regional differences between Chengdu-Chongqing and the Shandong Peninsula, between Chengdu-Chongqing and central and southern Liaoning, and between the Yangtze River Delta and Chengdu-Chongqing fluctuated less. The differences in air quality among most UAs showed a downward trend, among which Beijing-Tianjin-Hebei and Central Plains had the most significant decline, with an average annual decline of 10.42%. The differences between it and Guanzhong Plain, central and southern Liaoning and Guanzhong Plain, and Yangtze River Delta and Guanzhong Plain expanded, with an average annual growth rate of 9.54%, 8.33%, and 7.66%, respectively. Generally, there were specific differences in air quality among the ten UAs.

3.2.3. Overall Difference and Decomposition of Air Quality

Table 4 shows the overall Gini coefficient and decomposition results of air quality in the ten UAs. The table also demonstrates the source and contribution of air quality differences in each UA. The overall difference in air quality showed a fluctuating downward trend, and the overall difference narrowed, with a Gini coefficient of 0.1332–0.1077. The coefficient reached the maximum value of 0.1338 in 2015, fluctuated and decreased after 2015, increased slightly in 2019, and finally fell to the minimum value of 0.1077 in 2021, with a decrease of 2.99%. The contribution of inter-regional differences in air quality was much higher than that of intra-regional differences and intensity of transvariation, indicating that inter-regional differences were the primary source of overall differences, accounting for 71.4669–82.3444%. The contribution rate of the intensity of transvariation was between 12.1477 and 21.6388%, representing the second most prominent source of overall differences. The contribution of intra-regional differences was the lowest, ranging from 5.5079% to 6.8943%. Judging from the evolution trend of the sources of differences, the contribution of inter-regional differences stood at 71.4669% in 2014 and then showed a fluctuating upward trend until it reached 78.4832% in 2021, with an average annual growth rate of 1.347%, suggesting that the differences between UAs widened. The intensity of transvariation was used to identify the overlap between UAs. Its contribution fluctuated and declined from 21.6388% in 2014 to 15.3366% in 2021, with an average annual decrease of 4.80%, showing that the overlap between UAs weakened, and the difference in air quality decreased. The contribution of intra-regional differences did not witness significant change and remained around 6.5%. It decreased year by year from 2014 to 2020, with a slight increase in 2021, indicating that the overall air quality in the UAs was relatively stable.

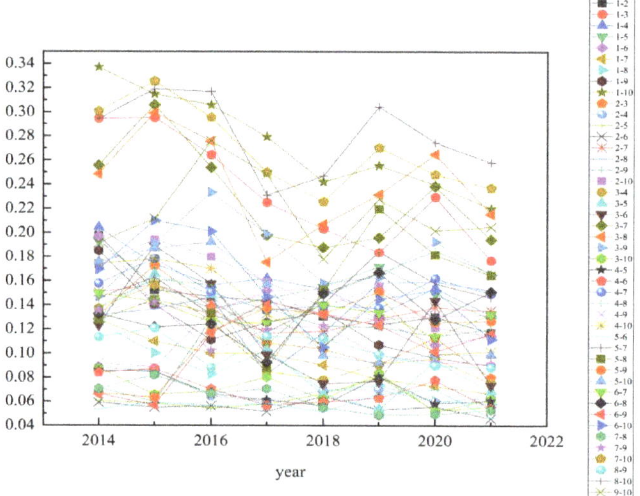

Figure 2. Evolution of Gini coefficients among the ten UAs.

Table 4. Dagum's decomposition of Gini coefficient of air quality in the UAs.

Year	Overall	Intra-Regional		Inter-Regional		Intensity of Transvariation	
		Source	Contribution Rate (%)	Source	Contribution Rate (%)	Source	Contribution Rate (%)
2014	0.1332	0.0091	6.8943	0.0944	71.4669	0.0286	21.6388
2015	0.1338	0.0089	6.6828	0.0990	74.5721	0.0249	18.7451
2016	0.1251	0.0076	6.0876	0.0953	76.5557	0.0216	17.3567
2017	0.1121	0.0072	6.4924	0.0812	72.9369	0.0229	20.5708
2018	0.1116	0.0072	6.5502	0.0803	72.6891	0.0229	20.7608
2019	0.1189	0.0071	6.0404	0.0932	78.9577	0.0177	15.0020
2020	0.1151	0.0063	5.5079	0.0942	82.3444	0.0139	12.1477
2021	0.1077	0.0066	6.1811	0.0840	78.4823	0.0164	15.3366
Mean	0.1197	0.0075	6.3046	0.0902	76.0006	0.0211	17.6948

3.3. Distribution Dynamics of Air Quality

The Gini coefficient presented the size and source of air quality differences in UAs, reflected the relative differences in air quality and could describe the dynamic changes in absolute air quality differences. Therefore, the kernel density estimation was adopted to analyze the distribution dynamics of the air quality of the ten UAs as a whole as well as each of them in terms of the distribution location, pattern, ductility, and polarization. Figure 3 presents the results of the KDE.

The distribution location reflects the air quality. From 2014 to 2021, the center of the kernel density distribution curve of the ten UAs as a whole and Beijing-Tianjin-Hebei, Pearl River Delta, the middle reaches of the Yangtze River, central and southern Liaoning, Shandong Peninsula, Guanzhong Plain, and Western Taiwan Strait move rightward, indicating that the air quality of the ten UAs as a whole and several individual UAs improved with an upward trend. The center of the kernel density distribution curve in Beijing-Tianjin-Hebei, Pearl River Delta, central and southern Liaoning, and the Shandong Peninsula moved rightward remarkably, showing an apparent upward trend in air quality. In contrast, the curve of the Yangtze River Delta, Chengdu-Chongqing, and Central Plains moved leftward, indicating that their air quality declined. Generally, the overall air quality of UAs improved, demonstrating that China's environmental protection has been effective.

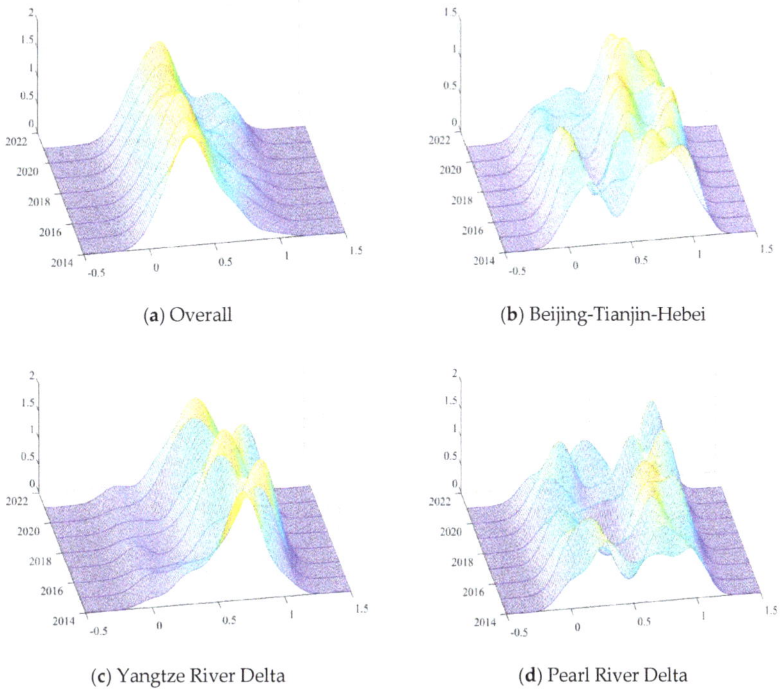

(**a**) Overall   (**b**) Beijing-Tianjin-Hebei

(**c**) Yangtze River Delta   (**d**) Pearl River Delta

Figure 3. *Cont.*

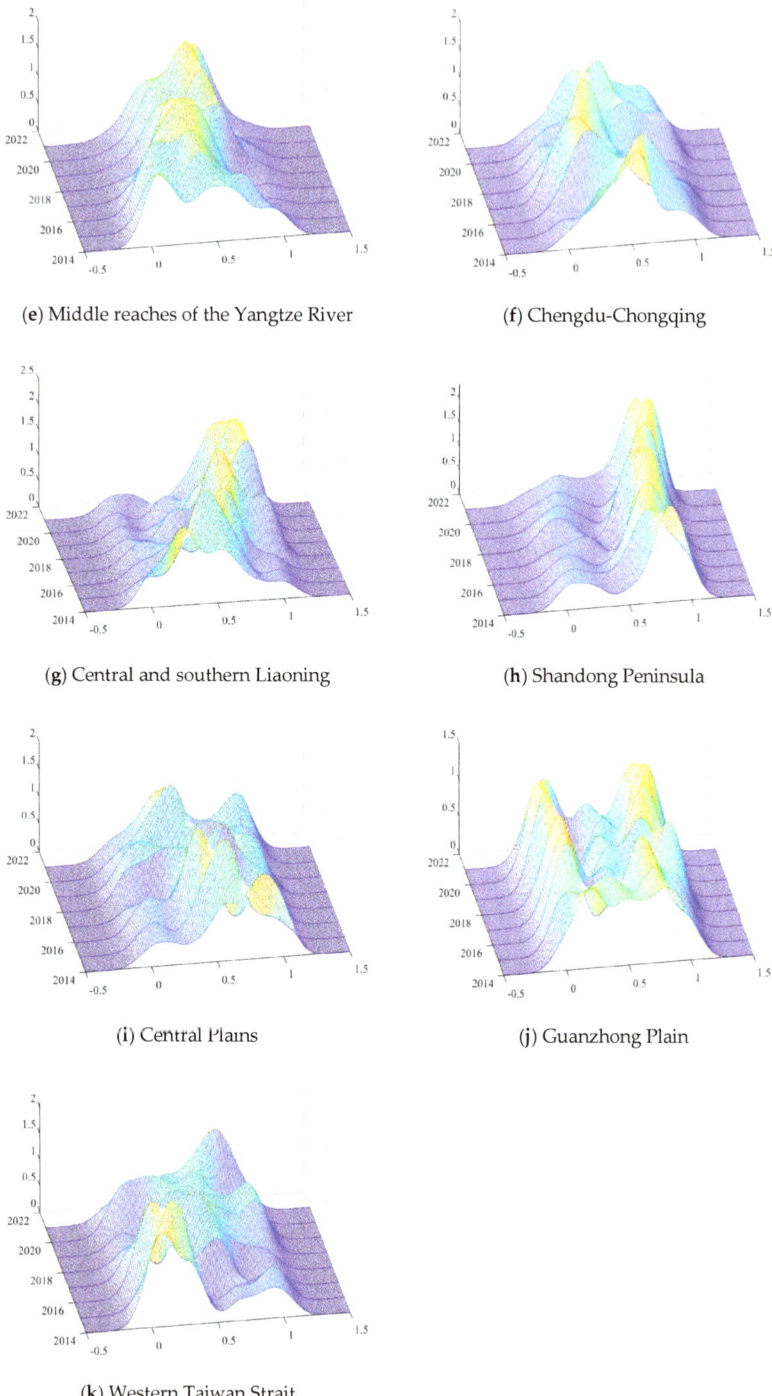

(**e**) Middle reaches of the Yangtze River   (**f**) Chengdu-Chongqing

(**g**) Central and southern Liaoning   (**h**) Shandong Peninsula

(**i**) Central Plains   (**j**) Guanzhong Plain

(**k**) Western Taiwan Strait

Figure 3. Distribution dynamics of air quality in UAs.

The distribution pattern reflects the spatial difference and polarization of air quality in each UA. Specifically, the height and width of the wave crests reflect the difference, and the number of wave crests reflects the polarization. From the distribution pattern, the overall main peak of the ten UAs showed a change of "fall-rise-fall-rise". The overall peak height decreased, and the width of the main peak continued to increase, indicating that the degree of internal dispersion was on the rise. The height of the main peak of Beijing-Tianjin-Hebei continued to rise, but the change in width was not prominent; the height of the Yangtze River Delta did not see much change, and the overall height showed a slight downward trend, with insignificant changes in width; the height of Pearl River Delta was rising, but the change in width was not noticeable; the height of the middle reaches of the Yangtze River increased, while the width decreased; the height of Chengdu-Chongqing decreased, but the width did not change significantly; the height of central and southern Liaoning increased, whereas the width was small; the height of Shandong Peninsula rose, but the width witnessed no remarkable change; the height of Central Plains fell, while the change in width was not apparent; the height of Guanzhong Plain increased, while the width did not change significantly; while the height of Western Taiwan Strait decreased, and the change of the width was reduced.

The distribution ductility reflects spatial differences in air quality. From the distribution ductility, Yangtze River Delta, central and southern Liaoning, and Central Plains showed a left-trailing phenomenon, while the ten UAs as a whole and some individual UAs displayed a right-trailing phenomenon. The air quality difference within the ten UAs as a whole and Guanzhong Plain tended to widen, while other UAs showed a convergence trend, suggesting that the air quality gap within these UAs was constantly narrowing.

From the perspective of polarization characteristics, the kernel density distribution curves of the ten UAs as a whole and each individual UA consisted of double peaks or multiple peaks, indicating that the air quality was multi-polarized to some extent, although there were some differences. The ten UAs as a whole changed from a single peak to multiple peaks, gradually showing polarization, and the distance between the main peak and the side peaks was small, showing slight differences within the UAs. The Shandong Peninsula showed pronounced double peaks from 2014 to 2021, and the distance between the two peaks was close, indicating minor differences within the UAs. However, there was a significant difference between the heights of the main peaks and side peaks of the ten UAs as a whole and the Shandong Peninsula, and the air quality within the UA was significantly different. The phenomenon of multiple peaks was common in other UAs. The difference in the heights between the peaks in UAs, such as Beijing-Tianjin-Hebei, Pearl River Delta, the middle reaches of the Yangtze River, Chengdu-Chongqing, Central Plains, Guanzhong Plain, and Western Taiwan Strait, was relatively small, and the air quality gap within the UA was also small.

3.4. Convergence Analysis of Air Quality

3.4.1. σ-Convergence

Table 5 shows the σ-convergence results of the air quality of the ten UAs as a whole and each individual UA. The overall air quality variation coefficient of the ten UAs decreased from 0.2415 in 2014 to 0.1919 in 2018, slightly increased to 0.2102 in 2019, showed a downward trend from 2019, and finally decreased to 0.1919 in 2021. Among the UAs, the coefficient of variation of Guanzhong Plain fluctuated and increased from 0.1312 in 2014 to 0.1926 in 2021, with an average annual increase of 5.6%. The other UAs showed a fluctuating downward trend, among which Western Taiwan Strait saw the most remarkable decline, with an average annual decrease of 8.24%. Generally speaking, the coefficient of variation of air quality in Guanzhong Plain increased without σ-convergence; the variation coefficient of the ten UAs as a whole and each individual UA showed a downward trend, all with σ-convergence. The air quality of each urban agglomeration was balanced. The regional disparity within these UAs narrowed, and the air quality of each UA was even.

Table 5. σ-convergence Gini coefficient of air quality in the UAs.

Year	Overall	Beijing-Tianjin-Hebei	Yangtze River Delta	Pearl River Delta	The Middle Reaches of the Yangtze River	Chengdu-Chongqing	Central and Southern Liaoning	Shandong Peninsula	Central Plains	Guanzhong Plain	Western Taiwan Strait
2014	0.2415	0.2562	0.1192	0.1505	0.1825	0.1023	0.0926	0.1585	0.0801	0.1312	0.1327
2015	0.2394	0.2196	0.1021	0.1038	0.1803	0.1111	0.1000	0.1769	0.1156	0.1073	0.1314
2016	0.2221	0.1926	0.0975	0.1002	0.1292	0.0985	0.1025	0.1669	0.0731	0.2019	0.1049
2017	0.2033	0.1936	0.0958	0.1250	0.1037	0.1114	0.0964	0.1471	0.0860	0.2517	0.0884
2018	0.1981	0.1739	0.1144	0.1035	0.1148	0.0853	0.0989	0.1522	0.0331	0.2571	0.0677
2019	0.2102	0.1644	0.1114	0.1087	0.1403	0.0825	0.0804	0.1165	0.0438	0.2146	0.0533
2020	0.2044	0.1444	0.0840	0.1130	0.1185	0.0922	0.0878	0.1281	0.0471	0.1905	0.0571
2021	0.1919	0.1158	0.1020	0.1013	0.1212	0.0929	0.0657	0.1220	0.0623	0.1926	0.0727

3.4.2. β-Convergence

(1) Absolute β-Convergence

We screened the models by LM test, Hausman test, LR test, and Wald test, analyzed the absolute β-convergence of the air quality of the ten UAs as a whole and each individual UA, and determined the convergence analysis model accordingly. After testing, Beijing-Tianjin-Hebei, Pearl River Delta, Chengdu-Chongqing, Shandong Peninsula, and Central Plains were shown to require a return to the traditional convergence model (Table 6).

The ten UAs as a whole and the middle reaches of the Yangtze River partially passed the LM test, significantly passed the LR test and Wald test, and supported the SDM (this study did not report the LR test and Wald test for the SDM, which was available for retrieval). The spatial lag coefficient ρ of the explained variables was significantly positive at a 1% level, indicating that the improved air quality in surrounding UAs promoted that of the target UA. The convergence coefficient was significantly negative at a 1% level, suggesting that the overall air quality of the ten UAs had an absolute β-convergence trend. θ was significantly positive, which means that the improved air quality in surrounding UAs substantially promoted that of the target UA. These findings revealed that spatial spillover was one of the critical factors in improving regional air quality.

Beijing-Tianjin-Hebei, Pearl River Delta, Chengdu-Chongqing, Shandong Peninsula, and Central Plains passed the LR and Wald tests, and some results failed the significance tests. In addition, after running SEM and SAR for Beijing-Tianjin-Hebei, Pearl River Delta, Chengdu-Chongqing, and the Shandong Peninsula, ρ or λ were shown to be not significant, and the SAR and SEM were shown to require a return to the traditional convergence model.

The spatial error coefficient λ of central and southern Liaoning and Western Taiwan Strait was negative and passed the 1% significance test, showing a certain degree of negative spatial correlation, which means that a "core-periphery" pattern characterized air quality improvement. There was a competitive relationship between the cities within the UA, but the areas with relatively poor air quality were shown to consistently catch up, and the central city had a siphon effect. There was a positive spatial correlation between the Yangtze River Delta and Guanzhong Plain, suggesting that the improvement of surrounding air quality boosted the air quality of the target area.

The convergence coefficient β of the UAs as a whole and each individual UA was negative and passed the 1% significance test, with an absolute β-convergence trend. Combining the spatial correlation and the convergence coefficient β, it could be seen that the spatial correlation of the UAs as a whole and each individual UA was one of the factors leading to the absolute β-convergence. While there was a significant negative spatial correlation between the central and southern Liaoning and Western Taiwan Strait, the absolute β-convergence still existed, indicating that the behindhand cities within the UA had strong momentums to catch up with cities ahead of them, and the air quality between cities was increasingly competitive.

Table 6. Absolute β-convergence test of air quality in the UAs.

Area	Overall	Beijing-Tianjin-Hebei	Yangtze River Delta	Pearl River Delta	The Middle Reaches of the Yangtze River	Chengdu-Chongqing	Central and Southern Liaoning	Shandong Peninsula	Central Plains	Guan-Zhong Plain	Western Taiwan Strait
Model	Two-way fixed SDM	Two-way fixed OLS	Two-way fixed SAR	Two-way fixed OLS	Two-way fixed SDM	Two-way fixed OLS	Two-way fixed SAR	Two-way fixed OLS	Two-way fixed OLS	Two-way fixed SEM	Two-way fixed SAR
β	−0.6836 *** (−26.29)	−0.3396 *** (−5.31)	−0.6685 *** (−11.49)	−1.0544 *** (−8.83)	−0.7488 *** (−14.22)	−0.2769 *** (−4.48)	−0.8017 *** (−10.15)	−0.4120 *** (−5.90)	−0.6237 *** (−6.31)	−0.5600 *** (−6.25)	−0.6182 *** (−6.56)
θ	0.3602 *** (8.60)				0.5511 *** (5.55)						
ρ or λ	0.4251 *** (14.26)		0.4465 *** (6.67)		0.4797 *** (6.56)		−0.2383 ** (−2.17)			0.2631 * (1.87)	−0.2156 ** (−2.28)
R^2	0.2780	0.2677	0.1432	0.5954	0.3149	0.1742	0.1945	0.2562	0.3408	0.3961	0.4270
Log-L	1754.4811		365.5720		364.5399		192.8965			115.7514	133.9820
Space fixed effect	257.97 ***	52.64 ***	113.70 ***	45.53 ***	52.46 ***	42.52 ***	117.55 ***	58.27 ***	34.40 ***	33.13 ***	43.54 ***
Time fixed effect	520.53 ***	22.63 ***	99.68 ***	35.97 ***	116.56 ***	45.03 ***	67.93 ***	33.85 ***	21.88 ***	35.89 ***	23.65 ***
Hausman test	141.00 ***	8.72 ***	119.96 ***	25.79 ***	22.66 ***	5.78 **	111.93 ***	18.88 ***	9.61 **	46.11 ***	16.45 ***
LM (SAR)	788.946 ***	52.436 ***	229.000 ***	52.867 ***	197.990 ***	97.714 ***	71.577 ***	109.267 ***	52.647 ***	63.4000 ***	8.386 ***
R-LM (SAR)	0.274	2.040	0.100	13.129 ***	0.282	0.202	0.140	0.053	0.216	2.634	1.433
LM (SEM)	825.689 ***	51.994 ***	237.974 ***	40.230 ***	214.203 ***	104.723 ***	77.903 ***	110.588 ***	60.323 ***	60.781 ***	6.984 ***
R-LM (SEM)	37.016 ***	1.598	9.074 **	0.492	16.495 ***	7.212 ***	6.466 **	1.375	7.891 ***	0.015	0.031

***, **, and * indicate significance at the 1%, 5%, and 10% levels, respectively, with t value in brackets, the same below.

(2) Conditional β-Convergence

The convergence of air quality in UAs was further explored by considering five control variables: population, technological progress, government financial resources, economic development, and industrial structure. According to the convergence analysis steps, the conditional β-convergence analysis models of the ten UAs as a whole and each individual UA were respectively determined. After comparing the results of each model, the traditional convergence model was finally selected. The results are shown in Table 7. The conditional convergence coefficients of the air quality of the ten UAs as a whole and each individual UA were all negative and passed the 1% significance test, indicating a trend of conditional convergence.

The regression coefficients of the control variables showed that in the ten UAs, the two variables X4 and X5 passed the significance level test and had a significant impact on the conditional β-convergence of the AQI, while X1, X2, and X3 did not pass the significance level test and had no significant effect on the convergence of the AQI. The influence of X5 was significantly positive, suggesting that the industrial structure could improve the air quality in UAs but had a particular inhibitory effect on narrowing the AQI gap between UAs. The gap in industrial structure was the main factor for the gap in air quality between UAs.

For each UA, the control variables, such as X1, X2, X3, X4, and X5, had different effects on improving the air quality. For example, X3 had a significant positive impact on the air quality of the middle reaches of the Yangtze River. The improvement of X3 substantially promoted the air quality of the middle reaches of the Yangtze River, but it had an inhibitory effect on narrowing the gap within the UA. Therefore, to improve the air quality of the UA, it is necessary for government financial resources to play a role, to optimize the allocation of resources, and improve people's living standards. X3 had a significant negative impact on the air quality of the Yangtze River Delta and had a promoting effect on reducing the air quality gap in the UA. However, it had no significant impact on the air quality of other UAs, Yangtze River Delta, and the middle reaches of the Yangtze River. Therefore, it is necessary to utilize the government's financial resources further to improve resource allocation, optimize the environment, and pay attention to the impact on the air quality gap within the UA.

3.4.3. Club Convergence

Dagum's decomposition of the Gini coefficient revealed that the air quality difference between UAs was an essential source of the overall difference. We conducted a club convergence test based on UAs at different levels to further examine the air quality difference between UAs. The absolute and conditional β-convergence test results of air quality in UAs at the two levels are shown in Table 8. The absolute club convergence coefficients of UAs at different levels were significantly negative, and the air quality among UAs at the same level was characterized by absolute β-convergence. With development of the economy, the communication between UAs has become increasingly close. The significant positive spatial correlation between UAs at different levels strengthened their connection and enhanced the absolute β-convergence trend.

After adding the relevant control variables, the conditional β-convergence coefficients of AQI of UAs at the two levels were significantly negative, and the air quality within UAs at different levels had a conditional β-convergence trend. There were spatial correlations within the UAs at the first level, including Yangtze River Delta, Pearl River Delta, and Beijing-Tianjin-Hebei. The spatial effect accounted for the conditional β-convergence trend in UAs at the first level, indicating that the interaction between high-level UAs was tighter. The influence of each control variable in UAs at different levels was different. For example, economic development was not significant in the first layer but significantly positive in the second layer. Therefore, the impact of each variable on the air quality of UAs varied from region to region.

Table 7. Conditional β-convergence test of air quality in the UAs.

Area	Overall	Beijing-Tianjin-Hebei	Yangtze River Delta	Pearl River Delta	The Middle Reaches of the Yangtze River	Chengdu-Chongqing	Central and Southern Liaoning	Shandong Peninsula	Central Plains	Guanzhong Plain	Western Taiwan Strait
β	−0.8172 *** (−26.92)	−0.8026 *** (−7.86)	−1.0159 *** (−12.85)	−1.1815 *** (−9.65)	−1.0406 *** (−14.69)	−0.9857 *** (−9.44)	−0.6289 *** (−6.24)	−0.8213 *** (−8.31)	−0.9052 *** (−7.64)	−0.9685 *** (−7.39)	−0.8641 *** (−7.24)
X1	−0.0691 (−1.06)	0.0398 * (0.14)	−0.3591 *** (−3.57)	−0.2754 (−0.78)	−0.3830 * (−1.76)	0.0771 (0.32)	0.3026 (1.31)	−0.1425 (−0.49)	−0.4329 (−1.22)	0.0365 (0.12)	0.3157 (0.80)
X2	0.0097 (1.28)	−0.0413 (−1.37)	0.0281 (0.84)	0.0688 (1.40)	−0.0290 (−1.27)	−0.0397 * (−1.86)	0.0308 * (−0.92)	0.0126 (0.52)	0.0069 (0.18)	−0.0064 (−0.14)	−0.0265 (−1.32)
X3	−0.0344 (−1.44)	−0.1009 (−1.23)	−0.1551 ** (−2.47)	0.0478 (0.33)	0.1415 *** (2.69)	−0.0158 (−0.13)	−0.0502 (−0.92)	−0.2694 * (−1.89)	−0.0921 (−0.50)	0.0324 (0.28)	−0.0979 (−0.96)
X4	−0.2050 *** (−8.32)	−0.1186 (−1.19)	−0.3206 *** (−6.22)	−0.3020 * (−1.79)	−0.3785 *** (−5.81)	−0.3991 *** (−3.49)	−0.0209 (−0.23)	−0.0983 (−1.30)	−0.1807 (−0.99)	−0.3080 ** (−2.26)	−0.0129 (−0.16)
X5	0.3617 *** (10.88)	0.3079 *** (3.36)	0.2050 ** (2.09)	0.3428 (0.97)	0.2310 ** (2.13)	0.2527 (1.55)	0.2504 *** (2.97)	0.3874 *** (4.10)	0.0051 (0.03)	0.1234 (0.66)	0.0826 (0.47)
R^2	0.4386	0.4852	0.5270	0.6640	0.5802	0.5078	0.3816	0.4631	0.4690	0.4969	0.4950

*** , ** , and * indicate significance at the 1%, 5%, and 10% levels, respectively, with t value in brackets, the same below.

Table 8. Club convergence test of air quality in the UAs.

Area	Absolute β-Converg		Conditional β-Converg	
	First Layer	**Second Layer**	**First Layer**	**Second Layer**
Model	**Two-Way Fixed SEM**	**Two-Way Fixed SEM**	**Two-Way Fixed SDM**	**Two-Way Fixed SEM**
β	−0.6172 *** (−20.01)	−0.7019 *** (−17.09)	−0.6425 *** (−20.60)	−0.7154 *** (−17.52)
θ			0.2247 *** (4.10)	
ρ or λ	0.5159 *** (14.39)	0.2463 *** (5.11)	0.4839 *** (12.84)	0.2369 *** (4.85)
X1			0.08830 * (1.76)	0.0727 (0.89)
X2			−0.0139 ** (−2.01)	0.0056 (0.78)
X3			−0.0146 (−0.74)	−0.0204 (−0.70)
X4			0.0444 (1.38)	0.0861 *** (2.64)
X5			0.0525 (1.44)	−0.0207 (−0.48)
R^2	0.2503	0.3008	0.3135	0.2686
Log-L	1133.4696	669.7970	1151.7943	675.6161
Space fixed effect	154.00 ***	147.82 ***	95.42 ***	148.18 ***
Time fixed effect	272.98 ***	224.09 ***	297.57 ***	232.87 ***
Hausman test	284.23 ***	269.66 ***	60.98 ***	289.99 ***
LM spatial lag	639.470 ***	195.079 ***	642.784 ***	191.207 ***
Robust LM spatial lag	0.324	1.540	1.929	0.520
LM spatial error	686.177 ***	207.926 ***	677.778 ***	203.645 ***
Robust LM spatial error	47.031 ***	14.387 ***	36.924 ***	12.958 ***

***, **, and * indicate significance at the 1%, 5%, and 10% levels, respectively, with t value in brackets, the same below.

4. Conclusions and Policy Implications

4.1. Conclusions

This study measured the air quality of ten UAs in China based on the AQI from 2014 to 2021. We analyzed, estimated, and decomposed the regional differences using Dagum's decomposition of the Gini coefficient, described the distribution dynamics using kernel density estimation, and tested the σ-convergence, β-convergence, and club convergence. The conclusions drawn were as follows:

(1) According to the AQI from 2014 to 2021, there were spatiotemporal differences in the air quality of the ten UAs in China: from the scale of time, the AQI of each UA showed a downward trend over the years; from the perspective of space, the overall AQI of UAs was high in the north and low in the south. In other words, the air quality of China's UAs gradually improved, and the air quality in the south was better than that in the north.

(2) Dagum's decomposition of the Gini coefficient demonstrated no significant difference in air quality within each UA, but there was a particular gap in air quality between UAs. Among them, the air quality difference between Beijing-Tianjin-Hebei and Pearl River Delta was the largest; the overall air quality difference in the UAs showed a decreasing

trend; and the regional difference in air quality was the primary source of the overall difference in air quality.

(3) In the distribution dynamics of the air quality estimated by kernel density estimation, the overall distribution curve of the ten UAs moved rightward, and the air quality of the UAs improved to a certain extent. The changing trend of height and width of the ten UAs as a whole and each individually were inconsistent. Except for the left-tailing phenomenon in the Yangtze River Delta, central and southern Liaoning, and Central Plains, the ten UAs as a whole and the other UAs showed a right-tailing phenomenon, but the ductility of the distribution curves was different. The kernel density estimation curves of the ten UAs as a whole and each individually consisted of double or multiple peaks, indicating that air quality was multi-polarized to some extent.

(4) The convergence model demonstrated that except for Guanzhong Plain, the air quality of the UAs as a whole and each of the other UAs featured σ-convergence, and the ten UAs as a whole and each individually had absolute and conditional β-convergence. In addition, the air quality in UAs at different levels was characterized by club convergence.

In the future, spatial autocorrelation analysis and standard deviation ellipse can be used to systematically analyze the regional differences in air quality, and the spatial econometric model and Geodetector can be used to analyze the influencing mechanism of the air quality of UAs.

4.2. Policy Implications

(1) It is necessary to constantly improve the air quality of China's UAs. This study shows that the AQI of each UA in China displays a downward trend over time, indicating that the air quality of each UA has improved to some extent, but still needs to be further improved. Improving air quality is a response to environmental protection and green development nowadays. Since there is still extensive resource utilization in China's UAs, it remains necessary to establish the concept of green development, rationally allocate resources, improve energy efficiency, reduce pollution emissions in industrial production, and improve air quality in UAs by both the government and the market.

(2) For UAs with significant internal differences in air quality, their central cities should continue to improve resource allocation and utilization efficiency through the agglomeration effect, strengthen the radiation and driving effect within the UA, and improve the overall air quality of the UA. It is crucial for other UAs with shrinking differences to further achieve regional coordination and overall prosperity within each UA. The development of each UA should fully consider each city's resource endowment and location advantages to avoid further increases in air quality differences.

(3) The Gini coefficient shows that regional differences are the primary source of the overall differences in air quality. There is a significant difference between Beijing-Tianjin-Hebei and Pearl River Delta and between Beijing-Tianjin-Hebei and Western Taiwan Strait due to the geographical differences between the north and the south. Therefore, the overall AQI in China's UAs is high in the north and low in the south. To this end, Beijing-Tianjin-Hebei, Shandong Peninsula, and other northern UAs should strive to build ecological cities while continuing to develop their economies, and should continue to strengthen the control of air pollutant emissions and reduce energy consumption in industrial production activities. The extensive utilization of resources should be addressed by the rational allocation of resources to avoid waste of resources in the production process, establish a green and intensive development concept, improve the air quality of various UAs in China, and vigorously promote the construction of an ecological culture and the overall air quality of UAs. For southern UAs with better air quality, it is also essential to build effective channels for cooperation and communication between UAs so that UAs with poor air quality can extensively learn from the regions with better air quality.

(4) There are absolute and conditional β-convergence trends in China's UAs, indicating that the differences in air quality among China's UAs are narrowing. China has a vast territory with different resource endowments and economic development levels, and it

is not easy to achieve a balanced development that is entirely undifferentiated in terms of air quality. However, to improve the air quality of UAs as a whole, it is still necessary to improve the air quality in cities with poor air quality, strengthen the concept of green development, and protect the atmospheric environment. It is also essential to strengthen the overall coordination of the region and promote the overall improvement of air quality in China's UAs. For Beijing–Tianjin–Hebei, Yangtze River Delta, and Pearl River Delta, the three most developed UAs in China, it is necessary to further improve the quality of development to drive national economic growth and improve air quality. The Yangtze River Delta and Chengdu Chongqing need to drive the central and western parts of China, respectively, to improve air quality. The other five regional UAs need to integrate green and innovative elements into the development process, and pay attention to the coordination of economic growth and environmental protection.

In short, the air quality of China's UAs needs to be further improved. The central government should play a crucial role in addressing the current air pollution problems in UAs. Legal and administrative means can be used to strengthen the supervision and management of air pollution, and green finance can also be used to encourage energy conservation, environmental protection, and technological development. Like all industrialized countries, China's industrialization is accompanied by air pollution. In the future, UAs need to promote cleaner production methods, energy-saving and environmental protection technologies, transform those industries with high pollution and high energy consumption, and further improve the air quality of China's UAs.

Author Contributions: Conceptualization, K.L.; methodology, K.L.; software, Y.X.; validation, K.L.; formal analysis, K.L.; investigation, Y.X.; resources, K.L.; data curation, Y.X.; writing—original draft preparation, Y.X.; writing—review and editing, K.L.; visualization, Y.X.; supervision, K.L.; project administration, K.L.; funding acquisition, K.L. All authors have read and agreed to the published version of the manuscript.

Funding: This research was funded by the National Natural Science Foundation of China, Grant No. 72004124.

Institutional Review Board Statement: Not applicable.

Informed Consent Statement: Not applicable.

Data Availability Statement: The data presented in this study are available on request from the corresponding author. The data are not publicly available because research is ongoing.

Conflicts of Interest: The authors declare no conflict of interest.

References

1. Li, C.Y.; Zhang, Y.M.; Zhang, S.Q. Applying the Super-EBM model and spatial Durbin model to examining total-factor ecological efficiency from a multi-dimensional perspective: Evidence from China. *Environ. Sci. Pollut. Res.* **2022**, *29*, 2183–2202. [CrossRef]
2. Tu, Z.G.; Hu, T.Y.; Shen, R.J. Evaluating public participation impact on environmental protection and ecological efficiency in China: Evidence from PITI disclosure. *China Econ. Rev.* **2019**, *55*, 111–123. [CrossRef]
3. Feng, D.; Zhang, Y.Q.; Zhang, X.Y. Applying a data envelopment analysis game cross-efficiency model to examining regional ecological efficiency: Evidence from China. *J. Clean. Prod.* **2020**, *267*, 122031. [CrossRef]
4. Chen, S.Y.; Chen, D.K. Air Pollution, Government regulations and high-quality economic development. *Econ. Res. J.* **2018**, *53*, 20–34.
5. Wang, Z.Y.; Shi, X.Y.; Pan, C.H.; Wang, S.S. Spatial and temporal characteristics of environmental air Quality and its relationship with seasonal climatic conditions in Eastern China during 2015–2018. *Int. J. Environ. Res. Public Health* **2021**, *18*, 4524. [CrossRef] [PubMed]
6. Bandh, S.A.; Shafi, S.; Peerzada, M.; Rehman, T.; Bashir, S.; Wani, S.A.; Dar, R. Multidimensional analysis of global climate change: A review. *Environ. Sci. Pollut. Res.* **2021**, *28*, 24872–24888. [CrossRef]
7. Zhou, D.; Lin, Z.L.; Liu, L.M.; Liu, L.M.; Qi, J.L. Spatial-temporal characteristics of urban air pollution in 337 Chinese cities and their influencing factors. *Environ. Sci. Pollut. Res.* **2021**, *28*, 36234–36258. [CrossRef]
8. Dong, D.; Xu, B.; Shen, N.; He, Q. The adverse impact of air pollution on China's economic growth. *Sustainability* **2021**, *13*, 9056. [CrossRef]

9.   Zheng, B.L.; Liang, L.T.; Li, M.M. Analysis of temporal and spatial patterns of $PM_{2.5}$ in Prefecture-Level Cities of China from 1998 to 2016. *China Environ. Sci.* **2019**, *39*, 1909–1919. [CrossRef]
10. Li, M.S.; Zhang, J.H.; Zhang, Y.J.; Zhou, L.; Li, Q.; Chen, Y.H. Spatio-temporal pattern changes of ambient air PM_{10} pollution in China from 2002 to 2012. *Acta Geogr. Sin.* **2013**, *68*, 1504–1512. [CrossRef]
11. Xu, L.J.; Zhou, J.X.; Guo, Y.; Wu, T.M.; Chen, T.T.; Zhong, Q.J.; Yuan, D.; Chen, P.Y.; Ou, C.Q. Spatiotemporal pattern of air quality index and its associated factors in 31 Chinese provincial capital cities. *Air Qual. Atmos. Health* **2017**, *10*, 601–609. [CrossRef]
12. Tui, Y.; Qiu, J.X.; Wang, J.; Fang, C.S. Analysis of spatio-temporal variation characteristics of main air pollutants in Shijiazhuang City. *Sustainability* **2021**, *13*, 941. [CrossRef]
13. Chukwu, T.; Morse, S.; Murphy, R. Spatial analysis of air quality assessment in two cities in nigeria: A comparison of perceptions with instrument-based methods. *Sustainability* **2022**, *14*, 5403. [CrossRef]
14. Lin, X.Q.; Wang, D. Spatio-temporal variations and socio-economic driving forces of air quality in Chinese cities. *Acta Geogr. Sin.* **2016**, *71*, 1357–1371. [CrossRef]
15. Huang, C.H.; Liu, K.; Zhou, L. Spatio-temporal trends and influencing factors of $PM_{2.5}$ concentrations in urban agglomerations in China between 2000 and 2016. *Environ. Sci. Pollut. Res.* **2021**, *28*, 10988–11000. [CrossRef]
16. Platikanov, S.; Terrado, M.; Pay, M.T.; Soret, A.; Tauler, R. Understanding temporal and spatial changes of O_3 or NO_2 concentrations combining multivariate data analysis methods and air quality transport models. *Sci. Total Environ.* **2022**, *806*, 150923. [CrossRef]
17. Batterman, S.; Ganguly, R.; Harbin, P. High resolution spatial and temporal mapping of traffic-related air pollutants. *Int. J. Environ. Res. Public Health* **2015**, *12*, 3646–3666. [CrossRef]
18. Mendez-Espinosa, J.F.; Rojas, N.F.; Vargas, J.; Pachón, J.E.; Belalcazar, L.C.; Ramírez, O. Air quality variations in Northern South America during the COVID-19 lockdown. *Sci. Total Environ.* **2020**, *749*, 141621. [CrossRef]
19. Liu, H.M.; Fang, C.L.; Huang, J.J.; Zhu, X.D.; Zhou, Y.; Wang, Z.B.; Zhang, Q. The spatial-temporal characteristics and influencing factors of air pollution in Beijing-Tianjin-Hebei urban agglomeration. *Acta Geogr. Sin.* **2018**, *73*, 177–191. [CrossRef]
20. Wang, Z.S.; Li, Y.T.; Chen, T.; Zhang, D.W.; Sun, F.; Pan, L.B. Spatial-temporal characteristics of $PM_{2.5}$ in Beijing in 2013. *Acta Geogr. Sin.* **2015**, *70*, 110–120. [CrossRef]
21. Zhan, J.Y.; Huang, G.C.; Zhou, H.; Duan, W.S.; Wu, A.A.; Wang, W.J.; Li, T. Spatial and temporal distribution characteristics and factors of particulate matter concentration in North China. *J. Nat. Resour.* **2021**, *36*, 1036–1046. [CrossRef]
22. Arowosegbe, O.O.; Röösli, M.; Adebayo-Ojo, T.C.; Dalvie, M.A.; Hoogh, K. Spatial and temporal variations in PM10 concentrations between 2010–2017 in South Africa. *Int. J. Environ. Res. Public Health* **2021**, *18*, 13348. [CrossRef] [PubMed]
23. Farsani, M.H.; Shirmardi, M.; Alavi, N.; Maleki, H.; Sorooshian, A.; Babaei, A.; Asgharnia, H.; Marzouni, M.B.; Goudarzi, G. Evaluation of the relationship between PM_{10} concentrations and heavy metals during normal and dusty days in Ahvaz, Iran. *Aeolian Res.* **2018**, *33*, 12–22. [CrossRef]
24. Xie, J.Y.; Suh, D.H.; Joo, S. Dynamic analysis of air pollution: Implications of economic growth and renewable energy consumption. *Int. J. Environ. Res. Public Health* **2021**, *18*, 9906. [CrossRef] [PubMed]
25. Shi, T.; Liu, M.; Hu, Y.M.; Li, C.L.; Zhang, C.Y.; Ren, B.H. Spatiotemporal pattern of fine particulate matter and impact of urban socioeconomic factors in China. *Int. J. Environ. Res. Public Health* **2019**, *16*, 1099. [CrossRef] [PubMed]
26. Jing, B.; Ni, Z.Y.; Zhao, L.Y.; Liu, K. Does rural-urban income gap exacerbate or restrain air pollution. *China Popul. Resour. Environ.* **2021**, *31*, 130–138. [CrossRef]
27. Dinda, S.; Coondoo, D.; Pal, M. Air quality and economic growth: An empirical study. *Ecol. Econ.* **2000**, *34*, 409–423. [CrossRef]
28. Wu, Y.P.; Dong, S.C.; Song, J.F. Modeling economic growth and environmental degradation of Beijing. *Geogr. Res.* **2002**, *21*, 239–246. [CrossRef]
29. Peng, J.Y.; Zhang, Y.G.; Xie, R.; Liu, Y. Analysis of driving factors on China's air pollution emissions from the view of critical supply chains. *J. Clean. Prod.* **2018**, *203*, 197–209. [CrossRef]
30. Zhang, R.L.; Mi, K.N. Energy consumption, structural changes and air quality: Empirical test based on inter-provincial panel data. *Geogr. Res.* **2022**, *41*, 210–228. [CrossRef]
31. Tan, X.; Yu, W.; Wu, S. The impact of the dynamics of agglomeration externalities on air pollution: Evidence from urban panel data in China. *Sustainability* **2022**, *14*, 580. [CrossRef]
32. Ma, L.M.; Zhang, X. The spatial effect of China's haze pollution and the impact from economic change and energy structure. *China Ind. Econ.* **2014**, *4*, 19–31. [CrossRef]
33. He, L.Y.; Wu, M.; Wang, D.Q.; Zhong, Z.Q. A study of the influence of regional environmental expenditure on air quality in China: The effectiveness of environmental policy. *Environ. Sci. Pollut. Res.* **2018**, *25*, 7454–7468. [CrossRef] [PubMed]
34. Wang, F.; Yang, J.; Shackman, J.; Liu, X. Impact of income inequality on urban air quality: A game theoretical and empirical study in China. *Int. J. Environ. Res. Public Health* **2021**, *18*, 8546. [CrossRef] [PubMed]
35. Chen, B.; Yang, S.; Xu, X.D.; Zhang, W. The impacts of urbanization on air quality over the Pearl River Delta in winter: Roles of urban land use and emission distribution. *Theor. Appl. Climatol.* **2014**, *117*, 29–39. [CrossRef]
36. Bandeira, J.M.; Coelho, M.C.; Sá, M.E.; Tavares, R.; Borrego, C. Impact of land use on urban mobility patterns, emissions and air quality in a Portuguese medium-sized city. *Sci. Total Environ.* **2011**, *409*, 1154–1163. [CrossRef] [PubMed]
37. Sun, C.W.; Luo, Y.; Yao, X. The effects of transportation infrastructure on air quality: Evidence from empirical analysis in China. *Econ. Res. J.* **2019**, *54*, 136–151.

38. Guo, Y.J.; Zhang, Q.; Lai, K.K.; Zhang, Y.Q.; Wang, S.B.; Zhang, W.L. The impact of urban transportation infrastructure on air quality. *Sustainability* **2020**, *12*, 5626. [CrossRef]
39. Huang, Z.J.; Song, L.; Gao, B.Y.; Jiang, L. Industrial land transfer, industrial selection and urban air quality. *Geogr. Res.* **2022**, *41*, 229–250. [CrossRef]
40. Zhou, J.; Lan, H.; Zhao, C.; Zhou, J. Haze pollution levels, spatial spillover influence, and impacts of the digital economy: Empirical evidence from China. *Sustainability* **2021**, *13*, 9076. [CrossRef]
41. Jiang, L.; Zhou, H.F.; Bai, L. Spatial Heterogeneity Analysis of Impacts of Foreign Direct Investment on Air Pollution: Empirical Evidence from 150 Cities in China Based on AQI. *Sci. Geogr. Sin.* **2018**, *38*, 351–360. [CrossRef]
42. Wang, J.H.; Ogawa, S. Effects of Meteorological Conditions on PM2.5 Concentrations in Nagasaki, Japan. *Int. J. Environ. Res. Public Health* **2015**, *12*, 9089–9101. [CrossRef] [PubMed]
43. Wang, J.Y.; Zhang, H.R.; Zhao, X.W.; Ji, J.H.; Yang, S.Z. Variation of air quality index and its relationship with meteorological elements in Beijing from 2012 to 2015. *Meteorol. Environ. Sci.* **2017**, *40*, 35–41. [CrossRef]
44. Kang, H.Y.; Liu, Y.L.; Li, T. Characteristics of air quality index and its relationship with meteorological factors in key cities of Heilongjiang Province. *J. Nat. Resour.* **2017**, *32*, 692–703. [CrossRef]
45. Wang, T.H.; Du, H.D.; Zhao, Z.Z.; Zhou, Z.M.; Russo, A.; Xi, H.L.; Zhang, J.P.; Zhou, C.J. Prediction of the impact of meteorological conditions on air quality during the 2022 Beijing Winter Olympics. *Sustainability* **2022**, *14*, 4574. [CrossRef]
46. Tiziano, T.; Umberto, R. An analytical air pollution model for complex terrain. *Environmetrics* **1994**, *5*, 159–165. [CrossRef]
47. Yu, S.; Wang, C.; Liu, K.; Zhang, S.; Dou, W. Environmental effects of prohibiting urban fireworks and firecrackers in Jinan, China. *Environ. Monit. Assess.* **2021**, *193*, 512. [CrossRef]
48. Zhao, L.; Liu, C.; Liu, X.; Liu, K.; Shi, Y. Urban spatial structural options for air pollution control in China: Evidence from provincial and municipal levels. *Energy Rep.* **2021**, *7*, 93–105. [CrossRef]
49. Aleluia, R.; Drouet, L.; Dingenen, R.; Emmerling, J. Future global air quality indices under different socioeconomic and climate assumptions. *Sustainability* **2018**, *10*, 3645. [CrossRef]
50. Zhang, J.H.; Tian, P.L.; Liu, X.; Yang, Y.; Wang, K.; Wang, W.J. Effect of urban greening on air quality: Take 27 provincial capitals in China as an example. *Bull. Bot. Res.* **2019**, *39*, 471–480. [CrossRef]
51. Irga, P.J.; Burchett, M.D.; Torpy, F.R. Does urban forestry have a quantitative effect on ambient air quality in an urban environment? *Atmos. Environ.* **2015**, *120*, 173–181. [CrossRef]
52. Bai, L.; Jiang, L.; Zhou, H.F.; Chen, Z.S. Spatio-temporal characteristics of air quality index and its driving factors in the Yangtze River Economic Belt: An empirical study based on bayesian spatial econometric model. *Sci. Geogr. Sin.* **2018**, *38*, 2100–2108. [CrossRef]
53. Zhang, Y.Q.; Bash, J.O.; Roselle, S.J.; Shatas, A.; Repinsky, A.; Mathur, R.; Hogrefe, C.; Piziali, J.; Jacobs, T.; Gilliland, A. Unexpected air quality impacts from implementation of green infrastructure in urban environments: A Kansas City case study. *Sci. Total Environ.* **2020**, *744*, 140960. [CrossRef] [PubMed]
54. Zhou, Y.J.; Liu, H.L.; Zhou, J.X.; Xia, M. Simulation of the impact of urban forest scale on $PM_{2.5}$ and PM_{10} based on system dynamics. *Sustainability* **2019**, *11*, 5998. [CrossRef]
55. Dagum, C. A new approach to the decomposition of the Gini income inequality ratio. *Empir. Econ.* **1997**, *22*, 515–531. [CrossRef]
56. Xiong, H.H.; Lan, L.Y.; Liang, L.W.; Liu, Y.B.; Xu, X.Y. Spatiotemporal differences and dynamic evolution of $PM_{2.5}$ pollution in China. *Sustainability* **2020**, *12*, 5349. [CrossRef]
57. Zhang, Z.M.; Wang, X.Y.; Zhang, Y.; Nan, Z.; Shen, B.G. The over polluted water quality assessment of Weihe River based on Kernel Density Estimation. *Procedia Environ. Sci.* **2012**, *13*, 1271–1282. [CrossRef]
58. Liu, K.; Wang, X.Y.; Zhang, Z.B. Assessing urban atmospheric environmental efficiency and factors influencing it in China. *Environ. Sci. Pollut. Res.* **2022**, *29*, 594–608. [CrossRef]
59. Cui, E.; Ren, L.; Sun, H.Y. Evaluation of variations and affecting factors of eco-environmental quality during urbanization. *Environ. Sci. Pollut. Res.* **2015**, *22*, 3958–3968. [CrossRef]
60. Wang, W.T.; Zhang, L.J.; Zhao, J.; Qi, M.Q.; Chen, F.R. The effect of socioeconomic factors on spatiotemporal patterns of $PM_{2.5}$ concentration in Beijing–Tianjin–Hebei region and surrounding areas. *Int. J. Environ. Res. Public Health* **2020**, *17*, 3014. [CrossRef]
61. Zhou, L.; Zhou, C.H.; Yang, F.; Che, L.; Wang, B.; Sun, D.Q. Spatio-temporal evolution and the influencing factors of $PM_{2.5}$ in China between 2000 and 2015. *J. Geogr. Sci.* **2019**, *29*, 253–270. [CrossRef]
62. He, Z.F.; Guo, Q.C.; Liu, J.Z.; Zhang, Y.Y.; Liu, J.; Ding, H. Spatio-temporal variation characteristics of air pollution and influencing factors in Hebei province. *J. Nat. Resour.* **2021**, *36*, 411–419. [CrossRef]
63. Xiao, J.C.; Yuan, Z. *China's Top Ten Urban Agglomerations*; Economic Science Press: Beijing, China, 2009.
64. Yazdani, M.; Baboli, Z.; Maleki, H.; Birgani, Y.T.; Zahiri, M.; Chaharmahal, S.S.H.; Goudarzi, M.; Mohammadi, M.J.; Alam, K.; Sorooshian, A.; et al. Contrasting Iran's air quality improvement during COVID-19 with other global cities. *J. Environ. Health Sci. Eng.* **2021**, *19*, 1801–1806. [CrossRef] [PubMed]

 *sustainability*

Article

Do Lawsuits by ENGOs Improve Environmental Quality? Results from the Field of Air Pollution Policy in Germany

Fabio Bothner *, Annette Elisabeth Töller and Paul Philipp Schnase

Policy Research and Environmental Politics, FernUniversität in Hagen, 58084 Hagen, Germany;
annette.toeller@fernuni-hagen.de (A.E.T.); paul-philipp.schnase@fernuni-hagen.de (P.P.S.)
* Correspondence: fabio.bothner@fernuni-hagen.de

Abstract: It is generally assumed that in EU Member States the right of recognized environmental organizations (ENGOs) to file lawsuits under the Aarhus Convention contributes not only to a better enforcement of environmental law, but also to an improvement of environmental quality. However, this has not yet been investigated. Hence, this paper examines whether 49 lawsuits that environmental associations filed against air quality plans of German cities between 2011 and 2019 had a positive effect on air quality by reducing NO_2 emissions in the respective cities. Using a staggered difference-in-differences regression model, we show that, on average, lawsuits against cities' clean air plans have a negative effect on NO_2 concentration in these cities. In fact, the NO_2 concentration in cities sued by ENGOs decreased by about 1.31 to 3.30 $\mu g/m^3$ relative to their counterfactual level.

Keywords: air quality; air pollution policy; NO_2 concentrations; diff-in-diff-regression

Citation: Bothner, F.; Töller, A.E.; Schnase, P.P. Do Lawsuits by ENGOs Improve Environmental Quality? Results from the Field of Air Pollution Policy in Germany. *Sustainability* **2022**, *14*, 6592. https://doi.org/10.3390/su14116592

Academic Editors: José Carlos Magalhães Pires and Álvaro Gómez-Losada

Received: 21 April 2022
Accepted: 25 May 2022
Published: 27 May 2022

Publisher's Note: MDPI stays neutral with regard to jurisdictional claims in published maps and institutional affiliations.

1. Introduction

In 2006, Germany introduced a right of action for recognized environmental associations (environmental non-governmental organizations, ENGOs), by way of implementing the Aarhus Convention and Directive 2003/35/EC. This right of action, which was initially limited in its clout, was then successively expanded (as a result of rulings of the European Court of Justice) to a general right of action in environmental matters [1] (p. 6), [2]. The idea behind this (e.g., on the part of the European Commission) was to enable the associations to significantly contribute to improving the notoriously precarious application of (European) environmental law and to enhance environmental quality [3–5]. However, whether the lawsuits indeed improve both, the application of environmental law and environmental quality have not yet been investigated.

Clean air policy is a case in point for the frequently deficient application of European environmental law [6,7]. The Ambient Air Quality Directive of 2008 contains concentration thresholds for several pollutants, of which particulate matter (PM10) and nitrogen dioxide (NO_2) are the most important. In 2018, Germany (as one of 13 Member States) was sued by the European Commission for non-compliance with the limit values for highly harmful NO_2 emissions. Although the provisions of the directive were translated into the Federal Immission Control Act (BImSchG), in 2018 actual NO_2 concentrations were still above the limit value in 57 major German cities [8] (p. 24). Between 2011 and 2019, environmental non-governmental organizations filed 49 lawsuits before German administrative courts, challenging air quality plans for German cities as being inadequate. These 49 lawsuits represent "most likely cases" for the question of possible environmental effects in that the lawsuits in all cases decided by courts to date have been fully or substantially successful [9].

The problematic enforcement of the Air Quality Directive has been the subject of a number of publications, but none has linked the aspect of real pollution reduction with the legal actions taken by ENGOs to challenge the local air quality plans. In this paper, we attempt to make this connection by being the first to address the question of whether ENGO lawsuits have a positive effect on air quality by reducing NO_2 concentrations in German cities. The theoretical assumption is—in a nutshell—that the lawsuits should motivate political decision makers to—finally—adopt measures that have the potential to effectively improve air quality. This should be the case particularly because they want to avert driving bans for diesel cars that were looming as a consequence of court decisions. Those measures should improve air quality, even if not to the extent that the concentration limits can be complied with.

To answer this research question, we proceed as follows: In the next section, we briefly review the state of the research on the application of environmental law and air quality (Section 2). Next, we describe our case in more detail: the Air Quality Directive and the lawsuits against air quality plans (Section 3). In Section 4, we develop our theoretical argument, which is mainly based on a rational choice perspective. We continue with a description of our data set and methodological approach that is based on a staggered difference-in-differences (DiD) design, given that we have variation in treatment timing. This is followed by the presentation of the results (Section 6), which we subsequently discuss (Section 7). The paper ends with a conclusion (Section 8).

2. The State of Research

Improving the environmental condition or health protection through policies is far from trivial. It is politically difficult to adopt restrictive regulations, because there is often strong resistance from powerful addressees [10]. Even the adoption of restrictive regulations does not guarantee that the environmental condition will improve or that health burdens will be reduced because environmental policy measures, in particular, are often not or not fully implemented. This deficient implementation of an ever-increasing number of environmental policies has been the subject of implementation research for decades. Accordingly, the design of the policies themselves, the resources and willingness of the administration as well as the (lack of) interest or even resistance on part of the addressees are the major factors determining (or undermining) effective application [11–13]. Moreover, the process of implementation is usually inherently political and controversial [14].

In EU member states, a major share of national environmental policies is based on European policies. Studies investigating the implementation of environmental policies in the European Union mainly focus on the transposition of directives into national law [15,16], which, however, is a necessary, but not a sufficient condition for the effective application of these policies on the ground [6,15]. Whereas a lot of research is interested in different aspects of implementation and compliance [12] (p. 440), significantly fewer studies investigate the effects of the political measures on the quality of the environment, the so-called impact. For several environmental parameters studies conclude that political measures do have an effect in the intended direction [17,18]. For air quality, however, Knill et al. [19] conclude in their study of 24 OECD countries between 1976 and 2003 that there is no robust relationship between the regulatory density and intensity of environmental policies and the air quality parameters for the pollutants CO_2, SO_2, and NO_x [19] (p. 436). A recent study finds that, in 14 OECD countries between 1990 and 2014, command and control regulations "lead to a significant reduction in air pollutant emissions—but only when they are adequately executed and enforced" [13,20] (p. 227).

The implementation of the European Ambient Air Quality Directive adopted in 2008 has been examined for certain countries or municipalities, although all studies so far focus on the development of air quality plans rather than on compliance with the concentration limits (e.g., for the Netherlands Bondarouk and Liefferink [21] and Bondarouk et al. [22]). For Germany, an evaluation by the Federal Environment Agency (UBA) looks at air quality plans published between 2008 and 2012 and concludes that most of the examined German cities are still very far from complying with the concentration limits [23] (p. 135). Gollata and Newig [24] examine 137 air quality plans and find that the multi-level structure (i.e., the typical situation, where the federal government implements the European directive, but the states—in cooperation with the federal government—apply it) proves to be rather obstructive to goal achievement in clean air policy [24] (p. 1323). Lenschow et al. [25]

analyze the application of the directive in 12 cities in Poland, the Netherlands, and Germany, asking to what extent appropriate territorial levels were involved in the implementation [25] (p. 521). They conclude that "the German federal system tended to shift responsibility downwards without the necessary legal and financial backing" [25] (p. 530).

In addition, there are a number of studies with a technical background assessing the effect of different kinds of low emission zones (LEZ) on air quality levels [26–30]. For instance, Jiang et al. [28] investigate the development of NO_2 concentrations in German cities between 2002 and 2012, i.e., before and after the introduction of LEZs in 2008. They find no or only a small reduction effect on NO_2 [28] (pp. 3378–3379), which is not surprising since, at least in Germany, these early LEZs mainly aimed at reducing particulate matter emissions and did not specifically target the high NO_x-emitting diesel vehicles.

Green et al. [29] even note for the case of London that the congestion charge introduced in 2003 has led to varied but substantial reductions in three traditional pollutants, yet "a more robust countervailing increase in harmful NO_2 likely reflecting the disproportionate share of diesel vehicles exempt from the congestion charge" [29]. In Madrid, in contrast, considerable reductions in NO_2 pollution were achieved by implementing a tailored LEZ [30].

The introduction of the right of environmental associations to file suits is the subject of a range of studies. Most publications on Germany focus on the legal success of the lawsuits [1,31], whereas Töller [9] investigates the role of lawsuits in fostering driving bans (or the threat of driving bans) in German cities. Studies on other countries discuss, among other things, the role of lawsuits in the respective legal systems [32] and analyze the role of lawsuits as a legal opportunity structure for the strategic orientation of environmental associations [33].

The present paper thus fills an important research gap arising at the intersection of these discussions: the above-described debates on the determinants of air quality parameters, on the implementation of the Ambient Air Quality Directive, and on the right of environmental associations to file lawsuits have not yet been related to each other. This paper establishes this connection by investigating, for the first time, what effect the lawsuits of environmental associations have on the development of urban NO_2 concentrations using the example of Germany.

3. The (Deficient) Implementation of the Ambient Air Quality Directive in Germany and the Lawsuits

3.1. The Ambient Air Quality Directive in Germany

Nitrogen oxides (NO_x) are emitted in cities, especially by diesel vehicles [34,35]. According to the World Health Organization (WHO), they can cause considerable damage to human health and reduce life expectancy if present in high concentrations, for example in urban areas [36] (pp. 73–122). The European Ambient Air Quality Directive 2008/50/EC adopted in 2008 therefore contains, among other things, a concentration threshold for nitrogen dioxide (NO_2), which was translated into the Federal Immission Control Act (BImSchG) and the 39th Federal Immission Ordinance (BImSchV). Since 1st January 2010, the annual mean of NO_2 concentration may not exceed 40 µg/m^3 [37]. According to Art. 23 of the directive, EU Member States are obliged to measure NO_2 concentrations in urban areas and to develop air quality plans if the threshold is exceeded. These air quality plans must contain appropriate measures to limit the period during which the limit value is exceeded to the shortest possible period (Art. 23 (1) 2 of Directive 2008/50/EC). In Germany, the development of air quality plans is the responsibility of the states (Länder), which organize this task differently [24].

Even though Germany was granted several extensions of the implementation deadline, and the measured NO_2 concentrations in Germany displayed an overall decline [28] (p. 3378), many urban areas still failed to meet the threshold of 40 µg/m^3 in the annual mean. Among the 57 German cities that failed to comply with the NO_2 limit value in 2018, the exceedances in Stuttgart (71 µg/m^3), Darmstadt (67 µg/m^3), and Munich (66 µg/m^3)

were particularly high. In response to the continued violation of the threshold, in the summer of 2015, the European Commission sent the Federal Government a warning letter and eventually initiated an infringement procedure in May 2018 [38,39].

3.2. Lawsuits Filed by DUH

In 2006, Germany introduced a right of action under environmental law for recognized environmental associations based on the Aarhus Convention and Directive 2003/35/EC [1] (pp. 6–8), [5] (p. 351). However, this right initially remained limited in its scope and was only gradually expanded after various rulings of the European Court of Justice [1] (pp. 7–9). The more than 300 recognized environmental associations in Germany have been using their right of action, especially since 2013, albeit in different ways depending on the associations and the subject matter of the complaint [1,31]. Between 2017 and 2020, lawsuits against air quality plans were the second largest group of all ENGOs' lawsuits. Legally, the lawsuits filed by environmental associations have proven to be exceptionally successful, while the average success rate of administrative lawsuits (excluding asylum law) is about 12%, the success rate of lawsuits by ENGOs between 2017 and 2020 was 36.3% with major variation according to the issue at stake. Lawsuits against air quality plans have always been successful so far [40] (p. 54).

The Deutsche Umwelthilfe (DUH) is a rather atypical environmental organization with a small membership base, a high dependency on donations, and an early specialization in litigation. Of a total of 49 lawsuits against air quality plans of German municipalities, 47 were filed by the DUH, making it by far the most important organization in this field.

At the beginning of 2011, shortly after the transposition of the Ambient Air Quality Directive into national law, DUH started to sue state governments for non-compliance with the NO_2 concentration limits stipulated by the Ambient Air Quality Directive [9,41]. As shown in Figure 1, the DUH filed lawsuits in a total of 47 cases between 2011 and 2019. The strong increase in lawsuits in 2015 is likely to be due to the expiry of the extensions granted by the European Commission in January 2015, while the increase in 2018 seems to have been encouraged by the Federal Administrative Court ruling of February 2018 (see below).

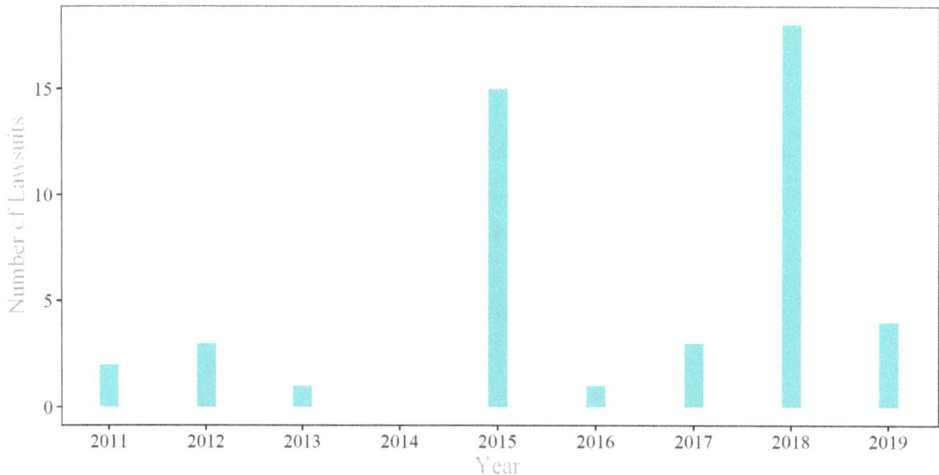

Figure 1. The 47 lawsuits by DUH against air quality plans 2011–2019.

3.3. The Court Decisions and Their Tangible Effects

To date, 21 court decisions have been adopted on these lawsuits by administrative courts—including three on a general note by the Federal Administrative Court in *Leipzig*. In all of the court decisions, the respective administrative courts held that the complaint was not only admissible, but also justified. This means that in all cases the respective air quality plan did not provide necessary measures to keep non-compliance with the NO_2 concentration limits to the shortest possible period, and that it had to be adjusted accordingly. In most cases this meant that driving bans for diesel cars had to be at least considered, if not adopted.

Beyond these fundamental commonalities, there are some differences in how straight-forward the courts argued regarding the necessity of adopting driving bans. In a first group of court decisions, the courts argued rather cautiously that driving bans could be a possible instrument for reaching compliance with the thresholds. A second group of court decisions took a more specific stance on driving bans. For instance, the Hamburg Administrative Court in 2014 argued that the city-state of Hamburg did not implement alternative measures successfully and thus would have a hard time in the future to justify that it did not adopt effective measures for economic, financial, or other reasons.

Finally, in a third group, courts considered driving bans to be inevitable in the respective case and sometimes even provided a precise timetable as to when driving bans should be introduced. The paradigmatic case is the decision of the Stuttgart Administrative Court, which decided in July 2017 that the air quality plan for Stuttgart was to be revised in a way that it contained the necessary measures to comply with the NO_2 concentration thresholds for the city of Stuttgart. The court found it doubtless that driving bans are suitable to achieve compliance with thresholds and that there is no other equivalent measure that would be less onerous. It held quite precisely that driving bans for cars with gasoline engines below Euro 3 and diesel cars below Euro 6 have to be considered. However, the state of Baden-Württemberg took the Stuttgart decision to the Federal Administrative Court. In February 2018, the Federal Administrative Court rejected the revision by and large. It accepted that a driving ban for diesel cars below Euro 6 and gasoline cars below Euro 3 would be the only effective measure, while also emphasizing that the principle of proportionality must be given adequate consideration, e.g., by introducing driving bans gradually and by establishing exceptions. After this judgement of the Federal Administrative Court, all but one of the subsequent court decisions argued for driving bans in this assertive way [9].

The fact that all courts have more or less explicitly ruled in favor of including diesel driving bans in air quality plans does not at all suggest that such driving bans have also been adopted for all these cities [9]. First, in many cases, the responsible states have appealed to the next higher instance or to the Federal Administrative Court, where possible. However, the higher-ranking courts have always upheld the essence of the decisions. Second, the introduction of driving bans requires the revision of air quality plans, which in most of the states is an elaborate and lengthy procedure between the state and the local level. Third, most states initially waited to implement what courts demanded, because they had doubts as to whether diesel driving bans were even legal, as they were not explicitly provided for in the legal regulations. This strategy proved to be invalid with the landmark decision by the Federal Administrative Court in February 2018 mentioned above. However, some state governments continued to ignore court decisions. Fourth, in 13 cases (in North Rhine-Westphalia), DUH and the state government agreed on settlements including a set of measures to comply with the threshold without adopting driving bans. Only in four cities (Hamburg, Stuttgart, Darmstadt, and Berlin) were driving bans for diesel cars imposed as a result of the court rulings [9].

4. Lawsuits and Their Effects from a Theoretical Point of View

If we assume that the lawsuits filed by ENGOs should result in more significant reductions in NO_2 concentrations than without the lawsuit, how can such an effect be theoretically conceived? What causal mechanisms might link the filing of a lawsuit to an improvement in air quality? While the expectation that the right of ENGOs to file lawsuits could contribute to improving the practical application of law in the EU is quite common in the literature [2,3] (p. 3), [5], the causal mechanism by which this should occur and how this is to affect environmental quality have not been elaborated further.

Rational choice institutionalism seems a useful approach for understanding the effects of institutions on agency, policies, and outcome parameters [42] (pp. 53–66). Looking at the NGOs' right of legal action from this perspective, lawsuits appear as procedures that may change the ways in which political decision makers perceive their preferences and accordingly chose their strategies. Political decision makers have to weigh the costs of high NO_2 concentrations against the costs and benefits of measures to reduce them. How could lawsuits affect this calculation? Plausibly, the filing of a lawsuit against a city's air quality plan would initially result in a broader public debate, both at the local and state level, on the problem of air quality, which until then was more a topic for expert circles. Thus, the problem cannot simply continue to be ignored. However, we can expect that in such a discussion, residents and local businesses affected by possible restrictions usually have a greater saying than those negatively affected by NO_2 concentrations, some of whom live in disadvantaged areas of the city and are often unaware of the negative impact of NO_2 on their health. As pointed out above, courts have from the outset ruled in favor of the environmental associations while signaling (albeit in varying degrees of concreteness) that diesel driving bans could be considered as ultima ratio. As a consequence, responsible administrations should be motivated to try to avert such driving bans. After all, driving bans would severely restrict citizens and the local business and lead to political resentment. Since in most states air quality plans are decided by state governments that ultimately depend on a parliamentary election, a loss of political confidence is feared. The political leaders therefore should be motivated by the lawsuits to adopt costly and unpopular alternative measures that would lead to a more effective reduction in NO_2 concentrations and thus make the imposition of driving bans unnecessary.

This motivation to decide on effective measures that are not driving bans is likely to grow over time with increasing levels of legal escalation (legal action, judgment, appeal, next-instance judgment, appeal for revision, and possibly even revision by the Federal Administrative Court) in the individual case. Albeit not each individual step would yield a quantifiable effect on air quality, all these measures over time can be assumed to have a negative effect on NO_x emissions in the respective urban area and on NO_2 concentrations, respectively (certainly, decreasing NO_x emissions does not necessarily translate in a linear way into lower NO_2 concentrations. Rather, there are intervening factors, such as the weather [30] (p. 6)).

Thus, it seems plausible that the lawsuits can have a negative effect on NO_2 concentrations, so that in the cities for which legal action is taken, NO_2 levels should decrease stronger after a lawsuit was filed than before and, additionally, be stronger than in cities where no lawsuit was filed. Moreover, due to the abovementioned logic of escalation, we also assume a more pronounced effect over time; hence, we formulate the following two hypotheses:

Hypothesis 1 (H1): The filing of a lawsuit against a city's air quality plan should have a negative effect on NO_2 concentrations in that city.

Hypothesis 2 (H2): This negative effect of lawsuits on NO_2 concentrations should increase over time, i.e., with the time that passes after the lawsuit was filed.

5. Materials and Methods

While the translation of the Air Quality Directive into national immission law came into force in January 2010, we set our investigation period from 2008 to 2019 (most recent available data when we conducted this research) in order to allow for a sufficient period of time prior to the filing of the first lawsuits in 2011, and thus to enable the consideration of a pre-treatment period for cases with early lawsuits as well. Moreover, 2019 is used as endpoint because the COVID-19 pandemic, which starts in early 2020, might have an impact on the results of the analysis.

To investigate whether the lawsuits have an effect on the NO_2 concentrations of German cities, we draw on data from the Federal Environment Agency's (UBA) annual evaluation of nitrogen oxide pollution [43], which includes the NO_2 monitoring stations operated by the states [44] (p. 6). The monitoring grid includes industrial, background, and traffic stations in urban, suburban, and rural areas with measurements available for the years from 2001 to 2019. For the purpose of our study, in a first step, all stations measuring background or industrial pollution are sorted out, and only those measuring traffic pollution remained. Second, according to our research question, only the stations that exceeded the limit of 40 $\mu g/m^3$ at least once during the investigation period are considered. This meant to include only those cities where a lawsuit could potentially be filed. The population of our study can thus be described as cities with traffic monitoring stations where the NO_2 limit value was exceeded at least once between 2008 and 2019.

Finally, we have to deal with problems of an unbalanced panel, which can be problematic for the DiD design we use in this study (an important issue for the validity of a DiD design is that the differences between the control and treatment groups are stable over time [45]. However, as the composition of the two groups changes across time due to missing data, this could bias the estimated results. This is rather unproblematic if the missing values are randomly distributed, which is not necessarily the case in our study). A special characteristic of the UBA's annual NO_2 measurement is that new monitoring stations are frequently set up in cities, whereas measurements at old stations are discontinued. The number of available monitoring stations therefore varies from year to year, and not all measurement series are complete during our investigation period. To handle this problem, stations for which no complete measurement series were available during the investigation period are sorted out.

Thus, our sample includes complete measurement series from 91 stations in 59 cities. From those 59 cities, 34 cities were never sued during the investigation period (non-treatment group), whereas in 25 cities, a lawsuit was filed at least at some point (treatment group). Furthermore, to overcome the problem that different stations located in the same city may correlate through their shared location, the values measured at stations within the same city are aggregated to a mean for that city. Our sample thus arrives at 59 cases, with annual measurements from 2008 to 2019 for every single case, resulting in a total of 708 observations in the sample (in addition, we run our model with a non-aggregated sample, i.e., we treat each station as a separate observation, using clustered standard errors at the city level. The results are similar to those for the aggregated sample and can be found in the Appendix A, see Figure A1 and Table A1).

Causally attributing policy measures and results in the real world is equally important and methodologically demanding [46] (pp. 37–47). As a result, there are different approaches to identify causal effects of policy "treatments". Quasi-experimental approaches in particular have become increasingly popular in recent years. A simple, but effective method for calculating the effects of (policy) measures are DiD regression models. While they are common in the field of political economy [47,48], they are also used in the context of air pollution measures [29,30]. In its simplest format, there are two groups and two time periods: a treated group (experimental group) and an untreated group (control group) [49] (p. 200). If, in the absence of treatment, the average outcomes of both groups follow parallel trends over time (parallel trend assumption), it is possible to calculate the average treatment effect (ATT) for the experimental group by comparing the average change in outcomes in the treated group with the average change in outcomes in the control group [49] (p. 200).

As already mentioned, our dataset contains monitoring stations in all German cities that have not complied with the limit value of 40 $\mu g/m^3$ in annual average for at least one time during the investigation period. While lawsuits were filed against some of these cities, there were no lawsuits in others. Thus, with a treated group (cities with lawsuits) and a non-treated group (cities without lawsuits) we have a good setting for a DiD regression.

However, unlike the simple DiD setting, there is variation in treatment timing in our study due to the lawsuits being filed in different cities at different times (see Figure 1). To date, it is common to use a two-way fixed effects regression model (TWFE) for analysis of groups with varying treatment timings. However, several recent studies indicate that the use of a TWFE in a staggered DiD, especially in the presence of effect heterogeneity, may cause problems that affect the estimation [49–51]. Although this does not have to result in complete design failure, some caution is needed when using a TWFE estimator to summarize treatment effects [51] (p. 255). Therefore, we rely on the model of Callaway and Sant'Anna [49]. Their approach not only allows us to estimate a treatment effect in the presence of effect heterogeneity and dynamic effects, but also proposes several ways to aggregate the ATT to answer different research questions. Especially for our study, which includes a rather small number of cases, it seems appropriate to use and interpret the aggregated ATTs, as they are more robust than the simple ATTs.

However, as with other DiD models, some basic assumptions must be made for the Callaway and Sant'Anna approach [49] (pp. 202–207). First, no unit is treated at the beginning of the observation period, and if a unit is treated, it remains treated until the end of the observation period. Second, there should be limited treatment anticipation, which is "likely to be the case when the treatment path is not a priori known and/or when units are not the ones who "choose" treatment status" [49] (p. 204). Third, the assumption of a parallel trend must hold at least under specific conditions. For our study, we assume that it holds even unconditionally. However, we may face the problem of non-random treatment in our study. This means that cities are not sued randomly but on certain characteristics. Although all cities in our data set exceeded the 40 $\mu g/m^3$ limit at least once and are therefore at risk of getting sued, it is not clear on what basis ENGOs like the DHU decide to sue cities. However, it does not seem far-fetched that cities with high NO_2 concentrations have a higher probability of being sued than cities with lower NO_2 concentrations. This assumption is supported by the fact that the cities sued have on average a significantly higher NO_2 concentration (see Figure 2). Fredrikson and de Oliveira [52] (p. 525) capture this problem by claiming: "with a non-random assignment to treatment, there is always the concern that the treatment states would have followed a different trend than the control states, even absent the reform". To address this issue, Frederikson and de Oliveira [52] propose to control for factors that lead to differences in time trends between groups. Normally this could be done by including control variables or by doing a matching procedure [52]. However, since in our case we assume that cities being sued mainly because of their NO_2 concentrations, including control variables seems not purposeful. Nevertheless, the fact that we have a variation in treatment timing allows us some sort of "matching" by using only treated cities and taking the "all not-yet-treated" cities as the control group [53]. Assuming that cities are sued because of specific characteristics (mainly the NO_2 concentration), the not-yet-treated cities seem to be a good control group since they should share the same characteristics as the already treated cities. To test for the effects of potential non-random treatment, we conduct our analysis once with the entire sample and once with the subsample of cities sued (the results for the subsample can be found in the Appendix A, see Figure A2 and Table A2).

6. Results

Figure 2 shows the mean NO_2 concentration in $\mu g/m^3$ from 2008 to 2019 for cities with a lawsuit against their air quality plan (treated group) and cities without a lawsuit against their air quality plan (not treated group). While treated cities have higher NO_2 concentration compared to untreated cities, the figure shows that NO_2 concentration follows a similar trend of decreasing values for both groups. This is true for the period before the first lawsuit in 2011, but also for the following years. For a simple DiD setting, (i.e., there is only one treatment time point) this would support the argument for maintaining the parallel trend assumption but against a significant effect of treatment. This is the case because the parallel trend is still intact until the end of the observation period, but for a substantial effect, we would expect the sued cities to reduce their NO_2 concentration to a greater extent than the non-sued cities. However, for our setting the interpretation is much more complex, since we have different treatment points (2011, 2012, 2015, 2017, 2018, and 2019). The variation of treatment timing causes different pre- and post-treatment periods. For example, cities that were sued in 2011 have a pre-treatment period of three years (2008, 2009, and 2010) and a post-treatment period of eight years (2012 to 2019). Cities sued in 2018, on the other hand, have a 10-year pre-treatment period but only one post-treatment year. Therefore, Figure 2 gives us an idea of the overall trend in urban NO_2 concentrations but cannot be used to assess the parallel trend assumption or the effect of treatment.

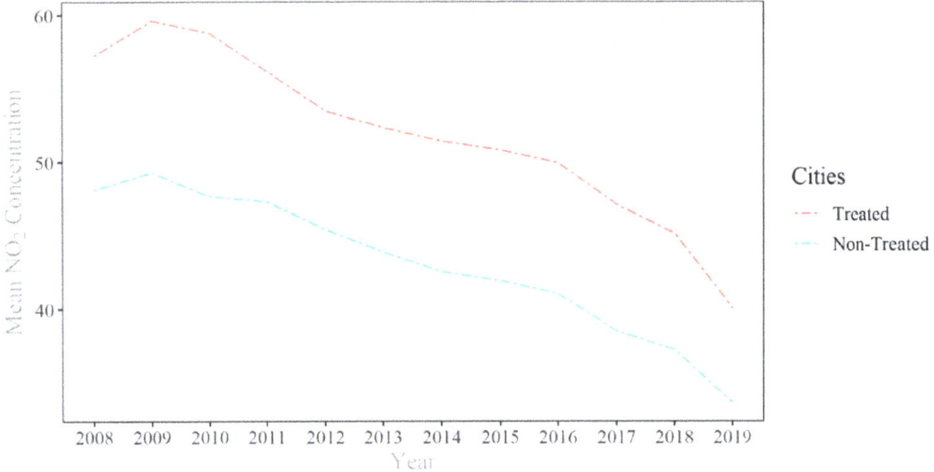

Figure 2. Mean NO_2 concentration in cities with and without lawsuit, 2008–2019.

Thus, as described in Section 5, we must use a staggered difference-in-differences design that allows us to test the assumption of a parallel trend and to calculate the effect of treatment even for different treatment time points. The most common way to do this is to use an event study plot (see Figure 3). The plot shows pre-treatment estimates that can be used as an indication about the parallel trend assumption as well as estimated post-treatment effects [49] (p. 218). The x-axis of the event study plot shows the years before and after treatment. In our case, the longest pre-treatment period is 10 years since our investigation starts in 2008, and 2019 is the last year in which cities are sued. The longest post-treatment period is 8 years since the first cities were sued in 2011 and the study period goes to 2019. The y-axis shows the partially aggregated effects of the treatment for both the pre- (red) and post-treatment period (blue) (The exact values are also shown in Table 1, line 2) In the pre-treatment period, we logically expect no effect of treatment, as there should be no significant difference between treated and non-treated cities. If this is the case, we can assume that the parallel trend assumption holds. A look at the pre-treatment period in Figure 3 shows that there is indeed no significant effect, which suggests that the parallel trend assumption holds for our case.

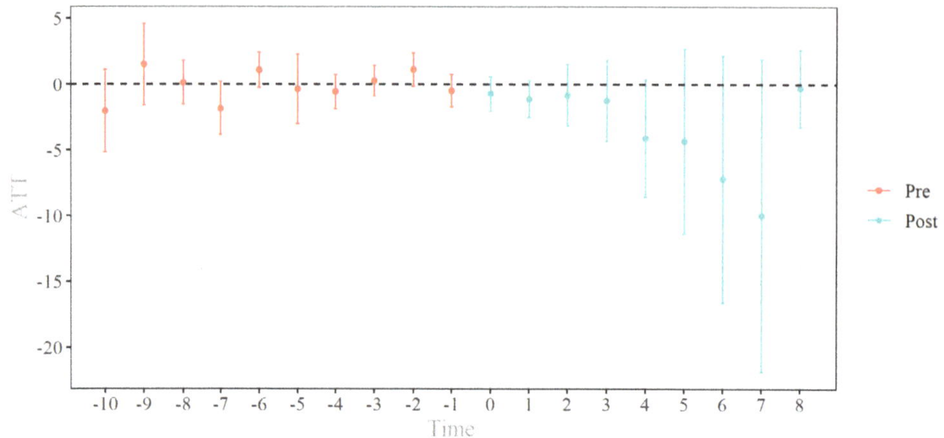

Figure 3. Event study plot.

Table 1. Aggregated treatment effect estimates.

	Partially Aggregated									Aggregated ATT
simple weighted average										−2.26 * (1.16)
Event study	$\underline{e = 0}$ −0.70 (0.49)	$\underline{e = 1}$ −1.10 (0.54)	$\underline{e = 2}$ −0.81 (0.89)	$\underline{e = 3}$ −1.24 (1.13)	$\underline{e = 4}$ −4.11 (1.70)	$\underline{e = 5}$ −4.32 (2.70)	$\underline{e = 6}$ −7.19 (3.51)	$\underline{e = 7}$ −9.95 (4.41)	$\underline{e = 8}$ −0.29 (1.14)	−3.30 ** (1.56)
Group-specific effects	$\underline{g = 11}$ −0.11 (0.79)	$\underline{g = 12}$ −7.10 ** (2.63)	$\underline{g = 13}$ −1.96 ** (0.66)	$\underline{g = 14}$	$\underline{g = 15}$ −1.80 (1.14)	$\underline{g = 16}$ −1.05 (0.55)	$\underline{g = 17}$ 2.25 ** (0.51)	$\underline{g = 18}$ 0.08 (0.66)	$\underline{g = 19}$ 0.47 (0.70)	−1.31 ** (0.54)

* Please note "e" indicates the effect after treatment, i.e., e = 1 reflects the effect 1 year after treatment. "g" indicates the effect for the observations treated in that year. For example, g = 11 reflects the effect for all units treated in 2011. *** $p < 0.01$, ** $p < 0.05$, * $p < 0.1$. For calculations we use the doubly robust approach instead of the outcome regression or inverse probability weighting. However, calculations with the outcome regression or inverse probability weighting show similar results and can be found in the Appendix A (see Figures A3 and A4, Tables A3 and A4). According to Callaway and Sant'Anna [49], all inference procedures use clustered bootstrapped standard errors at the city level (15,000 repetitions) and account for the autocorrelation of the data.

Regarding the post-treatment time, the plot indicates that the effect size becomes larger over time. However, we could not find a significant effect for the individual periods of the post-treatment period. The reason is that the group size varies and becomes smaller as more time passes, which is particularly problematic when the total number of observations is rather small, as is the case in our study [49] (p. 210). While at time zero we count every city that was sued, and at time seven we only count cities that were treated before 2013, as we can only observe the treatment effect after seven years for cities that were sued in 2011 and 2012. As Figure 1 shows, we have a large group of cities that were sued in 2015 and 2018. If we assume that the effect of being sued does not materialize immediately but rather requires a few years to take effect, this, in combination with the small number of observations, could explain why we do not find significant effects in the event study for the post-treatment periods.

As Callaway and Sant'Anna [49] point out, it seems more appropriate in such a setting to aggregate ATT into an overall effect of participating in the treatment. However, there are different methods for calculating the overall treatment effect, each with different advantages and disadvantages. Table 1 shows three ways to calculate such an overall effect, as well as the partially aggregated treatment effects required for this calculation. The simplest way is to estimate a weighted average across all groups and time points with weights proportional to group size (see Table 1, line 1). However, such an approach tends to overweight the effect of the early treated groups because we have more observations for them in the post-treatment period [49] (p. 212). Another approach is to use the ATTs estimated in the event study and aggregate them into an overall measure (see Table 1, line 2). In this case, the overall ATT is based on the average of the partially aggregated treatment effects of the post-treatment periods (e = 0 to e = 8). In contrast, Callaway and Sant'Anna [49] (p. 212) promote the idea to "first compute[s] the average effect for each group (across all time periods) and then averages these effects together across groups to summarize the overall average effect of participating in the treatment". Hence, the so-called aggregated group-specific ATT (see Table 1, row 3) is based on the aggregate average of the partially aggregated group-specific effects (g = 11 to g = 19). It can be interpreted similarly to the ATT in a classic two-group, two-period DiD design. For our study we calculate all three overall measures.

The simple weighted average shows a 2.26 $\mu g/m^3$ lower NO_2 concentration, while the aggregated event study average indicates a 3.30 $\mu g/m^3$ lower NO_2 concentration. The aggregated average effect of a lawsuit across all groups sued indicates a 1.31 $\mu g/m^3$ lower NO_2 concentration due to a lawsuit. All three aggregate ATT measures mostly paint the same picture, showing that a lawsuit against a city's air quality plan reduces NO_2 concentrations in that city.

7. Discussion

According to the aggregated group-specific ATT (see Table 1, line 3), the NO_2 concentration in cities that were sued by ENGOs decreased by roughly 1.31 $\mu g/m^3$ relative to their counterfactual level. How can this result be interpreted? Firstly, the estimated 1.31 $\mu g/m^3$ decrease should not be misinterpreted. It tells us that the NO_2 concentration in cities would be 1.31 $\mu g/m^3$ higher if there were no lawsuit against a city's air quality plan and not that the NO_2 concentration decreased by 1.31 $\mu g/m^3$ in absolute levels. Moreover, our results do not indicate that cities that have not been sued do not reduce NO_2 (see Figure 2), but rather suggest that sued cities reduce the NO_2 concentration to a larger extent. At first glance, however, it appears to be a rather small effect, but it should be noted that the aggregate group-specific ATT is an average effect that does not consider that later treated cities may experience a much smaller effect due to the time lag between the lawsuit and the adoption of measures against NO_2 concentration. This is due to the fact that the aggregate group-specific ATT weighs all groups equally, regardless of treatment duration and group size [49] (p. 210). For example, the treatment effect of the 2015 group (g = 15) is weighted equally with the treatment effect of the 2019 group (g = 19), even though the 2019 group is much smaller and experiences treatment for only one year. Thus, for our setting, it seems that the aggregated group-specific ATT underestimates the treatment effect. In contrast, the aggregate ATT of the event study (see Table 1, line 2) provides a measure of the mean effect of the treatment for the entire observation period. However, the aggregate ATT of the event study also does not take the group size into account. As already explained, in the event study setting, the group size becomes smaller with increasing length of exposure to the treatment (see Figure 3), which leads to a disproportional weighting of the effect for observations that receive the treatment very early (similar to the simple weighted average, but to a greater extent). Since Figure 3 indicates that, in our case, the treatment effect becomes stronger over time, this may lead to an overestimation of the aggregate ATT of the event study. Although the simple aggregated ATT (see Table 1, line 1) also overestimates the effect of the treatment, in our case, it seems to be the most appropriate measure for determining the treatment effect, as the other two measures either overestimate or underestimate the effect to a much greater extent.

Nevertheless, since all three measures are significant, we are confident that the first hypothesis is supported by our findings. In cities where legal action has been taken, NO_2 levels decrease more after the action than before the action and also more than in cities where no action has been taken. This assessment is supported by the findings of our subset analysis, in which we use only treated cities and use all the "not-yet-treated" cities as the control group (see Figure A1 and Table A1). We also find significant negative effects for the treatment. For the simple aggregated ATT, the coefficient is -2.44 and is significant at the 10% level. For the aggregated ATT of the event study the coefficient is also negative (-3.44) and significant even on a 5% level. Only for aggregated group-specific ATT is a smaller (-1.27) and no longer significant effect observed. However, this could be due to the now significantly reduced number of observations (300 to 709). Overall, the results of the subset analysis suggest that a treatment effect occurs even when the cities have similar NO_2 concentrations.

For our second hypothesis, however, the empirical evidence is not as clear as for the first hypothesis. Figure 3 and Table 1 indicate that the effect of a lawsuit increases over time but, as already mentioned, the partially aggregated effects of the event study are not significant due to the small numbers of observations within the groups (see Table 1, line 2). Further support for our hypothesis comes from the partially aggregated group-specific ATTs (see Table 1, line 3). The group-specific ATTs are estimated based on all observations within a group and across all post-treatment time points. For example, the 2012 group ATT is estimated based on all cities sued in 2012 over the years 2012 to 2019. Assuming that the effect takes time to become apparent or increases over time, early treated groups should have greater ATT. Overall, Table 1 shows that this is the case for the partially aggregated group-specific ATTs. As it shows, cities sued before 2016 ($g = 11$ to $g = 15$) tend to have large negative and significant effects, while groups treated after 2016 ($g = 16$ to $g = 19$) show small positive average treatment effects. In the case of the 2017 group ($g = 16$), there is indeed a significant positive effect, but this group is based on a single case, namely the city of Kiel, and should not be overinterpreted. Although the empirical evidence for the second hypothesis is not perfect, there is some evidence that the negative effect of lawsuits on NO_2 concentrations increases over time (this should not be misinterpreted as effect heterogeneity, since effect heterogeneity describes the phenomenon that different groups experience different treatment effect paths [50] (p. 193)).

8. Conclusions

The starting point of our paper is the question of whether lawsuits filed by ENGOs under the Aarhus Convention can lead to improvements in environmental quality, as assumed in the literature. For this question, the 49 lawsuits filed between 2011 and 2019 by ENGOs against the air quality plans for German cities represent a "most likely case" because they were exceptionally successful. Our theoretical argument is that the lawsuits should have motivated political decision makers to adopt more effective measures in order to avoid resorting to diesel driving bans. Those measures should have a negative impact on emissions and thus also on NO_2 concentrations. Indeed, the results of our DiD model suggest that sued cities have a 1.31 to 3.30 $\mu g/m^3$ lower NO_2 concentration relative to their counterfactual level. In addition, there is some evidence that lawsuits are not immediately effective, since the event study plot shows that the more time that passes after treatment, the larger the effects.

Our findings indicate that it is possible for lawsuits by ENGOs to lead to an improvement in air quality that would not have occurred without the lawsuit. However, it is still an open question of which actions and measures taken are exactly responsible for this improvement, i.e., which causal mechanisms connect the lawsuits with (improved) air quality and also what role they play, e.g., agency. With our DiD analysis, we applied a quantitative method to establish this causal connection as such. However, bringing light into the causal mechanisms will require further studies with research designs including qualitative methods, e.g., comparative case studies that look into what really happened in the affected cities after a lawsuit was filed, and which measures were adopted, causing a stronger decrease of NO_2 concentrations than we find elsewhere.

Furthermore, it is unclear how far this finding can travel. It seems at least doubtful that our findings are generalizable for all areas of environmental protection. It is important to note that ENGOs use the right to sue in a broad variety of areas with very different regulatory settings and conflict structures [40]. Besides air quality plans, ENGOs sue, for example, against the admission of wind energy plants [54], against water law permits, and against a variety of planning decisions. Success rates are lower than rates with air quality plans but still higher than on average of other administrative cases [40] (p. 54). In addition, in other areas besides air quality, it can be more ambivalent to determine what exactly improves environmental quality, especially in cases with conflicting environmental protection objectives. For instance, the ENGOs "Green League" and "NABU" are suing against the additional water pumping in the Grünheide area, which would be necessary due to the consumption of the recently completed Tesla Gigafactory [55]. Looking at lawsuits against the admission of wind energy plants, it is even debatable if the ENGOs' right to sue could harm specific environmental interests [54]. Thus, more research is needed on the effects of ENGOs' lawsuits in different areas.

Author Contributions: Conceptualization, A.E.T. and P.P.S.; methodology, F.B.; validation, A.E.T., P.P.S. and F.B.; formal analysis, F.B.; investigation, A.E.T. and P.P.S.; data curation, P.P.S.; writing—original draft preparation, A.E.T., P.P.S., and F.B.; writing—review and editing, A.E.T. and F.B.; visualization, F.B.; supervision, A.E.T.; funding acquisition, A.E.T. All authors have read and agreed to the published version of the manuscript.

Funding: This research received no external funding.

Institutional Review Board Statement: Not applicable.

Informed Consent Statement: Not applicable.

Data Availability Statement: Publicly available datasets were analyzed in this study. The data can be found here: https://www.umweltbundesamt.de/daten/luft/luftdaten/jahresbilanzen/eJxrWpScv9 B0UWXqEiMDQwMAMM8FtA== (accessed on 21 April 2022).

Acknowledgments: We are most grateful to Marcel Langner, UBA, Robin Kulpa, DUH, Andreas Hofmann, Leo Ahrens, Julian Erhardt, and Daniel Rasch for their very helpful comments on a previous version of this paper. All remaining errors are in our responsibility.

Conflicts of Interest: The authors declare no conflict of interest.

Appendix A

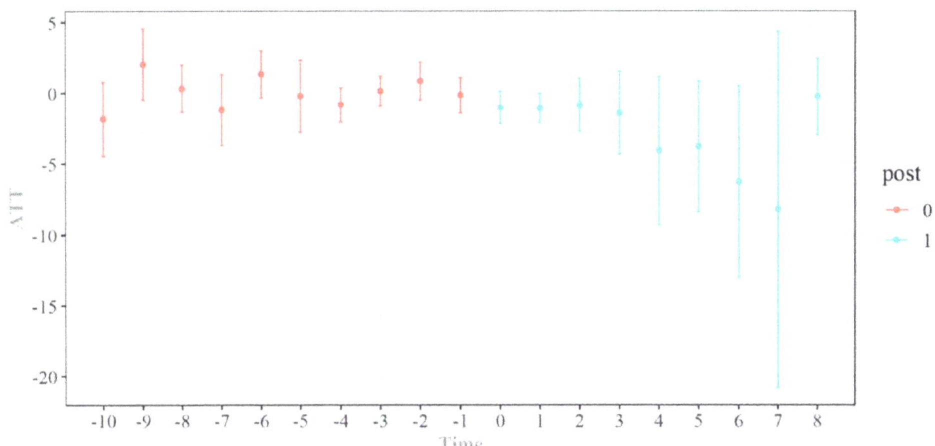

Figure A1. Event study plot of non-aggregated sample.

Table A1. Aggregated treatment effect estimates of non-aggregated sample *.

	Partially Aggregated									Aggregated ATT
simple weighted average										**−2.06** * (0.83)
Event study	$e = 0$ 1.02 (0.43)	$e = 1$ −1.05 ** (0.39)	$e = 2$ −0.84 (0.71)	$e = 3$ −1.40 (1.11)	$e = 4$ −4.05 (1.99)	$e = 5$ −3.75 (1.76)	$e = 6$ −6.24 (2.59)	$e = 7$ −8.20 (4.79)	$e = 8$ −0.24 (1.03)	**−2.98** ** (1.27)
Group-specific effects	$g = 11$ −0.11 (0.72)	$g = 12$ −7.40 ** (2.14)	$g = 13$ −2.21 ** (0.65)	$g = 14$	$g = 15$ −1.81 (1.22)	$g = 16$ −1.11 (0.54)	$g = 17$ 2.12 ** (0.47)	$g = 18$ −0.29 (0.54)	$g = 19$ 0.21 (0.75)	**−1.53** ** (0.55)

* Please note "e" indicates the effect after treatment, i.e., e = 1 reflects the effect 1 year after treatment. "g" indicates the effect for the observations treated in that year. For example, g = 11 reflects the effect for all units treated in 2011. *** $p < 0.01$, ** $p < 0.05$, * $p < 0.1$. According to Callaway and Sant'Anna, [49] all inference procedures use clustered bootstrapped standard errors at the city level (15,000 repetitions) and account for the autocorrelation of the data.

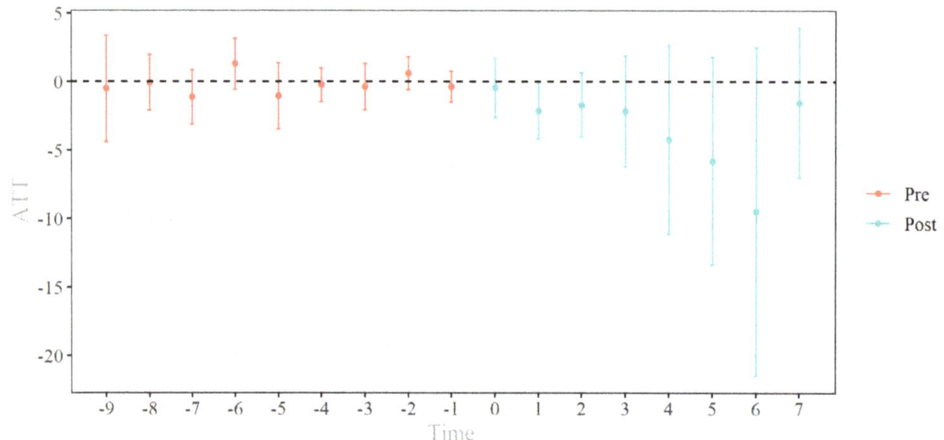

Figure A2. Event study plot (not-yet-treated vs. yet-treated).

Table A2. Aggregated treatment effect estimates (not-yet-treated vs. yet-treated) *.

	Partially Aggregated								Aggregated ATT
Simple weighted average									**−2.44** * (1.32)
Event study	$e = 0$ −0.43 (0.85)	$e = 1$ −2.13 ** (0.81)	$e = 2$ −1.69 (0.92)	$e = 3$ −2.14 (1.57)	$e = 4$ −4.19 (2.71)	$e = 5$ −5.77 (2.98)	$e = 6$ −9.46 (4.63)	$e = 7$ −1.50 (2.14)	**−3.41** ** (1.70)
Group-specific effects	$g = 11$ 1.10 (0.65)	$g = 12$ −6.79 ** (2.70)	$g = 13$ −3.35 ** (0.77)	$g = 14$	$g = 15$ −2.41 ** (0.96)	$g = 16$ −1.20 (0.55)	$g = 17$ 1.65 (1.22)	$g = 18$ 1.44 (2.16)	**−1.27** (0.54)

* Please note "e" indicates the effect after treatment, i.e., e = 1 reflects the effect 1 year after treatment. "g" indicates the effect for the observations treated in that year. For example, g = 11 reflects the effect for all units treated in 2011. *** $p < 0.01$, ** $p < 0.05$, * $p < 0.1$. For calculations we use the doubly robust approach instead of the outcome regression or inverse probability weighting. According to Callaway and Sant'Anna [49], all inference procedures use clustered bootstrapped standard errors at the city level (15,000 repetitions) and account for the autocorrelation of the data.

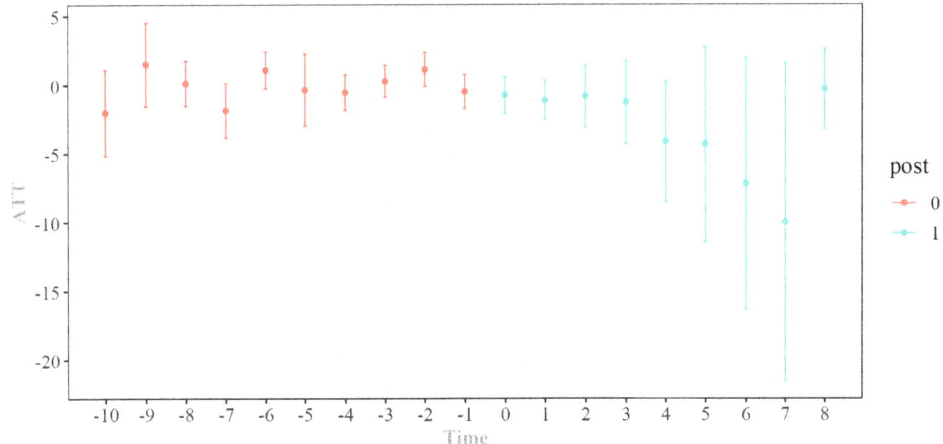

Figure A3. Event study plot based on outcome regression.

Table A3. Aggregated treatment effect estimates based on outcome regression*.

	Partially Aggregated									Aggregated ATT
simple weighted average										**−2.26 *** (1.16)
Event study	e = 0 −0.70 (0.50)	e = 1 −1.10 (0.55)	e = 2 −0.81 (0.88)	e = 3 −1.24 (1.15)	e = 4 −4.11 (1.68)	e = 5 −4.32 (2.71)	e = 6 −7.19 (3.51)	e = 7 −9.95 (4.43)	e = 8 −0.29 (1.13)	**−3.30 *** (1.56)
Group-specific effects	g = 11 −0.11 (0.72)	g = 12 −7.11 ** (2.97)	g = 13 −1.96 ** (0.66)	g = 14	g = 15 −1.80 (1.12)	g = 16 −1.05 (0.53)	g = 17 2.25 ** (0.51)	g = 18 0.08 (0.67)	g = 19 0.47 (0.68)	**−1.31 *** (0.54)

* Please note "e" indicates the effect after treatment, i.e., e = 1 reflects the effect 1 year after treatment. "g" indicates the effect for the observations treated in that year. For example, g = 11 reflects the effect for all units treated in 2011. *** $p < 0.01$, ** $p < 0.05$, * $p < 0.1$. According to Callaway and Sant'Anna [49], all inference procedures use clustered bootstrapped standard errors at the city level (15,000 repetitions) and account for the autocorrelation of the data.

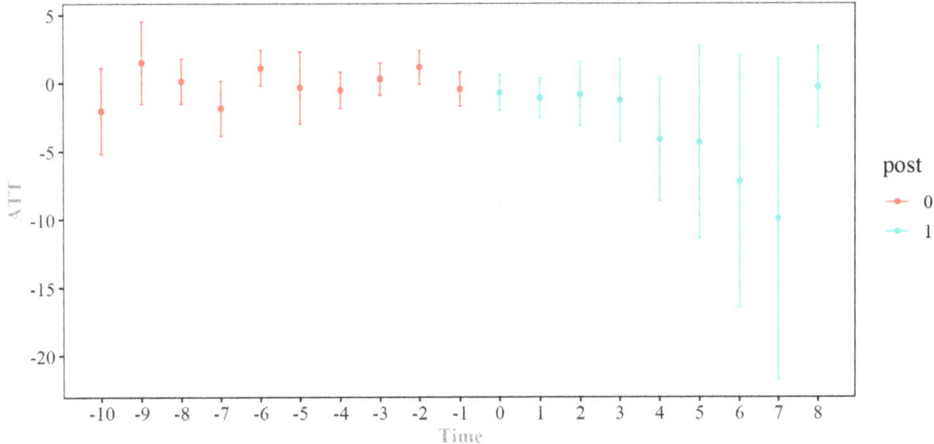

Figure A4. Event study plot based on inverse probability weighting.

Table A4. Aggregated treatment effect estimates based on inverse probability weighting*.

	Partially Aggregated									Aggregated ATT
simple weighted average										−2.26 * (1.16)
Event study	$e = 0$ −0.70 (0.50)	$e = 1$ −1.10 (0.55)	$e = 2$ −0.81 (0.88)	$e = 3$ −1.24 (1.15)	$e = 4$ −4.11 (1.71)	$e = 5$ −4.32 (2.69)	$e = 6$ −7.19 (3.51)	$e = 7$ −9.95 (4.47)	$e = 8$ −0.29 (1.13)	−3.30 ** (1.57)
Group-specific effects	$g = 11$ −0.11 (0.79)	$g = 12$ −7.11 ** (2.60)	$g = 13$ −1.96 ** (0.67)	$g = 14$	$g = 15$ −1.80 (1.12)	$g = 16$ −1.05 (0.54)	$g = 17$ 2.25 ** (0.52)	$g = 18$ 0.08 (0.65)	$g = 19$ 0.47 (0.69)	−1.31 ** (0.54)

* Please note "e" indicates the effect after treatment, i.e., e = 1 reflects the effect 1 year after treatment. "g" indicates the effect for the observations treated in that year. For example, g = 11 reflects the effect for all units treated in 2011. *** $p < 0.01$, ** $p < 0.05$, * $p < 0.1$. According to Callaway and Sant'Anna [49], all inference procedures use clustered bootstrapped standard errors at the city level (15,000 repetitions) and account for the autocorrelation of the data.

References

1. Schmidt, A.; Zschiesche, M. *Die Klagetätigkeit der Umweltschutzverbände im Zeitraum von 2013 bis 2016. Empirische Untersuchungen zu Anzahl und Erfolgsquoten von Verbandsklagen im Umweltrecht*; Sachverständigenrat für Umweltfragen: Berlin, Germany, 2018.
2. Sachverständigen Rat für Umweltfragen. *Verbandsklage Wirksam und Rechtskonform Ausgestalten: Stellungnahme zur Novelle des Umwelt-Rechtsbehelfsgesetzes*; Sachverständigenrat für Umweltfragen: Berlin, Germany, 2016.
3. European Commission. *Proposal for a Directive of the European Parliament and of the Council on Access to Justice in Environmental Matters*; COM/2003/0624 Final; European Commission: Brussels, Belgium, 2003.
4. Krämer, L. EU Enforcement of Environmental Laws: From Great Principles to Daily Practice–Improving Citizen Involvement. *Environ. Policy Law* **2014**, *44*, 247–271.
5. Hofmann, A. Left to interest groups? On the prospects for enforcing environmental law in the European Union. *Environ. Politics* **2019**, *28*, 342–364. [CrossRef]
6. European Commission. *Communication from the Commission to the European Parliament, the Council, the European Economic and Social Committee and the Committee of the Regions the EU Environmental Implementation Review: Common Challenges and How to Combine Efforts to Deliver Better Results*; COM/2017/063 Final; European Commission: Brussels, Belgium, 2017.
7. European Environmental Agency. *Air Quality in Europe–2019*; European Environmental Agency: Luxembourg, 2019.
8. Umweltbundesamt. Gesunde Luft. Schwerpunkt 1-2019. Available online: https://www.umweltbundesamt.de/sites/default/files/medien/2546/publikationen/sp_1-2019_web.pdf (accessed on 5 June 2020).
9. Töller, A.E. Driving bans for diesel cars in German cities: The role of ENGOs and Courts in producing an unlikely outcome. *Eur. Policy Anal.* **2021**, *7*, 486–507. [CrossRef]
10. Bernhagen, P. Air Pollution and Business Political Influence in Fifteen OECD Countries. *Environ. Plan. C Gov. Policy* **2012**, *30*, 362–380. [CrossRef]
11. Pressman, J.L.; Wildavsky, A. *Implementation: How Great Expectations in Washington Are Dashed in Oakland: Why It's Amazing That Federal Programs Work at All, This Being A Saga of the Economic Development Administration as Told by Two Sympathetic Observers Who Seek to Build Morals on a Foundation of Ruined Hopes*, 3rd ed.; University of California Press: London, UK, 1984; ISBN 0-52-005331-1.
12. Tosun, J. Environmental Monitoring and Enforcement in Europe: A Review of Empirical Research. *Environ. Pol. Gov.* **2012**, *22*, 437–448. [CrossRef]
13. Limberg, J.; Steinebach, Y.; Bayerlein, L.; Knill, C. The more the better? Rule growth and policy impact from a macro perspective. *Eur. J. Political Res.* **2021**, *60*, 438–454. [CrossRef]
14. Hupe, P.L.; Hill, M.J. 'And the rest is implementation.' Comparing approaches to what happens in policy processes beyond Great Expectations. *Public Policy Adm.* **2016**, *31*, 103–121. [CrossRef]
15. Jordan, A. The Implementation of EU Environmental Policy; A Policy Problem without a Political Solution? *Environ. Plan. C Gov. Policy* **1999**, *17*, 69–90. [CrossRef]
16. Börzel, T.A.; Buzogány, A. Compliance with EU environmental law. The iceberg is melting. *Environ. Politics* **2019**, *28*, 315–341. [CrossRef]
17. Le Quéré, C.; Korsbakken, J.I.; Wilson, C.; Tosun, J.; Andrew, R.; Andres, R.J.; Canadell, J.G.; Jordan, A.; Peters, G.P.; van Vuuren, D.P. Drivers of declining CO2 emissions in 18 developed economies. *Nat. Clim. Change* **2019**, *9*, 213–217. [CrossRef]
18. Steinebach. Water Quality and the Effectiveness of European Union Policies. *Water* **2019**, *11*, 2244. [CrossRef]
19. Knill, C.; Schulze, K.; Tosun, J. Regulatory policy outputs and impacts: Exploring a complex relationship. *Regul. Gov.* **2012**, *6*, 427–444. [CrossRef]
20. Steinebach, Y. Instrument choice, implementation structures, and the effectiveness of environmental policies: A cross-national analysis. *Regul. Gov.* **2022**, *16*, 225–242. [CrossRef]
21. Bondarouk, E.; Liefferink, D. Diversity in sub-national EU implementation: The application of the EU Ambient Air Quality directive in 13 municipalities in the Netherlands. *J. Environ. Policy Plan.* **2017**, *19*, 733–753. [CrossRef]

22. Bondarouk, E.; Liefferink, D.; Mastenbroek, E. Politics or management? Analysing differences in local implementation performance of the EU Ambient Air Quality directive. *J. Pub. Pol.* **2020**, *40*, 449–472. [CrossRef]

23. Diegmann, V.; Pfäfflin, F.; Wursthorn, H. *Bestandaufnahme und Wirksamkeit von Maßnahmen der Luftreinhaltung*; Umweltbundesamt: Dessau-Roßlau, Germany, 2014.

24. Gollata, J.A.; Newig, J. Policy implementation through multi-level governance: Analysing practical implementation of EU air quality directives in Germany. *J. Eur. Public Policy* **2017**, *24*, 1308–1327. [CrossRef]

25. Lenschow, A.; Becker, S.T.; Mehl, C. Scalar dynamics and implications of ambient air quality management in the EU. *J. Environ. Policy Plan.* **2017**, *19*, 520–533. [CrossRef]

26. Pestel, N.; Wozny, F. Health effects of Low Emission Zones: Evidence from German hospitals. *J. Environ. Econ. Manag.* **2021**, *109*, 102512. [CrossRef]

27. Gu, J.; Deffner, V.; Küchenhoff, H.; Pickford, R.; Breitner, S.; Schneider, A.; Kowalski, M.; Peters, A.; Lutz, M.; Kerschbaumer, A.; et al. Low emission zones reduced PM_{10} but not NO_2 concentrations in Berlin and Munich, Germany. *J. Environ. Manag.* **2022**, *302*, 114048. [CrossRef]

28. Jiang, W.; Boltze, M.; Groer, S.; Scheuvens, D. Impacts of low emission zones in Germany on air pollution levels. *Transp. Res. Procedia* **2017**, *25*, 3370–3382. [CrossRef]

29. Green, C.P.; Heywood, J.S.; Navarro Paniagua, M. Did the London congestion charge reduce pollution? *Reg. Sci. Urban Econ.* **2020**, *84*, 103573. [CrossRef]

30. Salas, R.; Perez-Villadoniga, M.J.; Prieto-Rodriguez, J.; Russo, A. Were traffic restrictions in Madrid effective at reducing NO_2 levels? *Transp. Res. Part D Transp. Environ.* **2021**, *91*, 102689. [CrossRef]

31. Lütkemeyer, E.; Hantsche, L.; Zschiesche, M. Der Ausbau der Windenergie unter den Bedingungen zunehmender gerichtlicher Auseinandersetzungen. In *Ökologische Debatte UfU Jahrbuch*, 1st ed.; Unabhängiges Institut für Umweltfragen e.V., Ed.; Unabhängiges Institut für Umweltfragen e.V.: Berlin, Germany, 2020; pp. 1–8.

32. Van den Broek, B.; Enneking, L. Public Interest Litigation in the Netherlands. A Multidimensional Take on the Promotion of Environmental Interests by Private Parties through the Courts. *ULR* **2014**, *10*, 77. [CrossRef]

33. Vanhala, L. Is Legal Mobilization for the Birds? Legal Opportunity Structures and Environmental Nongovernmental Organizations in the United Kingdom, France, Finland, and Italy. *Comp. Political Stud.* **2018**, *51*, 380–412. [CrossRef]

34. Umweltbundesamt. Sieben Fragen und Antworten zum Diesel. Available online: https://www.umweltbundesamt.de/themen/sieben-fragen-antworten-diesel (accessed on 18 May 2022).

35. Fayad, M.A.; AL-Ogaidi, B.R.; Abood, M.K.; AL-Salihi, H.A. Influence of post-injection strategies and CeO_2 nanoparticles additives in the C30D blends and diesel on engine performance, NO_X emissions, and PM characteristics in diesel engine. *Part. Sci. Technol.* **2021**, *39*. [CrossRef]

36. World Health Organisation. *Review of Evidence on Health Aspects of Air Pollution–Revihaap Project: Final Technical Report*; World Health Organisation: Genève, Switzerland, 2013.

37. European Parliament. European Council Directive 2008/50/EC of the European Parliament and the Council of 21 May 2008 on ambient air quality and cleaner air for Europe. *Off. J. Eur. Union* **2008**, *29*, 169–212.

38. European Commission. Aufforderungsschreiben—Vertragsverletzung Nr. 2015/2073. C (2015) 4009 Final. Available online: https://www.greenpeace.de/sites/default/files/media_type_pdf/2015_06_18_mahnschreiben_eu_kommission.pdf (accessed on 24 May 2022).

39. European Commission. EU-Kommission Verklagt Deutschland und Fünf Weitere Mitgliedsstaaten Wegen Luftverschmutzung. Available online: https://ec.europa.eu/germany/news/20180517-luftverschmutzung-klage_de (accessed on 6 June 2019).

40. Habigt, L.; Hamacher, L.; Tryjanowski, A.; Zschiesche, A.; Schmidt, A.; Heß, F.; Teßmer, D.; Franke, J.; Wozny, N.; Mareen, E.; et al. *Wissenschaftliche Unterstützung des Rechtsschutzes in Umweltangelegenheiten in der 19 in der Legislaturperiode*; Umweltbundesamt: Dessau-Roßlau, Germany, 2021.

41. Deutsche Umwelthilfe. Klagen für Saubere Luft–Hintergrundpapier. Deutsche Umwelthilfe e.V.: Radolfzell, Germany. Available online: https://www.duh.de/fileadmin/user_upload/download/Projektinformation/Ver-kehr/Feinstaub/Right-to-Clean-Air_Hintergrundpapier_D_Juli_2019.pdf (accessed on 11 October 2019).

42. Peters, B.G. *Institutional Theory in Political Science*, 4th ed.; Edward Elgar Publishing: Cheltenham, UK, 2019.

43. Umweltbundesamt. Jährliche Auswertung der Stickoxidwerte. Available online: https://www.umweltbundesamt.de/themen/luft/luftschadstoffe/stickstoffoxide (accessed on 11 June 2020).

44. Umweltbundesamt. Luftqualität 2019: Vorläufige Auswertung. Available online: https://www.umweltbundesamt.de/sites/default/files/medien/1410/publikationen/hgp_luftqualitaet2019_bf.pdf (accessed on 11 June 2020).

45. Wing, C.; Simon, K.; Bello-Gomez, R.A. Designing Difference in Difference Studies: Best Practices for Public Health Policy Research. *Annu. Rev. Public Health* **2018**, *39*, 453–469. [CrossRef]

46. Coglianese, C. *Measuring Regulatory Performance*; OECD: Paris, France, 2012.

47. Ahrens, L.; Bothner, F. The Big Bang: Tax Evasion After Automatic Exchange of Information Under FATCA and CRS. *New Political Econ.* **2020**, *25*, 849–864. [CrossRef]

48. Ahrens, L.; Hakelberg, L.; Rixen, T. A victim of regulatory arbitrage? Automatic exchange of information and the use of golden visas and corporate shells. *Regul. Gov.* **2020**. [CrossRef]

49. Callaway, B.; Sant'Anna, P.H. Difference-in-Differences with multiple time periods. *J. Econom.* **2021**, *225*, 200–230. [CrossRef]

50. Sun, L.; Abraham, S. Estimating dynamic treatment effects in event studies with heterogeneous treatment effects. *J. Econom.* **2021**, *225*, 175–199. [CrossRef]
51. Goodman-Bacon, A. Difference-in-differences with variation in treatment timing. *J. Econom.* **2021**, *225*, 254–277. [CrossRef]
52. Fredriksson, A.; de Oliveira, G.M. Impact evaluation using Difference-in-Differences. *RAUSP* **2019**, *54*, 519–532. [CrossRef]
53. Marcus, M.; Sant'Anna, P.H.C. The Role of Parallel Trends in Event Study Settings: An Application to Environmental Economics. *J. Assoc. Environ. Resour. Econ.* **2021**, *8*, 235–275. [CrossRef]
54. Töller, A. *Do ENGOs' Lawsuits against Wind Energy Plants Jeopardise the German "Energiewende"?* Working Paper; Fern Universität: Hagen, Germany, 2022.
55. Naturschutzbund Brandenburg. Verbände Gegen Erhöhte Wasserentnahme. Available online: https://brandenburg.nabu.de/umwelt-und-ressourcen/30964.html (accessed on 18 May 2022).

 *sustainability*

Article

Assessing the Impact of Local Policies on PM2.5 Concentration Levels: Application to 10 European Cities

Enrico Pisoni [1,*], Philippe Thunis [1], Alexander De Meij [2] and Bertrand Bessagnet [1]

[1] Joint Research Centre (JRC), European Commission, 21027 Ispra, Italy; philippe.thunis@ec.europa.eu (P.T.); bertrand.bessagnet@ec.europa.eu (B.B.)
[2] MetClim, 21025 Varese, Italy; alexander.de-meij@ext.ec.europa.eu
* Correspondence: enrico.pisoni@ec.europa.eu

Abstract: In this paper, we propose a methodology to evaluate the effectiveness of local emission reduction policies on PM2.5 concentration levels. In particular, we look at the impact of emission reduction policies at different scales (from urban to EU scale) on different PM2.5 baseline concentration levels. The methodology, based on a post-processing of air quality model simulations, is applied to 10 cities in Europe to understand on which sources local actions are effective to improve air quality, and over which concentration ranges. The results show that local actions are effective on low-level concentrations in some cities (e.g., Rome), whereas in other cases, policies are more effective on high-level concentrations (e.g., Krakow). This means that, in specific geographical areas, a coordinated approach (among cities or even at different administration levels) would be needed to significantly improve air quality. At last, we show that the effectiveness of local actions on urban air pollution is highly city-dependent.

Keywords: air quality; air quality plans; PM2.5 episodes

Citation: Pisoni, E.; Thunis, P.; De Meij, A.; Bessagnet, B. Assessing the Impact of Local Policies on PM2.5 Concentration Levels: Application to 10 European Cities. *Sustainability* **2022**, *14*, 6384. https://doi.org/10.3390/su14116384

Academic Editors: José Carlos Magalhães Pires and Álvaro Gómez-Losada

Received: 22 April 2022
Accepted: 21 May 2022
Published: 24 May 2022

Publisher's Note: MDPI stays neutral with regard to jurisdictional claims in published maps and institutional affiliations.

1. Introduction

Air quality in cities, in particular in relation to PM2.5, is an important topic indeed. It is well known that poor air quality is associated with impacts on human health, both in the short- and long-term [1]. Recent epidemiological evidence [2] motivates the need for accounting for fine particles in the scope of air quality regulations. While in the literature the topic of long-term exposure to air quality is well discussed (focusing on action plans to improve air quality in a time-frame of a few years), the issue of short-term exposure is more uncertain [3]), e.g., it is known that short-term action plans, triggered in cases of forecasted high pollution episodes, are not always effective [4]. In some cases (e.g., when limiting road traffic, or closing for few days an area to vehicles), only a limited impact on air quality concentrations is observed (in particular in terms of PM2.5 concentrations), and more research in this field is needed [5]. The difficulty arises from the fact that PM2.5 is produced through complex chemical reactions and dispersion/transport mechanisms, and from various sources, making it challenging to identify how to act to reduce concentration levels [6,7]. Modelling methodologies are often used to study the potential of local actions to reduce high pollution episodes in cities.

Usually, short-term action plans and the analysis of their effectiveness are based on model forecasts [8–12], differencing model simulations performed with and without a specific set of emission reduction measures. Even if is useful to predict the impact of a plan for a given episode, this information remains specific and does not provide insight on the impact of local strategies on air quality in general [13,14].

In this paper, we address the issue from another perspective. We use yearly modelling simulations, in which different sources are progressively switched off, to understand their cumulative impact on daily pollution concentrations. We then rank these impacts in terms of low and high concentration days, to see how policies would be effective in tackling different pollution ranges.

The sources considered for this analysis are the cities themselves; the urban areas in the neighborhood (to explore the benefit of coordinated urban actions); agriculture and the remaining anthropogenic and biogenic emissions, to quantify the added value of EU-wide policies.

In comparison to previous papers on this subject [15], the main difference lies in the use of daily PM2.5 values (not yearly averaged). In addition, in comparison to other similar analyses and visualization approaches (i.e., in CAMS annual source-receptor information, at policy.atmopshere.copernicus.eu), we provide here a combined view on sectors and geographical areas, with a method that is flexible to accommodate any possible geographical or sectoral category.

The model simulations are based on a full 100% switch-off of the sources of interest (city, agriculture . . .). While this choice allows assessing without any assumption of the full responsibility of the sources, it is not connected to real-life emission reductions that never reach this level. An important question is therefore to address whether we can interpolate these extreme (100% reduction) simulations to lower emission reduction intensities that are representative of local actions. We check this point by assessing the degree of nonlinearities of the model responses in the final part of this work. We also briefly assess the efficiency of local and large-scale actions with respect to other pollutants, such as NO_2 and O_3.

The proposed methodology is applied to 10 large European cities, to address the following questions:

- How much do the city, the surrounding urban areas, agriculture and the remaining EU emissions contribute to urban pollution?
- How do these contributions change across city?
- How do these contributions depend on specific days (characterised by low or high concentration values)?

The paper is structured as follows: in Section 2 we detail the methodology; in Section 3, the results (and the evaluation of the nonlinearities); and in Section 4 there is a discussion and conclusions.

2. Methodology

2.1. Modelling Set-Up

This work is based on a set of model simulations performed with the EMEP air quality model version 4.34 [16] for the entire year 2015, fed by the CAMS v2.2.1 emission inventory [17]. The emission inventory covers all PM2.5 precursor emissions (NOx, VOC, NH3, PPM, SO2) classifying emissions according to the GNFR (Gridded Nomenclature for Reporting) categories. More details on the emission inventories and sectoral split are provided in [18]. The modelling domain covers the whole Europe, with a spatial resolution of 0.1×0.1 degrees. The domain stretches from $-15.05°$ W to $36.95°$ E longitude and $30.05°$ N to $71.45°$ N latitude with a horizontal resolution of $0.1° \times 0.1°$ and 20 vertical levels, with the first level at about 45 m. The EMEP model uses meteorological data from the European Centre for Medium Range Weather Forecasting (ECMWF-IFS) for the meteorological year 2015. The temporal resolution of the meteorological input data is daily, with a 3 h timestep. The meteorological fields for EMEP are retrieved on a $0.1° \times 0.1°$ longitude latitude coordinate projection. A validation of the modelling application (checked against observations) is provided in [19].

2.2. Selection of Cities and Simulations

Figure 1 shows the 10 cities considered in this paper: Madrid, Paris, Lisbon, Milan, Rome, Berlin, Athens, Frankfurt, Krakow and Stockholm (for more information on the selected cities feature, please refer to the PM2.5 Urban Atlas, at https://publications.jrc.ec.europa.eu/repository/handle/JRC126221, accessed on 20 April 2022). These have been selected among large EU cities, to provide a balanced geographical coverage on the

EU territory. These cities are defined as a 'Functional Urban Area', i.e., composed of a city core and a commuting zone. The city core is the local administrative units, with a population density above $1500/km^2$ and a population above 50,000, where the majority of the population lives in an urban centre and the wider commuting zone consists of the surrounding travel-to-work areas where at least 15% of the employed residents work in the city.

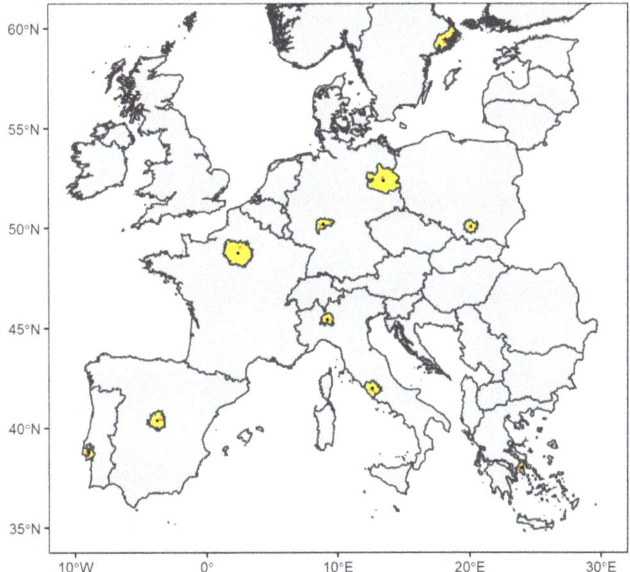

Figure 1. Map of Europe showing the 10 cities considered in this paper, focusing on the city areas (polygon, in yellow), the centroid of the polygon (black dots) and the highest concentration point (red dots), for each city.

The simulations performed on these 10 cities are designed with incremental emission reductions:
- The 'baseline' simulation (reference), considering the 2015 as reference year.
- The 'city scenario': in which all the emissions from the 10 cities are switched off. As cities are far away from each other, we assume that impacts from other cities on the city of interest are negligible.
- The 'urban scenario': in which emissions from all urban areas with a population $> 300/km^2$ are switched off, in addition to the 10 cities themselves. This scenario allows for estimating the additional benefit of reducing urban emissions around each city.
- The 'agriculture scenario': in which agricultural emissions are switched off, on top of the 'city' and 'urban' reductions. This is useful to evaluate the additional benefit on urban air quality of reducing one of the main emission sources in rural areas.
- The 'EU wide scenario': in which all-anthropogenic emissions remaining in Europe (here intended as the modelling domain) are switched off. This simulation is intended to assess the additional benefit from EU wide actions (the background contribution is derived by the baseline concentration simulation, summing up the components related to sea salt and dust).

Starting from these runs, the analysis in the following sections is then based on the 'relative potential' indicator computed with PM2.5 daily values, as:

$$\frac{\text{PM2.5}_{\text{BC}} - \text{PM2.5}_{\text{scenario}}}{\alpha \text{PM2.5}_{\text{BC}}} \times 100 \tag{1}$$

where $PM2.5_{BC}$ and $PM2.5_{scenario}$ represent the baseline and scenario PM2.5 daily concentrations; and α the emission reduction strength.

For this analysis, all aforementioned scenarios are based on a 100% emission reduction, i.e., a value of α equal to 1. To test the validity of interpolating these extreme emission reduction scenarios to moderate reduction values more representative of local action plans, we performed simulations (on a reduced set of cities) with emissions reduced by 10%, 20% and 50%. It means that, in those cases, the 'relative potential' indicator is then used with α values of 0.1, 0.2 and 0.5, respectively.

As mentioned above, we focus our analysis on daily averaged PM2.5 concentrations in 10 cities (see Figure 1). While the sources have been already defined (city, urban areas, agriculture and EU), we also need to define the receptor, i.e., the location where we quantify the different contributions. Two receptors are selected:

- The centroid of the city 'functional urban area';
- The location of the highest modelled concentration within the FUA (Functional Urban Areas [20]).

These 2 locations are selected to assess the spatial sensitivity of the contributions. In the final part of the paper, an additional point (the one with the highest value for 'highestConcentration $\times$ population') is also considered.

3. Results

We structure this section into three parts. In Section 3.1, we motivate the choice of PM2.5 as the most challenging pollutant in terms of selecting the appropriate time and scale for actions. In Section 3.2, we analyse the source contributions to PM2.5 daily values and assess how they vary in terms of city and type of episodes. Finally, in Section 3.3, we assess the robustness of the results, by quantifying the importance of the nonlinear effects. This step serves to estimate the possibility of interpolating our full emission reductions to lower reduction strengths more representative of practical policies. We also discuss the dependency of our results on the temporal dimension.

3.1. Why a Focus on PM2.5

It is important to stress that responses to local and wider-scale emission changes largely differ from one pollutant to the other. Although we focus on PM2.5 in this work, we also briefly address other pollutants, in particular O_3 and NO_2. Figure 2 compares the local (red) and EU wide responses (blue) obtained from EMEP simulations, for NO_2, PM2.5 and O_3. To obtain the local responses, emissions have been totally switched off over the functional urban area (FUA) of each of the ten cities. Responses are then expressed as relative fractions of the base case concentrations at the city centre location (see Section 2). EU responses represent the fraction of the base case concentration reduced when all emissions in Europe are switched off. For NO_2, almost 100% of the concentration can be explained by local emissions. This proportion lowers to 50% or less for PM2.5, and to 25% or less for O_3. While the remaining fraction of PM2.5 can be related to EU wide emissions, this is not the case for O_3, for which about half of the base case concentration is related to background values that do not depend on EU emissions (over the time scale considered, here 1 year). The remaining O_3 levels [21] depend on other factors such as contributions of methane [22], hemispheric transport [23], stratospheric intrusion [24], etc.

Figure 2 (showing, for different cities and scenarios, the 'relative potential' as previously introduced in Section 2) clearly highlights the need for a pollutant specific strategy, itself varying across cities. More specifically, Figure 2 shows how policies (at different levels) can improve air quality differently, depending on the pollutant and city considered.

Indeed, while O_3 is mostly driven by large-scale processes and NO_2 by local processes, PM2.5 shows a mixed behaviour, therefore more challenging to translate in terms of policy. This complex behaviour calls for a focused analysis for this pollutant.

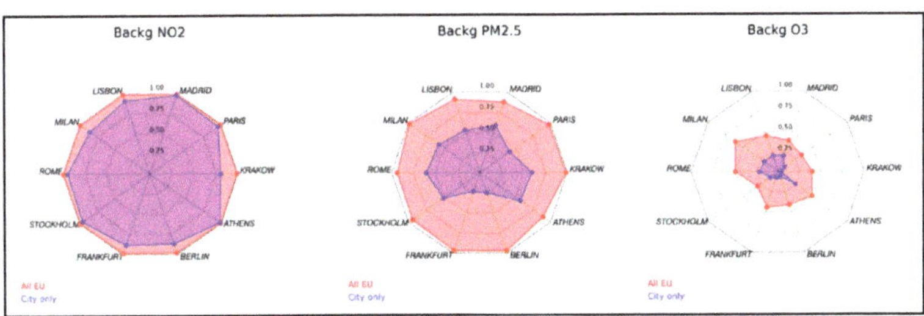

Figure 2. Comparison of EMEP responses (relative potential) to local (blue) and EU reductions (red) for yearly averaged PM2.5, yearly averaged NO_2 and daily 8 h max (summertime O_3). Emission reduction impacts are assessed at the city location where the maximum base case concentration is modelled.

3.2. Contribution to PM2.5

3.2.1. Yearly Average Results

Now focusing on PM2.5, Figure 3 shows the impacts of different sources (city, urban, agriculture remaining EU and natural) on the daily PM2.5 value for the 10 considered cities, ranked in terms of concentration bins. The receptor location for this analysis is the FUA 'centroid'.

In more detail, in Figure 3 (and in the following figures), the 'labels/colours' represent:

- City: represents the case reducing emissions in the 10 cities;
- AllCities: represents the case reducing all urban areas emissions;
- AllCities_Agri: represents the case reducing all urban areas emissions and agricultural emissions;
- AllEu: represents the case reducing all anthropogenic emissions;
- Natural: represent the remaining concentrations.

To better explain these results, let us take, e.g., the case of Milan, in Figure 3. The x-axis shows the baseline concentrations bins for a given city, with a first bin from 2.96 to 12.2 $\mu g/m^3$ and a final bin from 45.9 to 152 $\mu g/m^3$. Bins are created by splitting daily PM2.5 values in five equal quantiles for each city. In the case of Milan, 20% of the days have concentrations falling between 2.96 and 12.2 $\mu g/m^3$, 20% of the days between 45.9 and 152 $\mu g/m^3$, and so on. The y-axis identifies the relative contributions to the concentration. For low concentration values (lower bin), up to 50% of the pollution might be reduced by acting on local emissions (Milan city, violet), and an additional 25% from other urban emissions (in blue), while the remaining contributions depend on additional reductions achieved on agriculture and at the EU level. Finally, a small contribution is 'natural'. The same analysis holds for each bin and for each city. For the 'highest concentration bin' (concentrations between 45.9 and 152 $\mu g/m^3$), the picture is a bit different with a lower city contribution (around 40%), while the 'urban' and 'agricultural' emissions have more importance. For the city of Milan, these findings lead to an important message: for high-level concentrations, it is not sufficient to act at the city level (Milan), and cooperation with other cities is advisable. As cities usually implement short-term action plans in isolation when a pollution episode occurs, this brings attention to the potential added value for integrated short-term action plans, at least for some cities.

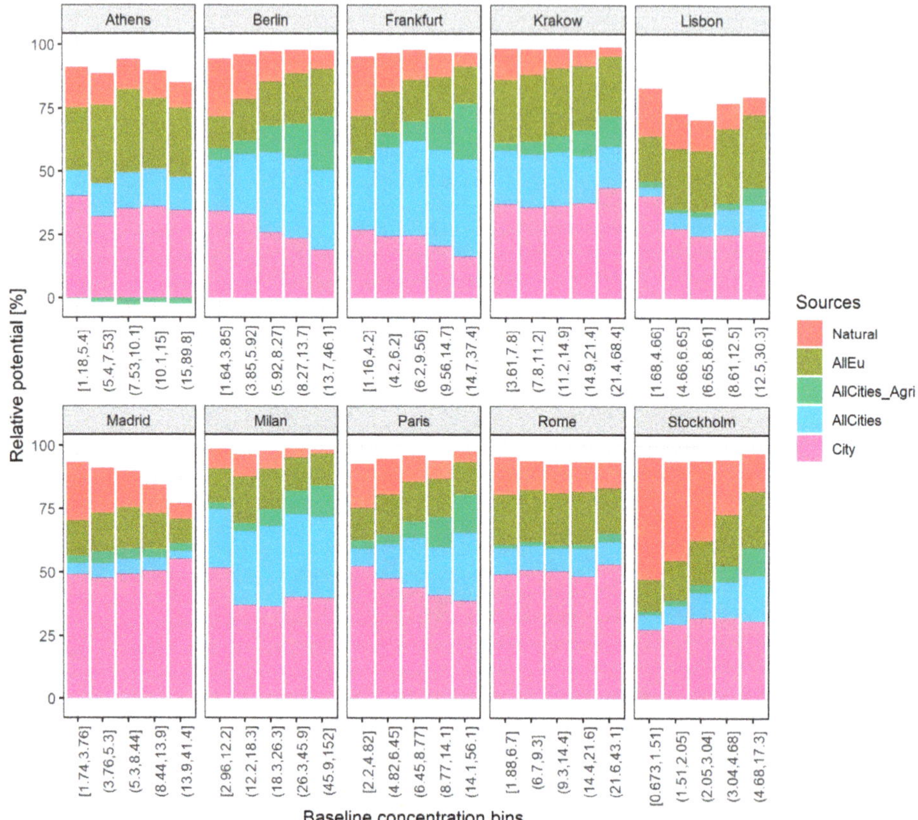

Figure 3. Impact of different sources (city, all cities, all cities and agricultural emissions, all domain, natural) on daily PM2.5 value for the 10 considered cities, classified in terms of baseline concentration bins. The results refer to the 'centroid' point, as defined in Figure 1.

Among the 10 cities, we identify three groups in terms of behaviour:

- Cities in which the 'city' impact gets less important when concentrations increase. Berlin, Frankfurt, Milan, Paris, Lisbon. For these cities, local plans are not sufficient to abate high pollution episodes;
- Cities in which the 'city' impact is independent of the concentration level: Athens, Rome and Stockholm. Local plans will have the same efficiency regardless of the concentration levels;
- Cities in which the 'city' impact becomes more important when concentration levels increase: Krakow and Madrid. Local plans are then more effective during high concentration episodes.

This analysis clearly shows that the effectiveness of local actions on pollution episodes is city-dependent.

If we focus on the highest concentration bin, only Madrid and Rome can half their concentrations through local actions. For other cities, actions need to be coordinated with other cities or combined with actions at a larger scale (EU) involving other sectors (e.g., agriculture) to be effective.

In the case of Athens, we also note a small increase (negative values) in PM2.5 concentrations when agricultural emissions are switched off on top of the city's reductions.

This signal is very small, and can be explained by some nonlinearities of the system, that become visible when 'piling up the results' of the different performed incremental runs.

For some cities (Lisbon, Madrid, Athens), part of the pollution remains uncontrollable despite all considered actions. For these cities, initial and boundary conditions (dust, but also sea salt for Lisbon and Athens) play an important role, mainly when pollution levels are high; this is also possibly due to the fact that these cities are closer to the boundaries of the modelling domain.

3.2.2. Yearly vs. Seasonal Results

Figure 4 proposes a similar analysis to Figure 3, but with the results aggregated seasonally. As expected, the highest concentration bins mainly occur during winter in most cities (e.g., Berlin, Frankfurt, Krakow, Madrid, Milan, Rome). However, in other cities, this is less contrasted (Athens, Lisbon and Stockholm) and autumn is also a relevant season for high pollution episodes. For Paris, spring is very important in terms of high pollution episodes due to ammonium nitrate formation enhanced by fertilizer spreading in neighbouring regions and countries [25].

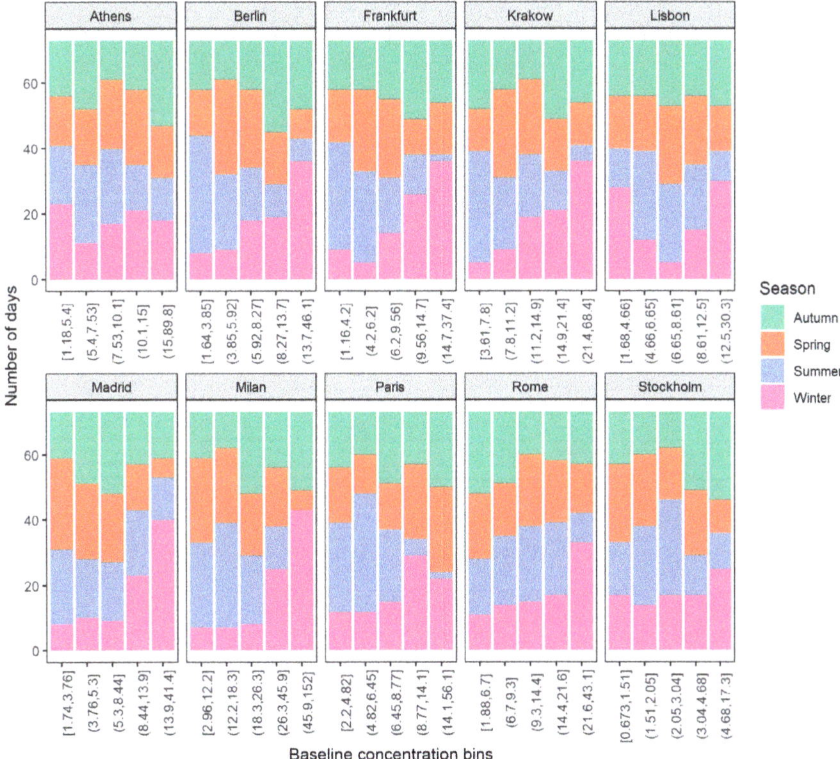

Figure 4. Classification of the PM2.5 daily values, per city, baseline concentration bin and season. Results refer to the 'centroid' point as defined in Figure 1.

3.2.3. Centroid vs. Hot-Spot Receptor

Figures 5 and 6 show similar results to Figures 3 and 4, but for the second receptor choice corresponding to the 'highest concentration point' within the FUA.

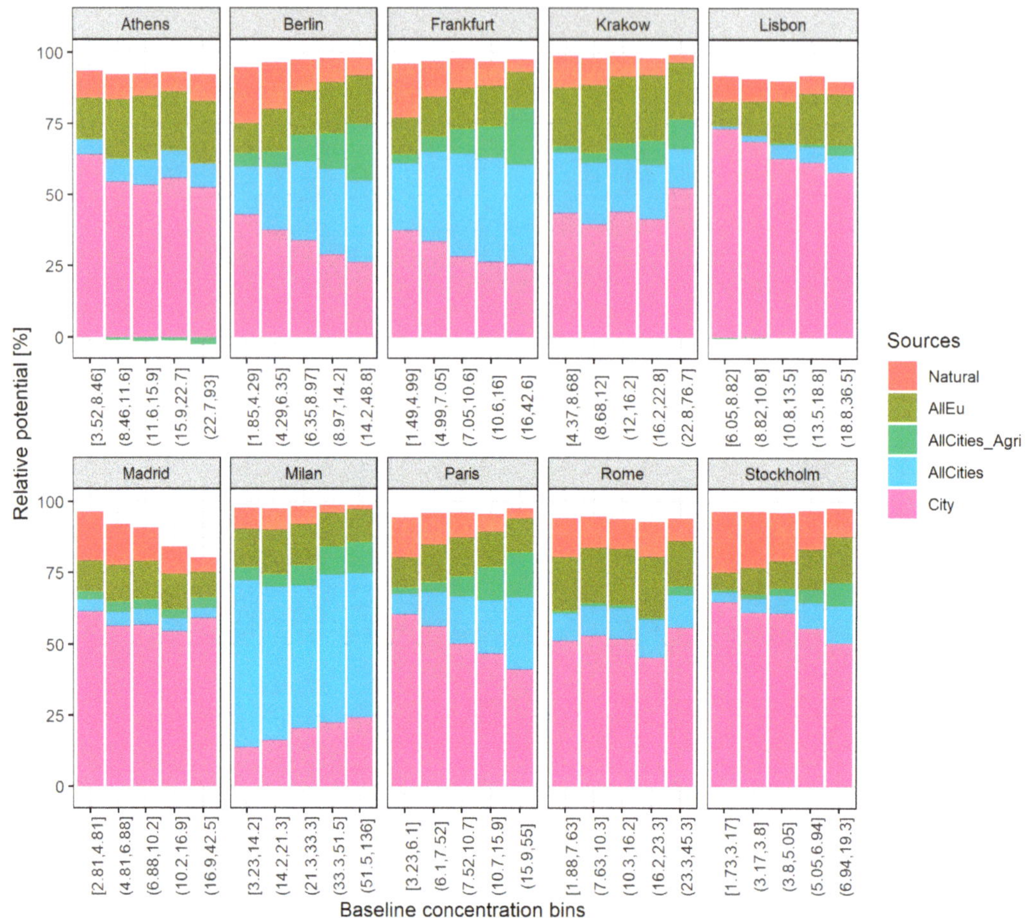

Figure 5. Impact of different sources (city, all cities, all cities and agricultural emissions, all domain, natural) on daily PM2.5 value for the 10 considered cities, classified in terms of baseline concentration bins. The results refer to the 'highest concentration' point, as defined in Figure 1.

With the focus on the 'highest concentration point', Lisbon and Stockholm can do much more with local actions, reducing up to 50% of their concentrations (while for the 'centroid' receptor, contributions were limited to roughly 25%). For both cities, the impact slightly decreases when considering higher concentration bins, but still remain high. Milan, on the contrary, shows an opposite behaviour with a city contribution that is reduced significantly to about 25% for the highest concentration bin, in contrast to 40–50% when considering the centroid. This is explained by the PPM emissions around the two points, that are 2.6 higher when focusing on the centroid, in contrast to the emissions close to the highest concentrations point; this emission issue is then causing different behaviours in the two cases, with higher local contributions when considering the centroid of the city (see Supplementary Materials for this analysis). Southern cities like Lisbon, Madrid and Athens display a larger 'remaining' contribution (i.e., from the boundary conditions mainly driven for particulate matter by Saharan desert dust). This is particularly noteworthy for Madrid during the high pollution episodes.

These results show the importance of the receptor point for the analysis of the policy impact as stressed in [26].

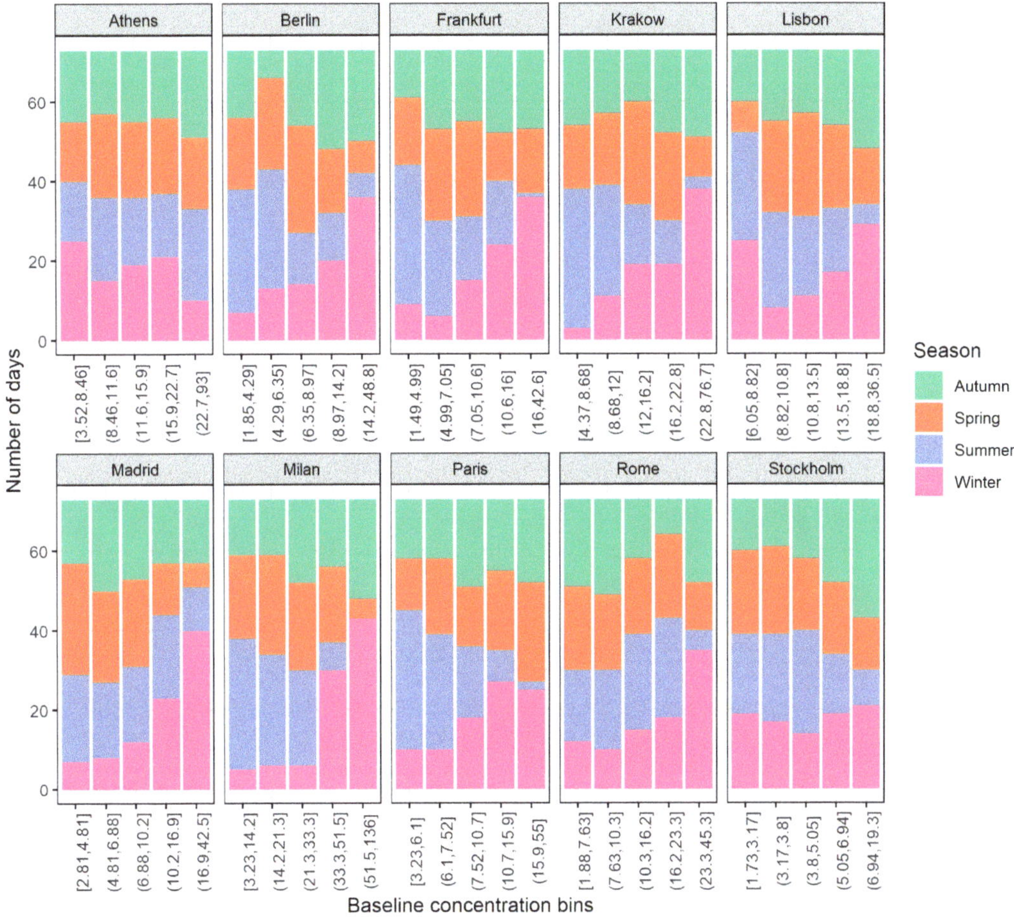

Figure 6. Classification of the PM2.5 daily values, per city, baseline concentration bin and season. Results refer to the 'highest concentration' point as defined in Figure 1.

3.3. Nonlinearities and Temporal Trends Assessment

Our methodology to evaluate the contribution of different sources on concentrations is based on full emission reductions (i.e., 100%) for all source areas (i.e., city, urban, agriculture …).

To evaluate the applicability of our approach to low and moderate (more realistic) emission reductions, we check here the importance of nonlinearities. We perform simulations with intermediate intensity emission reductions.

We simulated scenarios in which city emissions are reduced by 10%, 20% and 50% in three cities), and computed the associated relative potentials for the yearly, monthly and daily values. In the case of linear behaviour, relative potentials remain constant with the emission reduction intensity.

For the three considered cities (London, Madrid, Paris), the yearly average relative potential is shown in Figure 7, for emission reductions of 10% (RP10), 20% (RP20), 50% (RP50) and 100%. There are not big differences for yearly averages, meaning that concentration changes due to 100% emission reductions (potential 100) are similar to twice the concentration changes due to 50% emission reductions (all 'dots' overlap), regardless of the receptor choice (the same stands when rescaling the cases at 20% or 10% reductions, that is to say rescaling RP20 and RP10).

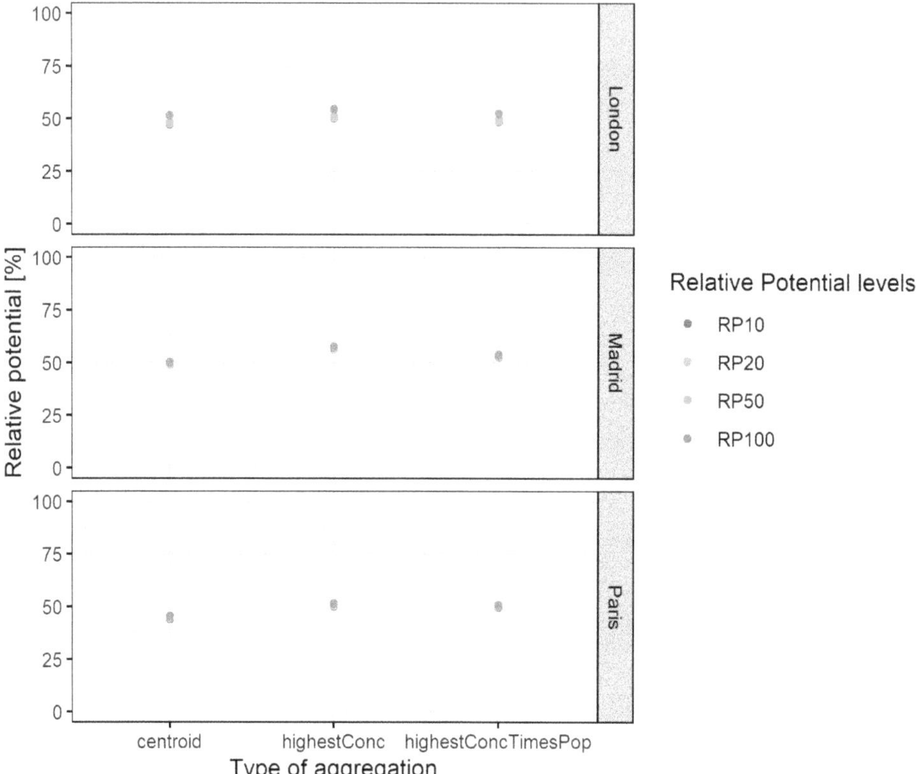

Figure 7. Nonlinearities for yearly values, considering different emission reduction scenarios at 10% (RP10), 20% (RP20), 50% (RP50) and 100% (RP100). RP stands for relative potentials. The terms 'centroid', 'highestConc' and 'highestConcTimesPop' refer to different points considered for the analysis, as the polygon centroid, the point with higher concentrations and the point with a higher value of concentration times population.

On top of this, we also checked temporal trends, and possible related nonlinear features. For monthly values (Figure 8), relative potentials obtained for reductions ranging from 10% to 100% remain very similar, emphasizing the robustness of the approach. It is interesting to note the strong variability of the monthly trends in Madrid, moving from 75% (in winter) to 25% (in summer), while for London and Paris, the patterns are quite flat. The monthly trends in London and Paris are similar, with a limited variability; on the contrary, Madrid shows a more pronounced variability. This means that, focusing on plans that could be seasonally adapted would be of interest in Madrid, but not necessarily in the other two cities.

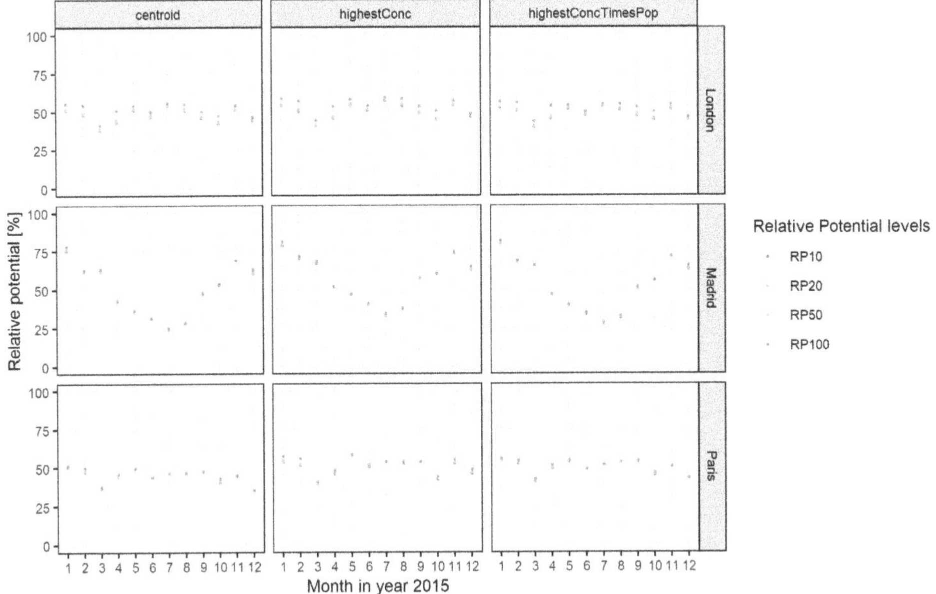

Figure 8. Nonlinearities and related pattern for monthly values.

Even for daily values (Figure 9), the 100% relative potential (y-axis) remains comparable to the potentials obtained for the 10%, 20% and 50% emission reductions (x-axis), at
least for these three cities. In a previous work [27], the nonlinear signal was more visible
than here. In that work [27], reductions were completed one pollutant at a time and on
larger regional domains; in that case, the nonlinear trends were appearing at monthly, and
increasing at daily, frequencies. In this study, the low nonlinear signal is probably lower
due to the fact that reductions are performed on a smaller domain (city scale), reducing all
pollutants at the same time.

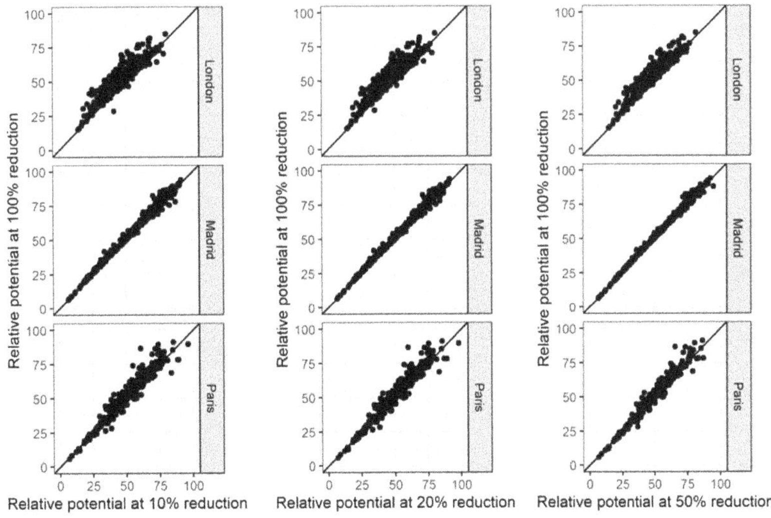

Figure 9. Daily values patterns.

4. Conclusions

In this work, we proposed a methodology to evaluate, for different cities, the effectiveness of local actions to reduce PM2.5 concentrations. The focus is on PM2.5, as it is the most complex pollutant to be managed at local scale, but also the one leading to the largest burden on human health [15].

The traditional approach to evaluate the effectiveness of local actions is to use models in the scenario mode and assess what would happen with the implementation of local action plans, during specific short periods. In this work we use a different approach, consisting of running a set of simulations, on top of the baseline, and switching off different sources of emissions to understand how different cities react to different 'sources' of emission reductions, depending on the pollution level. So, the focus of this study is not on a specific episode, but more in general on the effectiveness of measures at different concentration levels.

The results (with an analysis of 10 cities) show that the concentration changes (linked to the considered scenarios) differ depending on the city; in particular, we can have cities where local actions are very effective for a high level of concentrations, and other cities that show an opposite behaviour. The analysis performed for two different receptor points within the city domain ('centroid' vs. 'highest concentration' point) showed that policy interventions can also vary significantly, depending on where the analysis is made (i.e., the selected receptor).

The generalisation of the findings to more realistic emission reductions has been tested by looking at how results change when modifying the level of emission reductions. The conclusion is that the methodology is robust, i.e., that similar results would be obtained for lower levels of emission reductions (e.g., 10%, 20% or 50%).

The overall conclusion of this work is that cities can generally improve, significantly, the pollution level through local actions during high-level concentration episodes, for PM2.5 (and also NO_2). This is not the case for O_3, which exhibits a completely different pattern. However, for PM2.5, in many cases an approach that only looks at local policies is not sufficient, and an integrated approach (through coordinated actions with other urban areas and/or acting on rural area emissions, etc.) would be needed. Finally, we suggest that, prior to the selection of a final action plan, a sensitivity is performed for different receptor points within the domain of interest, to ensure the most efficient policy. Indeed, given the large observed differences in terms of the receptor, averaging the results in terms of the receptor would lead to masking these geographical specificities and to lowering the impact of local strategies. In terms of limitation, the main uncertainty of this study lies in the EU wide emissions that could differ from locally reported ones. This aspect can affect the quality of the results, even if the general methodology would remain valid.

Supplementary Materials: The following supporting information can be downloaded at: https://www.mdpi.com/article/10.3390/su14116384/s1, Figure S1. showing the primary PM yearly emissions for the 10 considered cities.

Author Contributions: E.P. coordinated the work and prepared the final version of the manuscript. P.T. conceived the initial idea for the paper, and contributed to the paper preparation. A.D.M. was in charge of the modelling needed for this work, and contributed to the general writing of the paper. B.B. contributed to the paper design and commented on the final version of the work. All authors have read and agreed to the published version of the manuscript.

Funding: This research received no external funding.

Data Availability Statement: Data for this publication are available at: https://doi.org/10.5281/zenodo.5513848.

Conflicts of Interest: The authors declare no conflict of interest.

References

1. Kim, K.-H.; Kabir, E.; Kabir, S. A review on the human health impact of airborne particulate matter. *Environ. Int.* **2015**, *74*, 136–143. [CrossRef] [PubMed]
2. Ohlwein, S.; Kappeler, R.; Kutlar Joss, M.; Künzli, N.; Hoffmann, B. Health effects of ultrafine particles: A systematic literature review update of epidemiological evidence. *Int. J. Public Health* **2019**, *64*, 547–559. [CrossRef] [PubMed]
3. Shah, A.S.V.; Lee, K.K.; McAllister, D.A.; Hunter, A.; Nair, H.; Whiteley, W.; Langrish, J.P.; Newby, D.E.; Mills, N.L. Short term exposure to air pollution and stroke: Systematic review and meta-analysis. *BMJ* **2015**, *350*, h1295. [CrossRef] [PubMed]
4. Lasry, F.; Coll, I.; Fayet, S.; Havre, M.; Vautard, R. Short-term measures for the control of ozone peaks: Expertise from CTM simulations. *J. Atmos. Chem.* **2007**, *57*, 107–134. [CrossRef]
5. Pisoni, E.; Christidis, P.; Thunis, P.; Trombetti, M. Evaluating the impact of "Sustainable Urban Mobility Plans" on urban background air quality. *J. Environ. Manag.* **2019**, *231*, 249–255. [CrossRef]
6. Pisoni, E.; Guerreiro, C.; Lopez-Aparicio, S.; Guevara, M.; Tarrason, L.; Janssen, S.; Thunis, P.; Pfäfflin, F.; Piersanti, A.; Briganti, G.; et al. Supporting the improvement of air quality management practices: The "FAIRMODE pilot" activity. *J. Environ. Manag.* **2019**, *245*, 122–130. [CrossRef]
7. Thunis, P.; Clappier, A.; Tarrason, L.; Cuvelier, C.; Monteiro, A.; Pisoni, E.; Wesseling, J.; Belis, C.; Pirovano, G.; Janssen, S.; et al. Source apportionment to support air quality planning: Strengths and weaknesses of existing approaches. *Environ. Int.* **2019**, *130*, 104825. [CrossRef]
8. Thunis, P.; Degraeuwe, B.; Pisoni, E.; Meleux, F.; Clappier, A. Analyzing the efficiency of short-term air quality plans in European cities, using the CHIMERE air quality model. *Air Qual. Atmos. Health* **2017**, *10*, 235–248. [CrossRef]
9. Borge, R.; Santiago, J.L.; de la Paz, D.; Martín, F.; Domingo, J.; Valdés, C.; Sánchez, B.; Rivas, E.; Rozas, M.T.; Lázaro, S.; et al. Application of a short term air quality action plan in Madrid (Spain) under a high-pollution episode—Part II: Assessment from multi-scale modelling. *Sci. Total Environ.* **2018**, *635*, 1574–1584. [CrossRef]
10. Akritidis, D.; Katragkou, E.; Zanis, P.; Pytharoulis, I.; Melas, D.; Flemming, J.; Inness, A.; Clark, H.; Plu, M.; Eskes, H. A deep stratosphere-to-troposphere ozone transport event over Europe simulated in CAMS global and regional forecast systems: Analysis and evaluation. *Atmospheric Chem. Phys.* **2018**, *18*, 15515–15534. [CrossRef]
11. Marecal, V.; Peuch, V.-H.; Andersson, C.; Andersson, S.; Arteta, J.; Beekmann, M.; Benedictow, A.; Bergstrom, R.W.; Bessagnet, B.; Cansado, A.; et al. A regional air quality forecasting system over Europe: The MACC-II daily ensemble production. *Geosci. Model Dev.* **2015**, *8*, 2777–2813. [CrossRef]
12. Adani, M.; Piersanti, A.; Ciancarella, L.; D'Isidoro, M.; Villani, M.G.; Vitali, L. Preliminary Tests on the Sensitivity of the FORAIR_IT Air Quality Forecasting System to Different Meteorological Drivers. *Atmosphere* **2020**, *11*, 574. [CrossRef]
13. Singh, V.; Carnevale, C.; Finzi, G.; Pisoni, E.; Volta, M. A cokriging based approach to reconstruct air pollution maps, processing measurement station concentrations and deterministic model simulations. *Environ. Model. Softw.* **2011**, *26*, 778–786. [CrossRef]
14. Agarwal, S.; Sharma, S.; Suresh, R.; Rahman, M.H.; Vranckx, S.; Maiheu, B.; Blyth, L.; Janssen, S.; Gargava, P.; Shukla, V.K.; et al. Air quality forecasting using artificial neural networks with real time dynamic error correction in highly polluted regions. *Sci. Total Environ.* **2020**, *735*, 139454. [CrossRef]
15. Thunis, P.; Degraeuwe, B.; Pisoni, E.; Trombetti, M.; Peduzzi, E.; Belis, C.A.; Wilson, J.; Clappier, A.; Vignati, E. PM2.5 source allocation in European cities: A SHERPA modelling study. *Atmos. Environ.* **2018**, *187*, 93–106. [CrossRef]
16. Simpson, D.; Benedictow, A.; Berge, H.; Bergström, R.; Emberson, L.D.; Fagerli, H.; Flechard, C.R.; Hayman, G.D.; Gauss, M.; Jonson, J.E.; et al. The EMEP MSC-W chemical transport model—Technical description. *Atmos. Chem. Phys.* **2012**, *12*, 7825–7865. [CrossRef]
17. Granier, C.; Darras, S.; van der Gon, H.D.; Doubalova, J.; Elguindi, N.; Galle, B.; Gauss, M.; Guevara, M.; Jalkanen, J.-P.; Kuenen, J.; et al. *The Copernicus Atmosphere Monitoring Service Global and Regional Emissions (April 2019 Version)*, Report April 2019 version; Copernicus Atmosphere Monitoring Service: Reading, UK, 2019. [CrossRef]
18. Kuenen, J.; Dellaert, S.; Visschedijk, A.; Jalkanen, J.-P.; Super, I.; van der Gon, H.D. CAMS-REG-v4: A state-of-the-art high-resolution European emission inventory for air quality modelling. *Earth Syst. Sci. Data* **2022**, *14*, 491–515. [CrossRef]
19. Thunis, P.; Crippa, M.; Cuvelier, C.; Guizzardi, D.; de Meij, A.; Oreggioni, G.; Pisoni, E. Sensitivity of air quality modelling to different emission inventories: A case study over Europe. *Atmospheric Environ. X* **2021**, *10*, 100111. [CrossRef]
20. Dijkstra, L.; Poelman, H.; Veneri, P. The EU-OECD definition of a functional urban area. In *OECD Regional Development Working Papers*; No. 2019/11; OECD Publishing: Paris, France, 2019. [CrossRef]
21. Itahashi, S.; Mathur, R.; Hogrefe, C.; Napelenok, S.L.; Zhang, Y. Modeling stratospheric intrusion and trans-Pacific transport on tropospheric ozone using hemispheric CMAQ during April 2010—Part 2: Examination of emission impacts based on the higher-order decoupled direct method. *Atmospheric Chem. Phys.* **2020**, *20*, 3397–3413. [CrossRef]
22. Fiore, A.M.; West, J.J.; Horowitz, L.W.; Naik, V.; Schwarzkopf, M.D. Characterizing the tropospheric ozone response to methane emission controls and the benefits to climate and air quality. *J. Geophys. Res.* **2008**, *113*, 8307. [CrossRef]
23. Derwent, R.G.; Utembe, S.R.; Jenkin, M.E.; Shallcross, D.E. Tropospheric ozone production regions and the intercontinental origins of surface ozone over Europe. *Atmos. Environ.* **2015**, *112*, 216–224. [CrossRef]
24. Butler, T.; Lupascu, A.; Coates, J.; Zhu, S. TOAST 1.0: Tropospheric Ozone Attribution of Sources with Tagging for CESM 1.2.2. *Geosci. Model Dev.* **2018**, *11*, 2825–2840. [CrossRef]

25. Bessagnet, B.; Beauchamp, M.; Guerreiro, C.; de Leeuw, F.; Tsyro, S.; Colette, A.; Meleux, F.; Rouïl, L.; Ruyssenaars, P.; Sauter, F.; et al. Can further mitigation of ammonia emissions reduce exceedances of particulate matter air quality standards? *Environ. Sci. Policy* **2014**, *44*, 149–163. [CrossRef]
26. Thunis, P.; Clappier, A.; de Meij, A.; Pisoni, E.; Bessagnet, B.; Tarrason, L. Why is the city's responsibility for its air pollution often underestimated? A focus on PM2.5. *Atmos. Chem. Phys.* **2021**, *21*, 18195–18212. [CrossRef]
27. Thunis, P.; Clappier, A.; Pisoni, E.; Degraeuwe, B. Quantification of non-linearities as a function of time averaging in regional air quality modeling applications. *Atmos. Environ.* **2015**, *103*, 263–275. [CrossRef]

Article

A Real-Time Comparison of Four Particulate Matter Size Fractions in the Personal Breathing Zone of Paris Subway Workers: A Six-Week Prospective Study

Rémy Pétremand [1], Guillaume Suárez [1], Sophie Besançon [2], J. Hugo Dil [3] and Irina Guseva Canu [1,*]

1. Department of Occupational and Environmental Health, Center of Primary Care and Public Health (Unisanté), University of Lausanne, Epalinges, 1066 Lausanne, Switzerland; remy.petremand@chuv.ch (R.P.); guillaume.suarez@unisante.ch (G.S.)
2. Régie Automne de Transport Parisien (RATP), 75012 Paris, France; sophie.besancon@ratp.fr
3. Institute of Physics, Ecole Polytechnique Fédérale de Lausanne (EPFL), 1015 Lausanne, Switzerland; hugo.dil@epfl.ch
* Correspondence: irina.guseva-canu@unisante.ch

Citation: Pétremand, R.; Suárez, G.; Besançon, S.; Dil, J.H.; Guseva Canu, I. A Real-Time Comparison of Four Particulate Matter Size Fractions in the Personal Breathing Zone of Paris Subway Workers: A Six-Week Prospective Study. *Sustainability* **2022**, *14*, 5999. https://doi.org/10.3390/su14105999

Academic Editors: José Carlos Magalhães Pires and Álvaro Gómez-Losada

Received: 19 April 2022
Accepted: 13 May 2022
Published: 15 May 2022

Publisher's Note: MDPI stays neutral with regard to jurisdictional claims in published maps and institutional affiliations.

Abstract: We developed a Bayesian spline model for real-time mass concentrations of particulate matter (PM10, PM2.5, PM1, and PM0.3) measured simultaneously in the personal breathing zone of Parisian subway workers. The measurements were performed by GRIMM, a gravimetric method, and DiSCmini during the workers' work shifts over two consecutive weeks. The measured PM concentrations were analyzed with respect to the working environment, the underground station, and any specific events that occurred during the work shift. Overall, PM0.3 concentrations were more than an order of magnitude lower compared to the other PM concentrations and showed the highest temporal variation. The PM2.5 levels raised the highest exposure concern: 15 stations out of 37 had higher mass concentrations compared to the reference. Station PM levels were not correlated with the annual number of passengers entering the station, the year of station opening or renovation, or the number of platforms and tracks. The correlation with the number of station entrances was consistently negative for all PM sizes, whereas the number of correspondence concourses was negatively correlated with PM0.3 and PM10 levels and positively correlated with PM1 and PM2.5 levels. The highest PM10 exposure was observed for the station platform, followed by the subway cabin and train, while ticket counters had the highest PM0.3, PM1, and PM2.5 mass concentrations. We further found that compared to gravimetric and DiSCmini measurements, GRIMM results showed some discrepancies, with an underestimation of exposure levels. Therefore, we suggest using GRIMM, calibrated by gravimetric methods, for PM sizes above 1 μm, and DiSCmini for sizes below 700 nm.

Keywords: occupational exposure; Bayesian spline model; time-series; public transport; particulate matter; inhalation

1. Introduction

Particulate matter (PM) is a common proxy indicator for air pollution. It consists of a complex mixture of solid and liquid particles of organic and inorganic substances suspended in the air. PM affects more people than any other pollutant [1]. Short-term exposures to coarse (PM10, i.e., particles with an average aerodynamic diameter < 10 μm) and fine (PM2.5, i.e., <2.5 μm) particles are clearly associated with all causes of cardiovascular, respiratory, and cerebrovascular mortality [2], while long-term PM exposure increases the risk of mortality from cardiovascular disease, respiratory disease, and lung cancer [3]. These associations persist below the exposure level outlined in the 2006 WHO guideline [4]. This led the WHO to reduce the recommended maximum annual average exposure level for PM2.5 from 10 μg/m^3 to 5 μg/m^3 and for PM10 from 20 μg/m^3 to 15 μg/m^3 [5]. The recommended maximum 24-hour average exposure was reduced from

$25~\mu g/m^3$ to $15~\mu g/m^3$ for PM2.5 and from $50~\mu g/m^3$ to $45~\mu g/m^3$ for PM10 [5]. Regarding levels of smaller particles (PM1 and PM0.1, i.e., <1 μm and 100 nm, respectively) that are beyond the guideline exposure levels, increasing epidemiological evidence suggests an association between short-term exposures and negative impacts on cardiorespiratory health, as well as the health of the central nervous system [6].

Epidemiological and toxicological studies show varying types and degrees of health effects related to PM, suggesting a role for both the chemical composition (such as transition metals and combustion-derived primary and secondary organic particles) and physical properties (size, shape, and surface area) along with concentration [7–17]. Yet, the research in this field is limited and controversial, particularly when comparing the results from epidemiological and experimental (in vivo and in vitro) studies [18]. This is especially true for PM in underground subway systems, where PM concentrations can be significantly higher than outdoors and have a very specific physio-chemical composition and size distribution [19–21]. Ultrafine particles (UFP) are the strongest contributor to subway pollution when the particle number concentration is used as the exposure metric [22]. Because of this size distribution and a highly ferruginous composition, along with the presence of trace metals (Mg, Al, Si, Ti, V, Cr, Mn, Ni, Cu, Zn, Ba, and Pb) [23,24], subway PM generates more reactive oxygen species (ROS) and oxidative-stress-related outcomes compared to other PM [18,25]. A comprehensive assessment of individual exposure to subway PM, particularly the finest size fractions, is urgently warranted in order to identify the sources and factors that contribute to high PM levels in individual subway stations and lines [26,27]. While the potential health impacts of subway PM on workers and/or commuters remain uncertain, exposure assessment studies are essential for risk assessments and exposure control interventions.

A Franco-Swiss epidemiological research project called "ROBoCoP" (Respiratory disease Occupational Biomonitoring Collaborative Project) was launched at the Parisian urban transport company (RATP) to address this issue [28]. Within this project, a 6-week longitudinal study was conducted among RATP workers to measure their personal exposures in terms of particle number and particle mass concentration using direct-reading instruments along with standardized gravimetric analysis [29]. The application of a Bayesian spline method to the collected UFP number measurements and contextual data enabled estimations of the differences in UFP exposure between subway professionals, stations, and various locations [22]. The developed model proved informative for documenting the magnitude and variability of UFP exposure and for understanding exposure determinants.

In the present study, we build on this existing work and aim to show that it can also be applied to other size fractions and measurement techniques, thereby demonstrating the general applicability of the model. Therefore, we apply a Bayesian spline method to the real-time mass concentrations of PM10, PM2.5, PM1, and PM0.3 measured simultaneously by an optical particle counter, and we analyze the exposure profiles and determinants of these aerosol fractions in the personal air samples of Parisian subway workers. In addition to covering the acute problem of PM exposure, our study also has a more general perspective. It gives insight into the limitations and intercomparability of different measurement techniques and, as such, can help design future studies.

2. Materials and Methods

2.1. Data Collection

Data were collected in the frame of a longitudinal pilot study dedicated to a comprehensive exposure assessment, as described in the study protocol [28]. We focused primarily on subway line 7. Line 7 entered into operation in 1910 and crosses Paris from the northeast to the southeast following a slightly curved route. This entirely underground line is one of the longest (22 km and 38 stations) and busiest (more than 136 million yearly passengers) in the Parisian subway network.

Nine subway professionals, who primarily work on line 7, were included from three different occupations: station agents (n = 3), locomotive operators (n = 3), and security guards (n = 3). Their tasks and exposure results were described in detail elsewhere [28,29]. The data collection lasted for a total duration of 6 weeks (from 7 October to 19 November 2019, i.e., 2 weeks per type of subway professional). RATP safety regulations do not allow any RATP professionals to wear any equipment other than what is used for their regular work. Therefore, airborne PM were collected as close as possible to the worker's personal breathing zone (PBZ) with appropriate equipment carried by two or three RATP technicians who job-shadowed RATP workers for their entire 6–8-hour shifts.

For the continuous measurement of airborne particles, we used the portable GRIMM Aerosol Spectrometer and Dust Monitor (GRIMM Aerosol Technik, Ainring, Germany) Model 1.109, which is considered a research-grade device of moderate cost [30–32]. The measuring principle of this model is based on light scattering off single particles with a semiconductor laser as a light source. Model 1.109 possesses 31 size channels for measuring particle size distribution within the range of 0.25 to 20 μm, recorded every 5 min (with a time resolution of 6 s). For each size fraction, the mass concentration is estimated. For these reasons, GRIMM Model 1.109 is considered suitable for aerosol research and the compilation of occupational health data [30–32]. As GRIMM results are not in compliance with European standards for PM10 and PM2.5, we complemented their measurement by standard gravimetric analysis (EN 12341). For this, the sampling train was equipped with a filter (PTFE Membrane Filters (37 mm), Sigma-Aldrich, Molsheim, France) in a cassette holder (Personal Impactor H-PEM, BGI, USA) connected to a cyclone and attached with flexible tubing to a pump (GilAir Plus, Sensidyne, Germany) operating at 4 L/min. Moreover, we used the particle counter "DiSCmini" (Testo, Monchaltorf, Suisse) to measure particles from 10 to 700 nm, yielding particle number concentration ($\#/cm^3$) and lung-deposited specific area (LDSA) (recorded every 6 s; time resolution of 1 s). In addition to instrumental PM measurements, technicians documented every participant's location and event that occurred during his/her work shift in a standardized activity logbook.

2.2. Data Management

The PM records and activity logbooks were processed as follows. First, we defined the time-series from daily collected PM measurements, and each time-series corresponded to a complete 6-hour work shift, linked with an activity logbook. The calibration of GRIMM measurements with reference to PM10 or PM2.5 levels is recommended, although not yet standardized [33,34]. In this study, we applied the most recent method [26] to standardize the PM10 or PM2.5 time-series using the temperature and relative humidity measurements as well as the gravimetric concentrations of PM10 or PM2.5. We tested several regression functions and determined that the power function had the highest R^2 fit to both the PM2.5 and PM10 gravimetric concentrations (Supplementary Figure S1). However, given the absence of measurement standards for the calibration of PM1 and PM0.3, we also used non-calibrated, raw time-series data to compare the aerosol dynamics according to the size fraction. To analyze their variation, three independent variables were extracted from activity logbooks: *Station*, *Environment*, and *Event*, along with their corresponding timing and duration. The *Event* variable documented the events that occurred during the recording (e.g., exposure to tobacco smoke, intervention on ticket distributor, subway cabin heater, train passing), as previously described [22]. The *Environment* variable defines the type of locality, or setting, that the participant was located in or visited during his/her work shift (e.g., sampling room, cloakroom, ticket-counter, underground corridor, subway platform). The variable *Station* corresponds to the participant's location in the subway rail network. When traveling underground between two subway stations on the same line, the *Station* variable was set to Tunnel for all corresponding time points.

In order to better understand the PM variations between subway stations, additional variables were retrieved from RATP records, namely the year of station opening, the year of the last station renovation, the annual number of passengers entering the station in 2019, and the number of ventilators per station, as well as the factors contributing to the natural station ventilation, such as station topography, the number of entrances, the number of correspondence concourses, the number of platforms and tracks, and the type of station design. The latter was assessed according to [35] and completed to account for local architectural particularities, with seven types of station design (coded from A to G) overall.

2.3. Statistical Analysis

To explore the association between PM mass concentration and independent variables, we developed a Bayesian spline model, as previously described [22]. We fitted four separate models that considered the log10 transformed PM10, PM2.5, PM1, and PM0.3 time-series as dependent variables and *Station*, *Environment*, and *Event* as independent variables, with an inter-day-specific intercept absorbing the corresponding job random effect:

$$Y_{ir} \sim N\left\{\mu_i + XStation_{ir}^T\alpha + XEnvironement_{ir}^T\beta + XEvent_{ir}^T\gamma + \zeta_i^T\mathbf{b}(t_{ir}),\ \sigma^2(Day_i)\right\}. \tag{1}$$

The models were fitted within a Bayesian framework and strictly validated using the "When to worry and how to Avoid the Misuse of Bayesian Statistics" (WAMBS) checklist [36]. This validation consisted of a convergence check for all 117 parameters (i.e., 45 stations, 8 locations, 10 events, 24 inter-day-specific intercepts, 24 inter-day-specific variances, and the 3 intercept and 3 variance parameters of the job random effect) using Gelman and Rubin convergence diagnostics and by visualizing the trace and density plots of all coefficients except the many ζ coefficients. A sensitivity analysis was performed on the prior distribution of α, β, and γ coefficients by varying the standard deviation from 5 to 3 or 10. In addition, we checked for large degrees of autocorrelation in the Markov chain using autocorrelation plots with lags varying from 1 to 20. Finally, we conducted a posterior predictive checking step by predicting the particle number concentration for complete time-series using the input data and then comparing it with the observed particle number concentrations.

Moreover, we explored the station-related variables described above as explanatory factors of β coefficients obtained by modeling. For this, we assessed the Pearson correlation coefficients. Given the exploratory and not hypothesis-driven analytical framework adopted here, we applied no correction for multiple comparisons.

All data management and statistical analyses were performed using the R software package (version 3.6.2). In the Bayesian inference step, we used the R2Jags library and JAGS standard software with the model described in Bayesian inference Using Gibbs Sampling (BUGS) format (10.16909/DATASET/28, File S1).

3. Results

3.1. Descriptive Results

Figure 1 summarizes daily PM mass concentrations for the four size fractions measured by GRIMM over six weeks in the PBZ of subway workers, stratified by their job. The values of the daily PM mass concentration per size fraction can be consulted at the Unisanté Research Data Repository (10.16909/DATASET/28, Excel File 1). It is worth noting that no real-time measurement records were available for two days out of ten for every job. This was due either to the difficulties in GRIMM use in the first two days of the field campaign (despite the fact that all RATP technicians were trained and well-experienced), uncontrolled disruptions in the dust monitor's functioning, or an unexplained stoppage of measurement recording. According to the available records, station agents had the most precise estimates of their daily individual exposure (appearing with the narrowest confidence intervals around the central estimates in Figure 1), regardless of the PM size fraction. This was in line with our previous study and was due to the relative stability of their tasks and the fixed nature of their workplace (in the ticket counter or its surrounding environment) [22]. As expected, the PM0.3 daily concentration was more than an order of magnitude lower compared to the other PM fractions, and this difference was consistent over time and across jobs. This is explained by the fact that the ultrafine particle contribution to the PM mass is very limited [37]. The variability in the daily mass concentration was the highest for PM0.3, followed by PM1. For PM2.5 daily mass concentration, the variability was the lowest overall; however, this was not consistent, as for some days PM10 mass concentration showed less variability than PM2.5. Overall, PM10 exposure, estimated as mass concentration, was predominant over that of the other PM size fractions in all jobs, closely followed by PM2.5 (Figure 1).

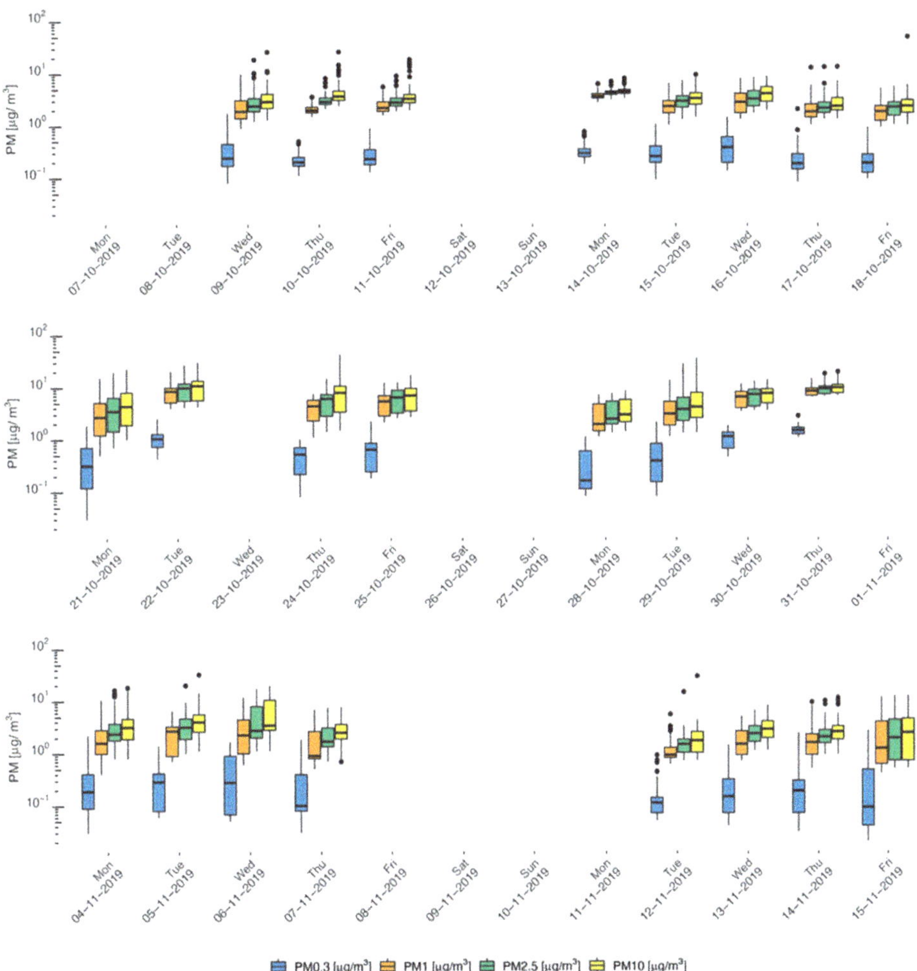

Figure 1. PM daily concentrations measured by GRIMM in the personal PBZs of Parisian subway workers: station agents (**top**), locomotive operators (**middle**), and security guards (**bottom**).

The correlations between gravimetric and real-time PM2.5 and PM10 measurements were rather fair (Supplementary Material, Figure S1), and this result is in line with previous studies of calibration issues [34,38–40]. When comparing the PM2.5 and PM10 concentrations before and after calibration (10.16909/DATASET/28, Excel Files 2 and 3, respectively), the geometric means estimated based on raw real-time records appeared unrealistically low (Excel File 3, sheet "GM_GSD_day"). Moreover, even after calibration, the real-time measurement of PM2.5 and PM10 underestimated the personal exposure in locomotive operators and security guards when measured as gravimetry mass [29], and this underestimation seems more important for PM10 levels. For instance, the geometric mean for PM10 of locomotive operators was 4.24 μg/m^3 before calibration, 88.12 μg/m^3 after calibration

(10.16909/DATASET/28, Excel Files 2 and 3, sheet "GM_GSD_job"), and 188.50 μg/m^3 when measured by the gravimetric method [29].

Figure 2 illustrates the integration of the contextual information collected through the daily activity logbooks to explain the variation in personal PM concentrations over the workers' work shifts. For the sake of clarity and comparability, we plotted data collected during the last day of the first week of exposure monitoring in each job, corresponding to the middle of the monitoring period. Professionals' work shifts usually started and ended in the sampling room at "Porte de la Villette" (Figure 2B,C), where the PM mass concentration is very low compared to other environments. On the 11 October 2019, all station agent PM records corresponded to the PM mass concentration in the ticket counter situated at the station Corentin Carriou (Figure 2A). The only recorded event this day was the opening of the ticket counter door, which consistently increased the PM mass concentration of all PM sizes fractions, particularly for PM10. It is remarkable that all PM size fraction mass concentrations evolved almost in parallel, although the increases in PM0.3 observed in locomotive operators and security guards (Figure 2B,C) were greater than those of other PM.

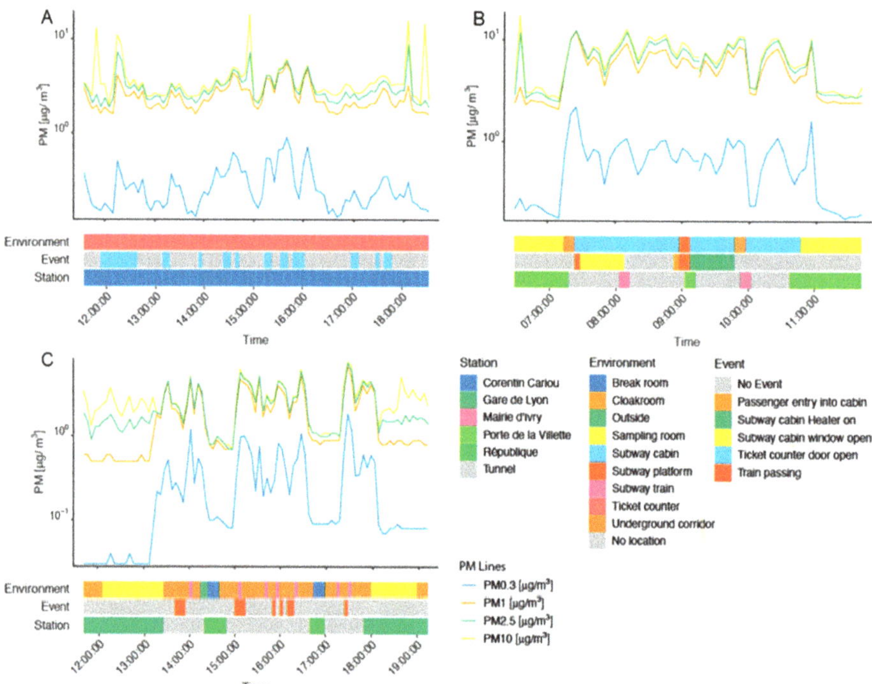

Figure 2. Real-time measurements of PM mass concentrations over the work shift in the PBZ for station agents (**A**) (11 October 2019), locomotive operators (**B**) (25 October 2019), and security guards (**C**) (7 November 2019).

The first PM0.3 peak in Figure 2B is particularly large, corresponding to the walk from the sampling room via the underground corridor before reaching the train and entering the cabin. Opening the window during the initial phase of driving the train seemed to decrease the PM mass concentration of all size fractions, while putting the heater on increased all types of PM in a very similar way. For the security guard, PM concentrations changed with environment, and every passing train event appeared to be followed by a peak in all PM size fractions, particularly for PM0.3 (Figure 2C). The shape of PM variation in this illustrative time-series clearly requires a model supporting the non-stationarity autocorrelation while taking into account different fixed effect variables (*Station*, *Environment*, and *Event*), and it confirms the relevance of the Bayesian spline model.

3.2. Bayesian Modeling Results

The fitting of the Bayesian spline model to the personal PM time-series resulted in a good mixing behavior in Markov chains. The model validity was supported by low Gelman and Rubin convergence diagnostics and autocorrelations, conducted in accordance with the WAMBS checklist [36]. Based on the visual examination of the trace and density plots of coefficients, we identified no conditions invalidating our models. Furthermore, we found that modifying the prior distribution for different parameters did not impact the estimation of the posterior distribution, thus demonstrating the robustness of the model. All of the estimated parameters of this model are available from the Unisanté Research Data Repository (10.16909/DATASET/28, Excel File 3, sheets "alpha_station", "beta_environments", and "gamma_event"). The model coefficients obtained when fitting the model to calibrated data were very similar (10.16909/DATASET/28, Excel File 2).

Figure 3, using the last example of the security guards' exposure monitored on the 7 November 2021, shows that the prediction of the PM mass concentration by the Bayesian spline model overlaps reasonably well with the observed values for all size fractions. Visually, the fit accuracy looks similar across size fractions, where some of the highest observed concentration peaks are above the model prediction curve. This is particularly the case for PM0.3. Although satisfactory according to the WAMBS guidelines [36], this model fit is worse than the fit obtained in our previous study on PM0.3 number concentration measured by DiSCmini [22]. This is due to the fact that fewer data were recorded by GRIMM, and thus available for model training, because of a lower time resolution of measurement recording compared to DiSCmini (5 min versus 6 s) and six days with missing GRIMM records.

Figure 4, panel A represents the posterior distribution of the estimated coefficients for every subway station along line 7. The coefficients are expressed as a fold change ($10^{coefficient}$) with respect to the reference station, Porte de la Villette, for each size fraction. The personal mass concentrations measured at this station were 0.31 ± 2.86 µg/m^3 for PM0.3, 2.71 ± 2.00 µg/m^3 for PM1, 3.32 ± 2.00 µg/m^3 for PM2.5, and 3.91 ± 2.03 µg/m^3 for PM10. The calibrated values for PM2.5 and PM10 were 70.08 ± 1.52 µg/m^3 and 82.66 ± 1.85 µg/m^3, respectively.

In line with the results of descriptive analysis, the coefficients corresponding to the different PM size fractions are rather similar at most subway stations (Figure 4A). However, a closer look reveals that PM2.5 raises the highest exposure concern, with 15 stations out of 37 showing significantly increased mass concentrations compared with the reference station. The coefficients corresponding to these PM2.5-polluted stations have credible intervals above 1. Twelve of these stations also had significantly higher PM1 mass concentrations compared to the reference. Regarding PM10, only 7 stations out of 37 had significantly higher mass concentrations, and an additional 4 stations presented an increase with borderline credibility. Finally, with respect to PM0.3, eight stations were significantly more polluted than the reference station. It is noteworthy that the magnitude of change in PM mass concentration between stations was not high and was rarely greater than 10% (Figure 4A). In this respect, the most polluted stations were Villejuif-Léo Lagrange, with the highest levels of PM10, PM2.5, and PM0.3; Louis Blanc, with the second-highest mass concentrations of PM0.3 and PM10; and Villejuif-Louis Aragon, with the highest PM1 and the second-highest PM2.5 and PM0.3 mass concentrations. Further analysis of the station characteristics revealed no correlation with the annual number of passengers entering the station, the year of station opening or renovation, or the number of platforms and tracks. In contrast, we observed a consistent negative correlation with the number of entrances for all size fractions (-0.11 for PM0.3, -0.15 for PM1, -0.24 for PM2.5, and -0.06 for PM10). The number of correspondence concourses for other stations was negatively correlated with PM0.3 and PM10 (-0.07 and -0.03, respectively) and positively correlated with PM1 and

PM2.5 (0.34 and 0.06, respectively). Station types C and F were associated with the highest levels of PM in all size fractions, whereas high PM0.3 concentrations were also observed in type A stations and PM10 in type G stations (results not shown). These findings are in line with the previously reported data suggesting the importance of general (natural) ventilation, which can be determined by the station architecture and topography [35]. Regarding the latter, the correlation coefficients between PM coefficients and minimal station altitude decrease with increasing PM size (0.28 for PM0.3, 0.19 for PM1, 0.06 for PM2.5, and 0.08 for PM10).

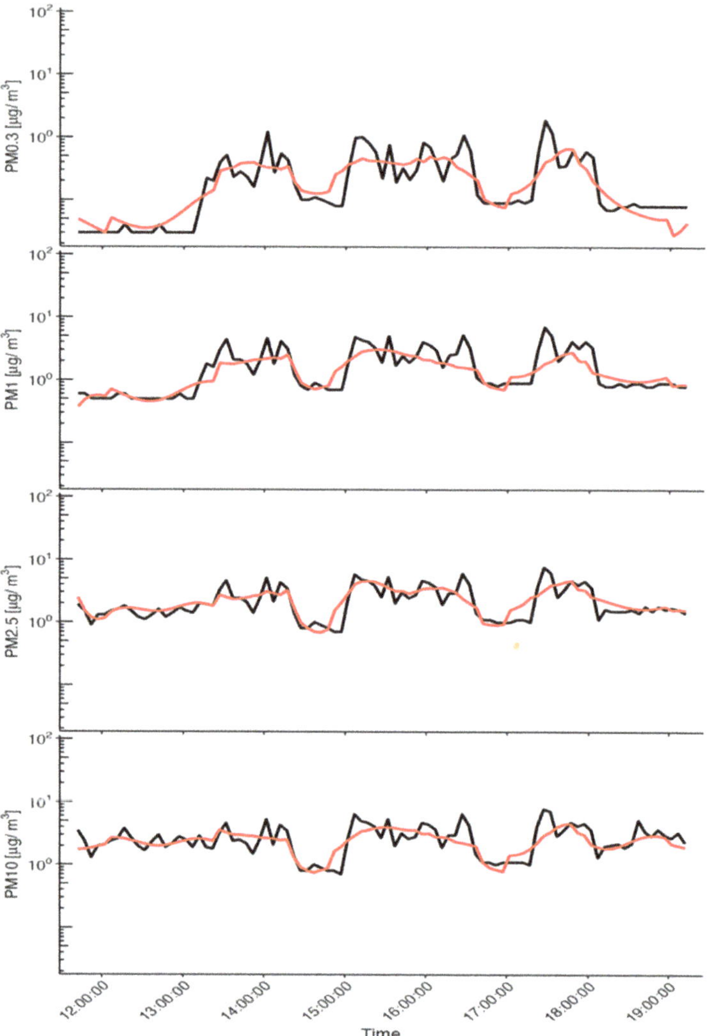

Figure 3. Prediction of the Bayesian spline model. Visualization of the Bayesian spline model prediction (red curve) for the personal PM10, PM2.5, PM1, and PM0.3 mass concentrations of the security guards on the 7 November 2019.

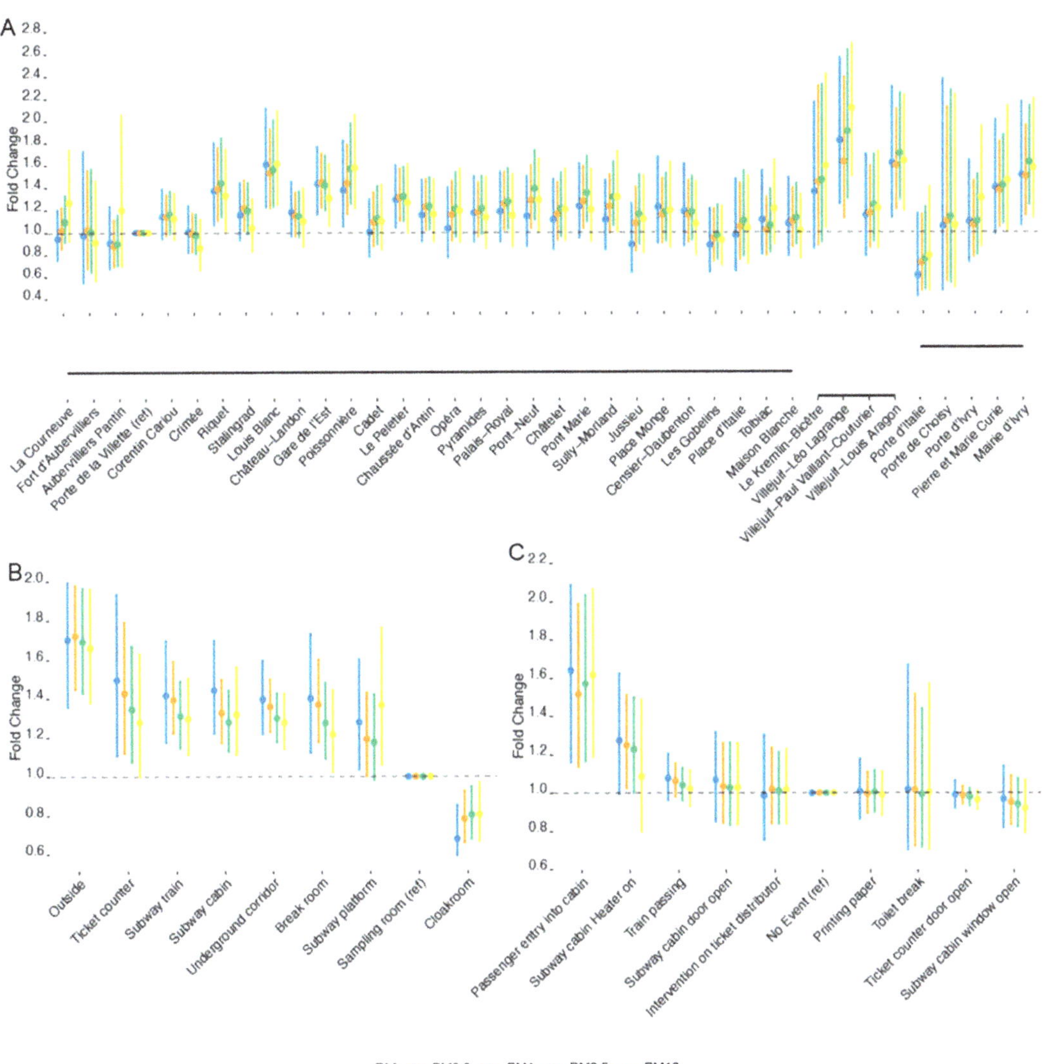

Figure 4. PM mass concentration variation on subway line 7 for *Stations* (**A**), *Environments* (**B**), and *Events* (**C**). The plots represent the posterior distribution of the coefficient transformed as fold change ($10^{coefficient}$) for every category of the studied factors. The bar is the 95% credible interval, and the point is the median of that distribution. The bottom sub-panel in panel (**A**) represents the subway line 7. The discontinuity shown at the *"Maison Blanche"* station corresponds to two embranchments, one towards *"Villejuif-Louis Aragon"* and the second towards *"Mairie d'Ivry"*. The reference category is noted with (ref).

Figure 4B shows that, with the exception of the cloakroom, all studied environments had higher PM exposure compared to the sampling room, although the highest levels of PM were measured outdoors. The highest PM10 exposure was observed at the station platform, followed by the subway cabin and train, while ticket counters had the highest PM0.3, PM1, and PM2.5 mass concentrations. Regarding the effect of studied events on the PM level (Figure 4C), only the event called "Passenger entry into cabin" was associated with a significant increase in the mass concentration of all size fractions. Indeed, despite the name, this event corresponds to opening the train cabin door to enter the cabin or have a study technician bring in some equipment; passengers, who should not disturb the locomotive operator, did not enter. Turning the heater on was also associated with a concentration increase, but only for fine and ultrafine particles. The increase in PM mass concentration when or after a train was passing or a cabin door was opened was small and not statistically significant.

3.3. Comparison of GRIMM and DiSCmini Results

It is noteworthy that in our previous study of UFP number concentration, we observed the opposite effect for the train passage, which lowered the UFP number concentration measured with DiSCmini [22]. The other increases in particle number concentration were identified in the same environments as in this study, although with different absolute concentrations. For instance, the subway platform was found to be more polluted with UFP than the ticket counter, identified as the second most UFP-polluted environment in that study, while the outdoor UFP number was not significantly higher than the reference UFP concentration [22]. In order to investigate these discrepancies, we compared GRIMM and DiSCmini measurement results in terms of particle number concentrations. For this, we integrated the GRIMM number concentration values—from 0.25 to 0.70 μm—in the particle size range corresponding to the DiSCmini operating range (0.01 to 0.7 μm) (Supplementary Material, Figure S2, 10.16909/DATASET/28, File 4).

As illustrated in Figure 5, using the measurements corresponding to the mid-point days in the PM monitoring interval, the two devices reflect relatively similar patterns in concentration changes over time. Although the latter records more peaks, probably due to a better time resolution (lower averaging span), the increases in concentration recorded with GRIMM correspond to increases in DiSCmini measurement results. Quantitatively, the particle number concentrations measured by the DiSCmini are three orders of magnitude higher than the ones measured by the GRIMM. This observation can be explained by the size distribution of the analyzed environmental aerosols, where the contribution of ultrafine particles (<100 nm) to the overall number concentration is particularly high [18,21,24]. Burkart et al. showed that, due to the number size distribution of urban environment aerosols, GRIMM (model 1.109) only measures a tiny part (6%) of the total particle number determined via a Vienna-type differential mobility analyzer (DMA) [41]. Although the particle number concentrations obtained with DiSCmini and GRIMM follow the same trends, particles with aerodynamic diameters below the GRIMM lower cut-off (0.25 μm) are expected to contribute largely to the aerosol distribution, thereby explaining the differences in Figure 5.

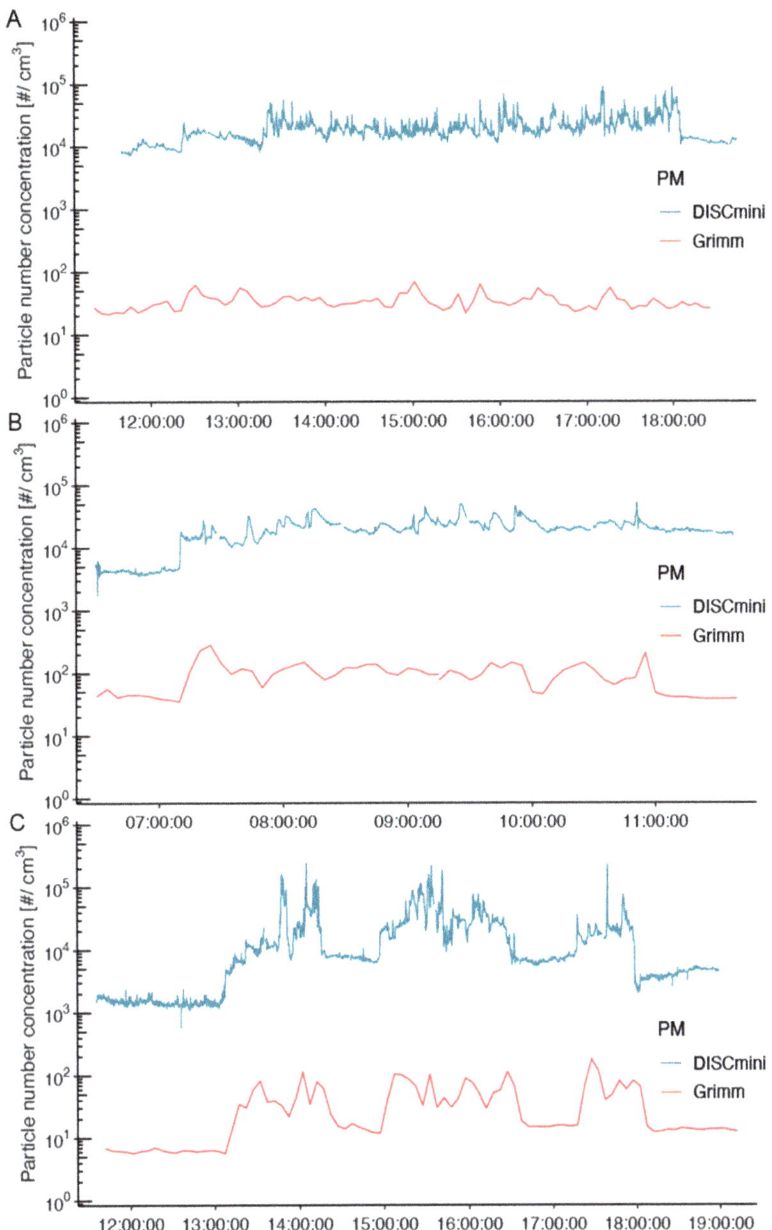

Figure 5. Total particle number concentration measured by DiSCmini for sizes from 0.01 to 0.7 μm (blue) and by GRIMM for sizes from 0.25 to 0.7 μm (red) in the personal breathing zone samples of Parisian subway workers. (**A**) Station agents (10 October 2019), (**B**) locomotive operators (25 October 2019), and (**C**) security guards (7 November 2019).

4. Discussion

4.1. Detection Physical Principles

Most of the commercially available direct-reading particle counters rely on a few physical principles that determine their strengths and limitations in a given context of use [39]. In a very generic manner, the use of instruments based on different physical principles implies questions to consider beyond the manual instructions: (i) which physical event generates a measurable change, and (ii) what is the nature of this change? In the case of particle counters, OPCs such as the GRIMM rely on light scattering properties inherent to particles in the micron-domain, with an exponential decrease in the scattering intensity with the particle diameter. Here, the instrument's lower cut-off diameter of about 0.25 μm is explained by the Mie diffusion domain at the laser wavelength. The scattering-based approach implies that the size determination is sensitive to a series of physical variables such as refractive index, density, and shape of particles. In addition, the conversion of the particle number distribution into mass concentration is done by applying a mathematical model with approximated values for particle density and morphology. Thus, in addition to the standard calibration (dolomite dust), the GRIMM instrument enables the possibility of adjusting the density variable through the correction-factor (C-factor) adapted to a particular aerosol [34]. Despite this measurement correction approach, one assumption still remains: namely, that the sensitive physical variables—density, refractive index, and shape— are unchanged during the period of measurement. For environmental, non-standardized aerosols, this assumption is even more problematic because a direct link between particle size and chemical composition, and thus refractive index and reflectivity, can be expected. An earlier comparison study performed in urban environments reports that GRIMM underestimates the mass concentration by about 20% when considering gravimetric data as reference and a comparable particle size range [41]. In accordance with this, in the present work, the GRIMM results underestimated PM2.5 and PM10 exposure as compared to the gravimetric method, and this was the case for both calibrated and non-calibrated comparisons. However, the positive effect of calibration on GRIMM measurements was clearly visible as it significantly reduced the gap with corresponding gravimetric data for both PM fractions.

In turn, personal monitors such as DiSCmini, Partector, or NanoTracer rely on the electrical measurement (current intensity) of the particle-borne charges resulting from the initial electrical diffusion charging of the aerosol. Since the electrical diffusion charging behavior of submicron particles is size-dependent, these devices provide quantitative information on both the particle number concentration and the modal diameter in the UFP domain (typically from 20 to 400 nm). The conversion of electrical mobility into aerodynamic diameter requires a series of approximations of the particle shape (spherical) and the size distribution (lognormal). Considering the ultrafine condensation particle counter (UCPC) as a reference instrument, Todea et al. showed that the DiSCmini overestimates the number concentration by about 30% in the aforementioned size range [42]. The same authors identify as interfering variables the presence of particles with sizes > 400 nm—discarded via the use of an impactor—and aggregates in the aerosol for which the charge is greater than for the equivalent non-aggregated particle. Similarly, Mills et al. reported a deviation from reference—here the scanning mobility particle sizer (SMPS)—for DiSCmini number concentration of about 21% in the case of polydisperse aerosols [43].

4.2. Qualitative Reliability of Measurements and Inter-Device Comparison Issues

It can be argued that the data obtained with the GRIMM and DiSCmini shown in Figure 5 follow a similar trend and that differences are due to the time resolution and different sensitivities to UFP for the two techniques. However, this cannot explain the opposite results for the "train passing" event obtained from the model, nor the quantitatively different response to other events and environments. A closer inspection of Figure 5 reveals some hints about the qualitative differences in the responses of both methods. For example, around 14h00 in Figure 5C, DiSCmini shows a dip where GRIMM shows a peak, similar to around 10h00 in Figure 5B. Such comparisons for both methods, including *Station, Location,* and *Event* indicators, can be found in the full online data set (10.16909/DATASET/28, File 4), and some representative mismatches are reproduced in Figure S3. Many occasions can be found where DiSCmini and GRIMM show opposite trends, and the measurements on 30 October 2019 even show a completely different shape. This clearly indicates a problem with the direct comparison of data from both methods.

Based on the qualitative consistency of the data, our opinion is that the results from DiSCmini are more representative of the real submicron PM concentration. This is based on the observation that all DiSCmini measurement sessions show a similar type of behavior with features clearly related to *Event*, *Location*, or *Station* variables, whereas the GRIMM data appear to show a less structured response. This is directly reflected in the Bayesian model fits and the uncertainties in the changes to all model parameters. Furthermore, the GRIMM results show almost identical curves, different by a multiplication factor, for all PM sizes (Figure 2). This suggests that results for the smallest particle range are strongly influenced by the presence of PM10 and PM2.5. These factors lead us to conclude that the GRIMM is not well suited to characterize PM concentrations for sizes below 700 nm in a typical underground aerosol environment with a mixture of particle sizes and compositions. Instead of a straightforward interpretation of the particle number concentration that considers the full operational capability of each instrument, the rational delimitation of the optimal particle size range for both the GRIMM and DiSCmini enables a useful co-deployment covering all the aerosol dimensions.

5. Conclusions

One of the aims of our study was to use the GRIMM as an intermediate link between the standardized gravimetric method for coarse and fine particles and the real-time measurement method for UFP using DiSCmini. Unfortunately, the large discrepancy in trends between GRIMM and DiSCmini for PM with sizes below 700 nm (the upper limit of the latter method) renders this approach unfeasible. Given the high quality of the Bayesian model fit for the DiSCmini, we have good confidence in this method for real-time measurements of UFP. However, an independent calibration would be needed to determine absolute particle concentrations [30,40]. This goes beyond the scope of this study, but one could envision combining DiSCmini measurements with scanning electron microscopy (SEM) or transmission electron microscopy (TEM) analysis on representative samples.

Another goal of this study was to independently measure the PM concentrations of four different size fractions in a real-time fashion as a function of various parameters applicable to workers in the Parisian subway. The GRIMM promises exactly this functionality; however, based on the results discussed above, it is clear that a single device is not suitable for this aim, and further instruments are needed. We suggest using a GRIMM, previously calibrated by gravimetric methods in the environment of interest, for PM sizes above 1 µm, and a DiSCmini for sizes below 700 nm. This co-deployment of DiSCmini and GRIMM in urban sites delivers valuable information on the dynamic evolution of the aerosol in terms of the number concentration and size distribution (or modal size), covering a large size domain from ultrafine to fine particles. Such dynamic information is particularly useful for identifying possible sources of emissions and for understanding the interventions/changes that govern the aerolics of the system.

Despite the above-discussed discrepancies between the measurement results of GRIMM and the gravimetric method on the one hand and GRIMM and DiSCmini on the other hand, we demonstrated the relevance of our Bayesian spline model for analyzing four time-series of PM concentrations according to subway stations, locations, and events. The strengths of this study lie in the assessment of personal (breathing zone) exposure using multiple devices and ad hoc modeling of exposure variations for four PM sizes simultaneously. This enabled us to evidence a differential exposure profile in terms of PM sizes in subway stations and workplaces, a singular behavior of UFP compared with fine and coarse particles. Further effort should be focused on the development and improvement of portable and affordable devices as well as calibration methods that provide reliable estimates for different exposure components in complex environments, such as underground railways.

Supplementary Materials: The following supporting information can be downloaded at: https://www.mdpi.com/article/10.3390/su14105999/s1, Figure S1: Regression analysis between the simultaneously collected gravimetric and GRIMM data; Figure S2: Particle number concentrations measured in personal samples; Figure S3: Further comparison between DiSCmini and GRIMM.

Author Contributions: G.S., S.B. and I.G.C., data collection; G.S. and S.B., lab analyses; R.P. and I.G.C., statistical analysis; R.P. and I.G.C., drafting the manuscript; G.S., J.H.D. and S.B., critically reviewing and expanding the manuscript. All authors have read and agreed to the published version of the manuscript.

Funding: This research was funded by the Swiss National Science Foundation (Grant N° IZ-COZ0_177067 "ROBoCoP").

Institutional Review Board Statement: The study protocols were approved by the French Personal Protection Committees South-Est II (N° 2019-A01652 55), Declaration of Conformity to the French National Commission for Computing and Freedoms (CNIL) N° 2220108.

Informed Consent Statement: Written consent to participate was obtained from all study participants.

Data Availability Statement: All data related to this study are available at 10.16909/DATASET/28.

Acknowledgments: This study would not have been possible without the support of N.B. Hopf, J.J. Sauvain, and M. Hemmendinger from Unisanté; T. Ben Rayana, G. Carillo, V. Jouannique, and A. Debatisse from RATP; and all RATP technicians and study participants.

Conflicts of Interest: The authors declare no conflict of interest.

References

1. WHO. Ambient (Outdoor) Air Quality and Health. 2021. Available online: https://www.who.int/news-room/fact-sheets/detail/ambient-(outdoor)-air-quality-and-health (accessed on 13 April 2022).
2. Orellano, P.; Reynoso, J.; Quaranta, N.; Bardach, A.; Ciapponi, A. Short-term exposure to particulate matter (PM10 and PM2.5), nitrogen dioxide (NO_2), and ozone (O_3) and all-cause and cause-specific mortality: Systematic review and meta-analysis. *Environ. Int.* **2020**, *142*, 105876. [CrossRef] [PubMed]
3. Chen, J.; Hoek, G. Long-term exposure to PM and all-cause and cause-specific mortality: A systematic review and meta-analysis. *Environ. Int.* **2020**, *143*, 105974. [CrossRef] [PubMed]
4. WHO. *Air Quality Guidelines for Particulate Matter, Ozone, Nitrogen Dioxide and Sulfur Dioxide. Global Update 2005. Summary of Risk Assessment*; WHO: Geneva, Switzerland, 2006.
5. WHO. *Global Air Quality Guidelines. Particulate Matter (PM2.5 and PM10), Ozone, Nitrogen Dioxide, Sulfur Dioxide and Carbon Monoxide*; WHO: Geneva, Switzerland, 2021.
6. Bencsik, A.; Lestaevel, P.; Guseva Canu, I. Nano- and neurotoxicology: An emerging discipline. *Prog. Neurobiol.* **2018**, *160*, 45–63. [CrossRef] [PubMed]
7. Guo, C.; Lv, S.; Liu, Y.; Li, Y. Biomarkers for the adverse effects on respiratory system health associated with atmospheric particulate matter exposure. *J. Hazard. Mater.* **2022**, *421*, 126760. [CrossRef] [PubMed]
8. Guo, Q.; Wang, X.; Gao, Y.; Zhou, J.; Huang, C.; Zhang, Z.; Chu, H. Relationship between particulate matter exposure and female breast cancer incidence and mortality: A systematic review and meta-analysis. *Int. Arch. Occup. Environ. Health* **2021**, *94*, 191–201. [CrossRef]
9. Kelly, F.J.; Fussell, J.C. Toxicity of airborne particles-established evidence, knowledge gaps and emerging areas of importance. *Philos. Trans. A Math. Phys. Eng. Sci.* **2020**, *378*, 20190322. [CrossRef]
10. Liang, Q.; Sun, M.; Wang, F.; Ma, Y.; Lin, L.; Li, T.; Duan, J.; Sun, Z. Short-term PM(2.5) exposure and circulating von Willebrand factor level: A meta-analysis. *Sci. Total Environ.* **2020**, *737*, 140180. [CrossRef]
11. Liang, S.; Zhang, J.; Ning, R.; Du, Z.; Liu, J.; Batibawa, J.W.; Duan, J.; Sun, Z. The critical role of endothelial function in fine particulate matter-induced atherosclerosis. *Part. Fibre Toxicol.* **2020**, *17*, 61. [CrossRef]
12. Lin, L.; Li, T.; Sun, M.; Liang, Q.; Ma, Y.; Wang, F.; Duan, J.; Sun, Z. Effect of particulate matter exposure on the prevalence of allergic rhinitis in children: A systematic review and meta-analysis. *Chemosphere* **2021**, *268*, 128841. [CrossRef]
13. Milici, A.; Talavera, K. TRP Channels as Cellular Targets of Particulate Matter. *Int. J. Mol. Sci.* **2021**, *22*, 2783. [CrossRef]
14. Ning, J.; Zhang, Y.; Hu, H.; Hu, W.; Li, L.; Pang, Y.; Ma, S.; Niu, Y.; Zhang, R. Association between ambient particulate matter exposure and metabolic syndrome risk: A systematic review and meta-analysis. *Sci. Total Environ.* **2021**, *782*, 146855. [CrossRef] [PubMed]
15. Sun, M.; Liang, Q.; Ma, Y.; Wang, F.; Lin, L.; Li, T.; Sun, Z.; Duan, J. Particulate matter exposure and biomarkers associated with blood coagulation: A meta-analysis. *Ecotoxicol. Environ. Saf.* **2020**, *206*, 111417. [CrossRef] [PubMed]

16. Wang, F.; Ahat, X.; Liang, Q.; Ma, Y.; Sun, M.; Lin, L.; Li, T.; Duan, J.; Sun, Z. The relationship between exposure to PM2.5 and atrial fibrillation in older adults: A systematic review and meta-analysis. *Sci. Total Environ.* **2021**, *784*, 147106. [CrossRef] [PubMed]
17. Zhu, H.; Wu, Y.; Kuang, X.; Liu, H.; Guo, Z.; Qian, J.; Wang, D.; Wang, M.; Chu, H.; Gong, W.; et al. Effect of PM(2.5) exposure on circulating fibrinogen and IL-6 levels: A systematic review and meta-analysis. *Chemosphere* **2021**, *271*, 129565. [CrossRef]
18. Loxham, M.; Nieuwenhuijsen, M.J. Health effects of particulate matter air pollution in underground railway systems—A critical review of the evidence. *Part. Fibre Toxicol.* **2019**, *16*, 12. [CrossRef]
19. Martins, V.; Moreno, T.; Minguillón, M.C.; Amato, F.; de Miguel, E.; Capdevila, M.; Querol, X. Exposure to airborne particulate matter in the subway system. *Sci. Total Environ.* **2015**, *511*, 711–722. [CrossRef]
20. Smith, J.; Barratt, B.; Fuller, G.; Kelly, F.; Loxham, M.; Nicolosi, E.; Priestman, M.; Tremper, A.; Green, D. PM2.5 on the London Underground. *Environ. Int.* **2019**, *134*, 105188. [CrossRef]
21. Wen, Y.; Leng, J.; Shen, X.; Han, G.; Sun, L.; Yu, F. Environmental and Health Effects of Ventilation in Subway Stations: A Literature Review. *Int. J. Environ. Res. Public Health* **2020**, *17*, 1084. [CrossRef]
22. Pétremand, R.; Wild, P.; Crézé, C.; Suarez, G.; Besançon, S.; Jouannique, V.; Debatisse, A.; Canu, I.G. Application of the Bayesian spline method to analyze real-time measurements of ultrafine particle concentration in the Parisian subway. *Environ. Int.* **2021**, *156*, 106773. [CrossRef]
23. Qiao, T.; Xiu, G.; Zheng, Y.; Yang, J.; Wang, L.; Yang, J.; Huang, Z. Preliminary investigation of PM1, PM2.5, PM10 and its metal elemental composition in tunnels at a subway station in Shanghai, China. *Transp. Res. D Transp. Environ.* **2015**, *41*, 136–146. [CrossRef]
24. Loxham, M.; Cooper, M.J.; Gerlofs-Nijland, M.E.; Cassee, F.R.; Davies, D.; Palmer, M.R.; Teagle, D.A.H. Physicochemical Characterization of Airborne Particulate Matter at a Mainline Underground Railway Station. *Environ. Sci. Technol.* **2013**, *47*, 3614–3622. [CrossRef] [PubMed]
25. Loxham, M.; Woo, J.; Singhania, A.; Smithers, N.P.; Yeomans, A.; Packham, G.; Crainic, A.M.; Cook, R.B.; Cassee, F.R.; Woelk, C.H.; et al. Upregulation of epithelial metallothioneins by metal-rich ultrafine particulate matter from an underground railway. *Met. Int. Biomet. Sci.* **2020**, *12*, 1070–1082. [CrossRef] [PubMed]
26. Luglio, D.G.; Katsigeorgis, M.; Hess, J.; Kim, R.; Adragna, J.; Raja, A.; Gordon, C.; Fine, J.; Thurston, G.; Gordon, T.; et al. PM2.5 Concentration and Composition in Subway Systems in the Northeastern United States. *Environ. Health Persp.* **2021**, *129*, 27001. [CrossRef] [PubMed]
27. Van Ryswyk, K.; Kulka, R.; Marro, L.; Yang, D.; Toma, E.; Mehta, L.; McNeil-Taboika, L.; Evans, G.J. Impacts of Subway System Modifications on Air Quality in Subway Platforms and Trains. *Environ. Sci. Technol.* **2021**, *55*, 11133–11143. [CrossRef]
28. Canu, I.G.; Hemmendinger, M.; Sauvain, J.J.; Suarez, G.; Hopf, N.B.; Pralong, J.A.; Ben Rayana, T.; Besançon, S.; Sakthithasan, K.; Jouannique, V.; et al. Respiratory Disease Occupational Biomonitoring Collaborative Project (ROBoCoP): A longitudinal pilot study and implementation research in the Parisian transport company. *J. Occup. Med. Toxicol.* **2021**, *16*, 1–11.
29. Canu, I.G.; Crézé, C.; Hemmendinger, M.; Ben Rayana, T.; Besançon, S.; Jouannique, V.; Debatisse, A.; Wild, P.; Sauvain, J.; Suárez, G.; et al. Particle and metal exposure in Parisian subway: Relationship between exposure biomarkers in air, exhaled breath condensate, and urine. *Int. J. Hyg. Environ. Health* **2021**, *237*, 113837. [CrossRef]
30. Giordano, M.R.; Malings, C.; Pandis, S.N.; Presto, A.A.; McNeill, V.; Westervelt, D.M.; Beekmann, M.; Subramanian, R. From low-cost sensors to high-quality data: A summary of challenges and best practices for effectively calibrating low-cost particulate matter mass sensors. *J. Aerosol Sci.* **2021**, *158*, 105833. [CrossRef]
31. Masic, A.; Bibic, D.; Pikula, B.; Blazevic, A.; Huremovic, J.; Zero, S. Evaluation of optical particulate matter sensors under realistic conditions of strong and mild urban pollution. *Atmos. Meas. Tech.* **2020**, *13*, 6427–6443. [CrossRef]
32. Kuula, J.; Mäkelä, T.; Aurela, M.; Teinilä, K.; Varjonen, S.; González, Ó.; Timonen, H. Laboratory evaluation of particle-size selectivity of optical low-cost particulate matter sensors. *Atmos. Meas. Tech.* **2020**, *13*, 2413–2423. [CrossRef]
33. GRIMM Aerosol Technik. *Portable Laser Aerosolspectrometer and Dust Monitor Model 1.108/1.109. Users Manual*; GRIMM Aerosol Technik: Ainring, Germany, 2010; p. 81.
34. Cheng, Y.-H.; Lin, Y.-L. Measurement of Particle Mass Concentrations and Size Distributions in an Underground Station. *Aerosol Air Qual. Res.* **2010**, *10*, 22–29. [CrossRef]
35. Reche, C.; Moreno, T.; Martins, V.; Minguillón, M.C.; Jones, T.; De Miguel, E.; Capdevila, M.; Centelles, S.; Querol, X. Factors controlling particle number concentration and size at metro stations. *Atmosph. Environ.* **2017**, *156*, 169–181. [CrossRef]
36. van de Schoot, R.; Depaoli, S.; King, R.; Kramer, B.; Märtens, K.; Tadesse, M.G.; Vannucci, M.; Gelman, A.; Veen, D.; Willemsen, J.; et al. Bayesian statistics and modelling. *Nat. Rev. Methods Prim.* **2021**, *1*, 1–26. [CrossRef]
37. Asbach, C.; Alexander, C.; Clavaguera, S.; Dahmann, D.; Dozol, H.; Faure, B.; Fierz, M.; Fontana, L.; Iavicoli, I.; Kaminski, H.; et al. Review of measurement techniques and methods for assessing personal exposure to airborne nanomaterials in workplaces. *Sci. Total Environ.* **2017**, *603–604*, 793–806. [CrossRef] [PubMed]
38. Cheng, Y.H. Comparison of the TSI Model 8520 and Grimm Series 1.108 portable aerosol instruments used to monitor particulate matter in an iron foundry. *J. Occup. Environ. Hyg.* **2008**, *5*, 157–168. [CrossRef] [PubMed]
39. Zuidema, C.; Stebounova, L.V.; Sousan, S.; Gray, A.; Stroh, O.; Thomas, G.; Peters, T.; Koehler, K. Estimating personal exposures from a multi-hazard sensor network. *J. Expo. Sci. Environ. Epidemiol.* **2020**, *30*, 1013–1022. [CrossRef] [PubMed]

40. Zuidema, C.; Stebounova, L.V.; Sousan, S.; Thomas, G.; Koehler, K.; Peters, T.M. Sources of error and variability in particulate matter sensor network measurements. *J. Occup. Environ. Hyg.* **2019**, *16*, 564–574. [CrossRef] [PubMed]
41. Burkart, J.; Steiner, G.; Reischl, G.; Moshammer, H.; Neuberger, M.; Hitzenberger, R. Characterizing the performance of two optical particle counters (Grimm OPC1.108 and OPC1.109) under urban aerosol conditions. *J. Aerosol Sci.* **2010**, *41*, 953–962. [CrossRef]
42. Todea, A.M.; Beckmann, S.; Kaminski, H.; Bard, D.; Bau, S.; Clavaguera, S.; Dahmann, D.; Dozol, H.; Dziurowitz, N.; Elihn, K.; et al. Inter-comparison of personal monitors for nanoparticles exposure at workplaces and in the environment. *Sci. Total Environ.* **2017**, *605–606*, 929–945. [CrossRef]
43. Mills, J.B.; Park, J.H.; Peters, T.M. Comparison of the DiSCmini aerosol monitor to a handheld condensation particle counter and a scanning mobility particle sizer for submicrometer sodium chloride and metal aerosols. *J Occup. Environ. Hyg.* **2013**, *10*, 250–258. [CrossRef]

 *sustainability*

Article

Surface Ozone Pollution: Trends, Meteorological Influences, and Chemical Precursors in Portugal

Rafaela C. V. Silva [1,2] and José C. M. Pires [1,2,*]

[1]   LEPABE—Laboratory for Process Engineering, Environment, Biotechnology and Energy,
     Faculty of Engineering, University of Porto, Rua Dr. Roberto Frias, 4200-465 Porto, Portugal;
     up201603609@edu.fe.up.pt
[2]   AliCE—Associate Laboratory in Chemical Engineering, Faculty of Engineering, University of Porto,
     Rua Dr. Roberto Frias, 4200-465 Porto, Portugal
*   Correspondence: jcpires@fe.up.pt; Tel.: +351-22-508-2262

Abstract: Surface ozone (O_3) is a secondary air pollutant, harmful to human health and vegetation. To provide a long-term study of O_3 concentrations in Portugal (study period: 2009–2019), a statistical analysis of ozone trends in rural stations (where the highest concentrations can be found) was first performed. Additionally, the effect of nitrogen oxides (NO_x) and meteorological variables on O_3 concentrations were evaluated in different environments in northern Portugal. A decreasing trend of O_3 concentrations was observed in almost all monitoring stations. However, several exceedances to the standard values legislated for human health and vegetation protection were recorded. Daily and seasonal O_3 profiles showed high concentrations in the afternoon and summer (for all inland rural stations) or spring (for Portuguese islands). The high number of groups obtained from the cluster analysis showed the difference of ozone behaviour amongst the existent rural stations, highlighting the effectiveness of the current geographical distribution of monitoring stations. Stronger correlations between O_3, NO, and NO_2 were detected at the urban site, indicating that the O_3 concentration was more NO_x-sensitive in urban environments. Solar radiation showed a higher correlation with O_3 concentration regarding the meteorological influence. The wind and pollutants transport must also be considered in air quality studies. The presented results enable the definition of air quality policies to prevent and/or mitigate unfavourable outcomes from O_3 pollution.

Keywords: cluster analysis; meteorological influence; Multiple linear regression; NO_x influence; Pearson's correlation; rural trends; surface ozone

Citation: Silva, R.C.V.; Pires, J.C.M. Surface Ozone Pollution: Trends, Meteorological Influences, and Chemical Precursors in Portugal. *Sustainability* 2022, 14, 2383. https://doi.org/10.3390/su14042383

Academic Editor: Alessandra De Marco

Received: 22 December 2021
Accepted: 17 February 2022
Published: 19 February 2022

Publisher's Note: MDPI stays neutral with regard to jurisdictional claims in published maps and institutional affiliations.

1. Introduction

With the constant increase of air pollutants emissions since the Industrial Revolution, an increase of ozone (O_3) near the Earth's surface has been observed [1–3]. This pollutant is a powerful oxidant, affecting human health by causing respiratory and cardiovascular diseases [4]. It also affects vegetation and ecosystems, leading to crop yield and biodiversity losses [5,6]. A total value of 769.2 billion USD loss equivalent to decreases in agricultural production and the occurrence of respiratory diseases and mortality could be ascribed to O_3 exposure in China [7].

O_3 is a secondary pollutant resultant of the reaction between nitrogen oxides (NO_x) and volatile organic compounds (VOCs) released to the atmosphere from natural and (mostly) anthropogenic activities [1]. The photochemical regime for surface ozone production determines the sensitivity of O_3 to anthropogenic sources. Usually, urban and suburban zones are VOC-limited due to higher levels of NO_x emissions, while less-populated zones (rural sites) are NO_x-limited [8,9]. Domínguez-López et al. [10] analysed the effect of NO_x (NO_2 and NO) concentrations in surface ozone at different locations through a spatial and temporal variation study. An opposite daily variance was observed between O_3, NO, and NO_2 concentrations in urban and suburban areas. Maximum ozone concentrations were observed in the early afternoon, while NO_x concentrations usually achieve two peaks (early morning and late afternoon). In rural sites, no hourly peak O_3 or NO_x was observed. Sun et al. [11] showed that rural sites presented O_3 levels (approximately) 30 µg m^{-3} higher than urban regions in the study period of 1990–2019. Therefore, rural and remote areas are the object of many studies, as the highest ozone levels are found there [12,13]. Several factors (such as higher average temperatures and usually lower NO_x emissions) lead to a more significant accumulation of this pollutant in these areas. There is also evidence of the influence of other chemical air pollutants in O_3 tropospheric levels, such as anthropogenic aerosols, that can affect O_3 photolysis rates directly (through earth radiation scattering) and indirectly (due to the formation of clouds), leading to a weaker surface insolation [14,15]. Li et al. [9] found that the increase of O_3 concentrations (predicted due to the decrease in NO_x emissions) was in fact generated by the abrupt decrease in $PM_{2.5}$, resulting in a reduction of aerosols as a sink of HO_2, stimulating the production of this secondary pollutant. O_3 is known for its complex formation, depending on many variables. In addition to chemical precursors, meteorological variables are also considered in studying and predicting ozone levels. In different parts of the globe, Fang et al. [16] and Afonso and Pires [17] reached the same results through correlation analysis: ozone shows a positive correlation with temperature, meaning it increases with the increase of this meteorological parameter, and the opposite with relative humidity, that shows a negative correlation with O_3. Other variables that can lead to the formation or elimination of surface O_3 are air pressure, wind speed, and direction. The same authors also concluded that O_3 has a negative correlation with local air pressure and a positive correlation with wind speed. Being a pollutant resultant of a radiation-induced chemical reaction, O_3 shows higher levels with clear skies, which is also related to the presence of anticyclones. The association of O_3 concentrations with different weather conditions is more detailed in Domínguez-López et al. [10].

In Portugal, surface O_3 pollution has been characterised in many studies [17–21], especially in the northern [17,22–24] and rural regions [13,25,26]. To complete the study area with more recent data and improve O_3 trends' understanding in Portugal, this study aims to (i) determine the evolution of O_3 levels in Portuguese rural stations, focusing on exceedances to legally imposed levels for human and vegetation protection; (ii) compare the relationship between the O_3 concentrations and one of its precursors (NO_x) for different environments; and (iii) evaluate the effect of meteorological conditions in ozone concentrations. This study presents a long-term characterisation (data from more than 10 years) of O_3 concentrations from different environments, including regions in which no similar study was performed yet. In addition, the statistical models were applied to explanatory variables that were selected based on the knowledge of the chemical reactions of O_3 production. This integration enables better performance in predicting O_3 concentrations, allowing, in advance, to prevent and/or mitigate possible unfavourable outcomes.

2. Materials and Methods

2.1. Ozone Concentration Analysis at Rural Stations

For a better understanding of the trends and levels of surface ozone at rural sites, the study period 2009 to 2019 was selected. Table 1 represents the geographical information of the 15 existing rural stations [27]. The minimum distance to the seashore was estimated using a tool available on Google Earth that allows measuring the distance between two points. Figure S1 presents the map with the geographical distribution of the rural sites.

The statistical analyses were performed only for O_3 concentrations recorded at stations with a monitoring efficiency higher than 75%. The temporal trend of O_3 concentrations was assessed by determining the annual average concentration for each station. The exceedances to all thresholds presented in the current legislation were determined: (i) alert (240 µm m^{-3}) and information (180 µm m^{-3}) threshold, (ii) target value for human protection (120 µm m^{-3}), and (iii) AOT_{40}, representative of vegetation protection, as a target value (8000 µg m^{-3} h) and a long-term goal (6000 µg m^{-3} h) [28]. The data collected at four rural sites from different regions in Portugal were then used to characterise the spatial variability of O_3 concentrations. Histograms with density functions and violin plots were plotted in Python.

Table 1. Geographical coordinates, altitude, and minimum distance to the seashore of the existing rural sites.

Monitoring Stations		Geographical Coordinates	Altitude (m)	Minimum Distance to the Seashore (km)
CH	Chamusca	39°21′15″ N 08°28′03″ W	143	58.5
CR	Cerro	37°18′45″ N 07°40′43″ W	300	21.1
DN	Douro Norte	41°22′17″ N 07°47′27″ W	1086	78.4
ER	Ervedeira	39°55′28″ N 08°53′34″ W	68	4.5
FA	Faial	38°36′18″ N 28°37′53″ W	310	0.7
FD	Fundão	40°13′59″ N 07°17′58″ W	461	133.6
FM	Fornelo do Monte	40°38′39″ N 08°06′00″ W	731	53.4
FP	Fernando Pó	38°38′14″ N 08°41′30″ W	57	25.2
LR	Lourinhã	39°16′48″ N 09°14′04″ W	143	7.6
ML	Minho-Lima	41°48′08″ N 08°41′38″ W	777	14.2
MOV	Montemor-o-Velho	40°12′08″ N 08°40′08″ W	14	17.9
MV	Monte Velho	38°04′37″ N 08°47′55″ W	53	1.2
SN	Sonega	37°52′16″ N 08°43′26″ W	235	1.4
ST	Santana	32°48′28″ N 16°53′11″ W	0	125.1
TR	Terena	38°36′54″ N 07°23′51″ W	187	125.1

Cluster analysis (CA) was applied to the rural stations to evaluate the representativeness of the stations with local O_3 measurement. This analysis aims to group monitoring sites in the same class/cluster according to the observed behaviour (daily concentration fluctuation) of collected O_3 data. A hierarchical clustering method was used, generating solutions with 1 to n clusters. Ward's minimum variance method was used to determine the cluster distance. A dendrogram (or tree diagram—graphical representation of hierarchical CA) was determined for each year. Based on the obtained dendrograms, a matrix of relative frequencies was used to pair each monitoring site in clusters, using measured O_3 concentrations. This matrix enables the definition of the number of different O_3 behaviours in the studied remote areas. Daily profiles of O_3 concentrations were determined for each group of monitoring sites and the daily distribution of the daily maximum values using Microsoft Excel Macros developed by the authors.

2.2. O_3 and NO_x Relationship

For an evaluation of the relationship between O_3 and its chemical precursor (nitrogen oxides, NO, and NO_2), urban (Ermesinde-Valongo, EV), suburban (Mindelo-Vila do Conde, MD), and rural (Douro Norte, DN) stations were selected (see Figure 1). Considering data availability (O_3 and its chemical precursors monitoring efficiency close or higher than 75% for the selected monitoring stations), the studied period was 2019. A correlation matrix (Pearson's Correlation—PC) was obtained with a O_3, NO_2, NO, and NO_2/NO ratio for each monitoring station. Multiple linear regression (MLR) was applied to better understand the NO_x influence in O_3 production. The method aims to develop a linear model to predict the output variable (Y) with several predictor ones (X_i), having each one a regression coefficient (b_i, $i = 1, n$) with b_0 being the Y-intercept, as shown in Equation (1).

$$Y = b_0 + \sum b_i X_i \tag{1}$$

The statistical significance of correlation coefficients and MLR parameters was evaluated by applying a t-test with a significance level of 5%.

Figure 1. Geographical distribution of O_3 monitoring stations EV (urban), MD (suburban), and DN (rural).

2.3. Meteorological Data and O_3 Relationship

A similar analysis was performed to assess the effect of meteorological conditions on O_3 concentrations using data collected at Pedras Rubras/Aeródromo EMA (Automatic Meteorological Station) and the O_3 of Meco-Perafita (MP) and VNTelha-Maia (VNT), which are both suburban stations (Figure 2). The data evaluated corresponds to 2015 (the year in which selected monitoring stations presented a monitoring efficiency close to or higher than 75%). The meteorological data evaluated are mean air pressure (P, in hPa), mean air temperature (T, in °C), relative humidity (RH, in %), wind direction (WD, in degrees), wind speed (WS, in m/s), and solar radiation (SR, in kJ/m^2). The data was provided by the national meteorological, seismic, and oceanographic organisation IPMA (Instituto Português do Mar e da Atmosfera, website: https://www.ipma.pt/en/index.html, accessed on 15 March 2021).

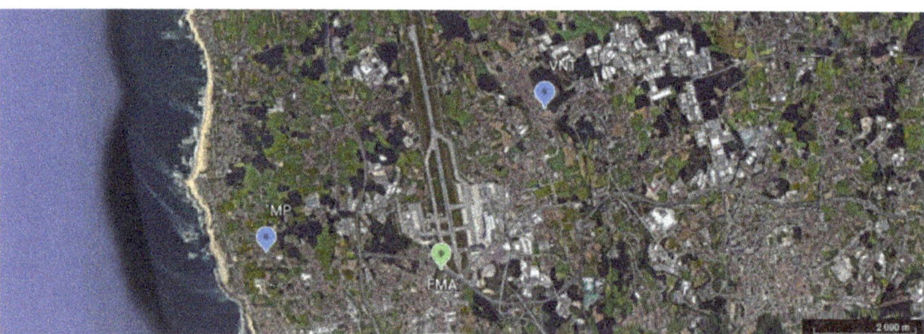

Figure 2. Geographical distribution of the meteorological station (EMA) and O_3 monitoring stations, MP and VNT.

The PC and MLR methods were applied to the meteorological parameters (except for wind data) and O_3 concentrations for annual quarters. A wind rose was developed to visualise the wind speed and direction registered, resorting to a MATLAB-coded program. To evaluate the wind speed and direction influence on ozone, violin plots were used to represent the frequency of O_3 concentration in each wind direction and speed interval considered.

3. Results and Discussion

3.1. Surface Ozone at Rural Stations

Table 2 presents the monitoring efficiency of each station during the study period. Figure 3 shows the annual average O_3 concentrations at each monitoring station in the studied period. The highest annual average O_3 concentrations were achieved at the DN, CR, SN, FA, FM, and ML sites. The highest values were observed for almost all stations during the period of 2001–2008 [13], except for the SN station, which shows a significant variance in mean values during the study period (Figure 3b). O_3 concentrations have been decreasing at most stations with rates between -0.12 and -1.35 µg m^{-3} year^{-1}. This behaviour has already been reported at rural Portuguese sites by Pires et al. [13] between 2001 and 2008. The exceptions to this trend are stations ML and SN, with high increase rates of 5.04 and 5.06 µg m^{-3} year^{-1}, respectively, and those with lower increase rates, the MOV (1.26 µg m^{-3} year^{-1}) and FA (0.06 µg m^{-3} year^{-1}) stations. Another study reported similar results for the MOV, FM, and ER stations between 2009 and 2011 [29]. Borrego et al. [22] detected a slight but significant decrease in ozone levels since 2006 in Portuguese rural stations. Yan et al. [30] reported decreases in rural European monitoring stations between 1995 and 2014. In the shorter period of 2000 to 2014, the decrease of O_3 concentrations was also registered by Chang et al. [31] and Proietti et al. [32] in Europe. The latter study reported a more significant decrease rate (-0.22 µg m^{-3} year^{-1}) in Mediterranean Europe.

Table 2. Air quality monitoring efficiency (in percentage) for rural stations.

	CH	CR	DN	ER	FA	FD	FM	FP	LR	ML	MOV	MV	SN	ST	TR
2009	94	0	84	96	44	99	95	99	94	89	94	0	99	0	97
2010	100	19	83	99	89	96	97	98	0	56	99	0	99	0	93
2011	100	40	86	100	98	99	94	96	98	67	100	0	100	0	92
2012	99	99	96	95	91	100	99	99	98	100	99	0	86	0	100
2013	99	87	94	90	98	100	98	93	89	97	92	0	88	0	100
2014	93	88	82	99	94	99	96	91	94	30	99	0	100	0	100
2015	99	97	27	99	99	84	90	100	98	12	99	96	57	0	94
2016	98	89	25	97	93	98	98	99	96	18	68	62	99	99	99
2017	98	88	54	99	100	99	74	98	98	0	100	0	93	100	99
2018	97	94	98	52	99	97	83	98	100	59	78	50	0	100	92
2019	91	95	90	0	100	92	94	100	94	59	0	0	0	100	91

Table 3 shows the exceedances to EU standard values for human protection (alert threshold, information threshold, and target value) during the study period. The alert threshold (240 µg m^{-3}) was exceeded in the DN, FM, and MOV stations. Apart from the TR, CR, ST, and FA stations, the information threshold (180 µg m^{-3}) was exceeded in all others. The highest number of exceedances was determined for the DN station. It is also the station with the highest number of exceedances in the period under study, followed by the FM station. The highest number of exceedances for both standard limits was determined in 2010. Regarding the target value for human health protection (120 µg m^{-3}), DN, CH, and FM were the three stations that recorded the most exceedances (in the respective order). In the Portuguese islands, exceedances to this limit were also observed. However, in ER, LR, SN, CR, TR, ST, and FA stations, exceedances were below 25 on average over 3 years, meaning that their O_3 concentrations were within the legislation requirements. However, none of the stations complied with the long-term objective imposed for human health protection.

Four stations were selected to deeply analyse the O_3 behaviour in different regions of Portugal (representing North, Centre, and South Regions and Islands). A histogram with frequency distribution for 2012 and violin plots representing monthly ozone concentrations are displayed for each selected station in Figure S3. The station located at a higher altitude (DN) presented the highest concentration levels, with more elevated minimum levels. The inner located station (FD) showed significant variation in the ozone concentration, with a maximum value not going far beyond 150 µg m^{-3} unlike DN, whose peak reached 250 µg m^{-3}. The other inland station, located further south (FP), showed similar behaviour to the FD station but reached a higher peak ozone level. All inland stations show peak values in summer, but higher mean concentrations in spring. This characteristic is known in northern hemisphere remote locations, where the peak concentrations can be related to a higher stratosphere-troposphere exchange or be enhanced by increased solar radiation after winter months with accumulated NO_x and VOC levels [33,34]. In the station at sea level (FA), the lowest ozone concentrations (not surpassing 120 µg m^{-3}) were observed. In all the mentioned stations, the maximum daily concentration occurs at 15 h–16 h, except for the FA station, which shows an almost non-existing daily variation of the concentration. A study conducted in southern Spain also presented that the higher O_3 values appear in high-altitude locations, usually closer to the sea [35]. DN station was the focus of other studies due to register the majority òf Portuguese legislated threshold exceedances. Carvalho et al. [25] attribute DN's high levels to long-range transport of pollutants from NE winds, and Borrego et al. [22] showed that background values contribute more than 50% to the local O_3 concentrations.

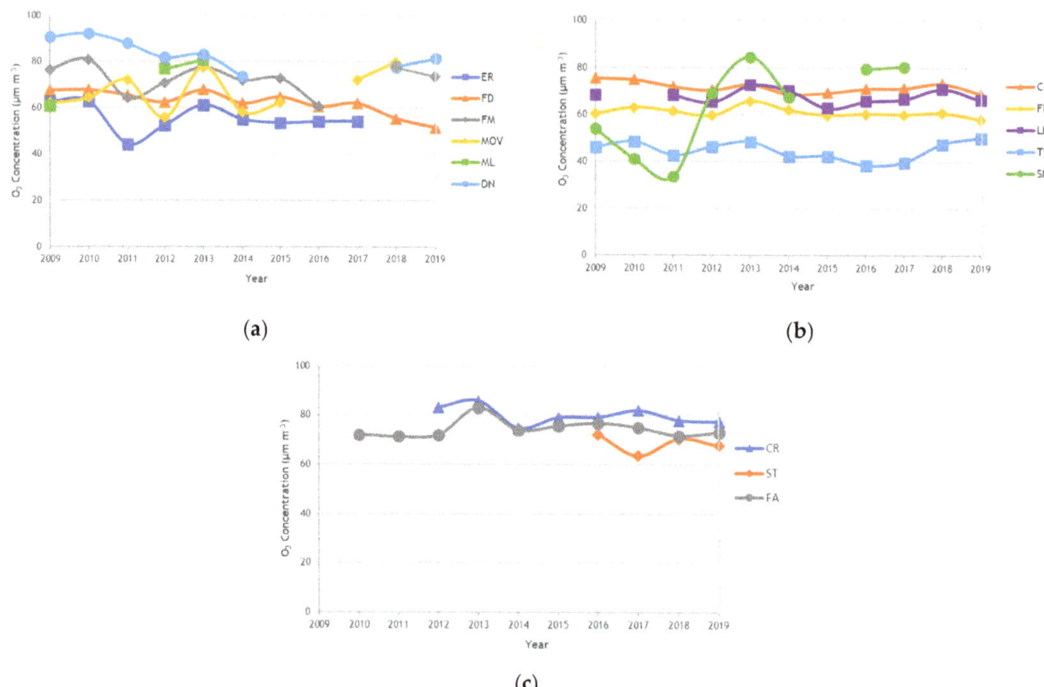

Figure 3. O_3 annual average concentration in each of Portugal's most important regions: (**a**) Regions of North and Centre; (**b**) Regions of Lisbon and Tejo Valley and Alentejo; (**c**) Regions of Algarve and Islands.

Table 3. Exceedances to O_3 alert threshold (AT), information threshold (IT), and target value (TV) for human health protection.

		CH	CR	DN	ER	FA	FD	FM	FP	LR	ML	MOV	SN	ST	TR
	AT	0	-	3	0	-	0	0	0	0	0	0	0	-	0
2009	IT	3	-	37	9	-	0	11	4	0	0	1	0	-	0
	TV	57	-	76	21	-	33	48	32	19	0	24	0	-	0
	AT	0	-	4	0	0	0	1	0	-	-	0	0	-	0
2010	IT	16	-	76	8	0	3	36	1	-	-	9	0	-	0
	TV	55	-	66	29	2	31	66	37	-	-	26	0	-	0

Table 3. *Cont.*

		CH	CR	DN	ER	FA	FD	FM	FP	LR	ML	MOV	SN	ST	TR
2011	AT	0	-	0	0	0	0	0	0	0	-	0	0	-	0
	IT	4	-	30	0	0	0	1	7	0	-	3	0	-	0
	TV	38	-	67	0	0	15	2	27	19	-	53	0	-	0
2012	AT	0	0	0	0	0	0	0	0	0	0	0	0	-	0
	IT	4	0	16	0	0	0	5	0	5	0	0	0	-	0
	TV	36	30	31	7	0	11	22	17	15	28	7	6	-	0
2013	AT	0	0	0	0	0	0	0	0	0	0	0	0	-	0
	IT	2	0	18	0	0	0	8	0	5	1	5	2	-	0
	TV	50	29	37	23	0	22	45	34	31	35	52	53	-	0
2014	AT	0	0	0	0	0	0	0	0	0	-	0	0	-	0
	IT	0	0	0	0	0	0	0	0	0	-	0	0	-	0
	TV	15	5	10	3	0	8	16	12	13	-	3	9	-	0
2015	AT	0	0	-	0	0	0	0	0	0	-	0	-	-	0
	IT	0	0	-	0	0	0	0	0	0	-	0	-	-	0
	TV	24	20	-	3	0	14	20	23	4	-	111	-	-	0
2016	AT	0	0	-	0	0	0	0	0	0	-	-	0	0	0
	IT	6	0	-	3	0	1	1	8	1	-	-	6	0	0
	TV	38	10	-	9	0	9	9	23	12	-	-	50	0	0
2017	AT	0	0	-	0	0	0	-	0	0	-	0	0	0	0
	IT	3	0	-	3	0	0	-	2	0	-	0	10	0	0
	TV	35	11	-	4	1	10	-	15	10	-	30	55	0	0
2018	AT	0	0	0	-	0	0	0	0	0	-	1	-	0	0
	IT	1	0	3	-	0	0	0	0	2	-	3	-	0	0
	TV	26	10	17	-	0	0	20	12	8	-	31	-	0	0
2019	AT	0	0	0	-	0	0	1	0	0	-	-	-	0	0
	IT	0	0	5	-	0	0	5	1	0	-	-	-	0	0
	TV	12	5	36	-	0	0	24	12	8	-	-	-	2	1

The EU legislation presents two different values (AOT_{40}) for the long-term goal and the target value for short-term exposition impact estimations for vegetation protection. Figure 4 presents the calculated AOT_{40} values at each monitoring site, showing the exceedances to these two parameters. Both legislated values were not surpassed at the TR, ST, and FA stations (the last two are in the Portuguese islands).

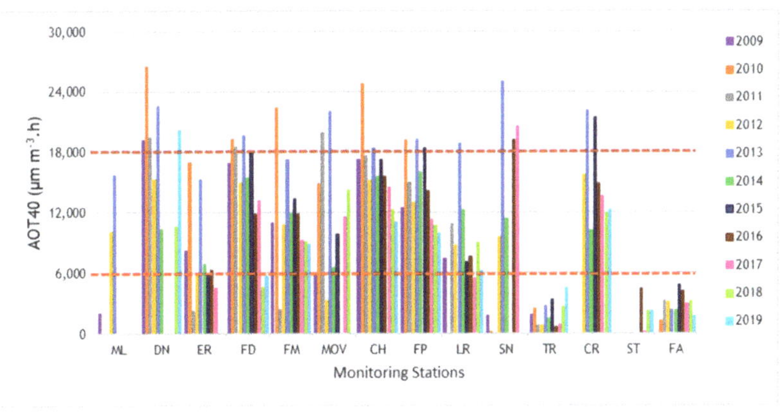

Figure 4. AOT_{40} levels in each station in the study period.

CA was performed to group monitoring stations according to O_3 data for each year. Figure 5 shows, as an example, the dendrogram obtained with data from 2012, forming three groups of stations (dendrograms achieved with the data collected in the remaining years are present in the Supplementary Material. The distribution of stations by these selected clusters was considered for a relationship analysis (Table S1). The frequency with which each pair of stations was grouped in the same class was determined, considering the minimum number of occurrences of stations, i.e., the minimum number of years in which the station is considered. For the final cluster formation, a minimum of 70% of correspondence between stations was considered, resulting in (i) FD, FP, ER, and TR; (ii) DN and FM; (iii) FA and ST; (iv) MOV and LR; (v) ML; (vi) CH; and (vii) SN and (viii) CR. The high number of clusters and the fact that half have only one station assigned are indicative of their importance. Pires et al. [13] also performed CA on Portuguese rural sites, obtaining similar grouping results.

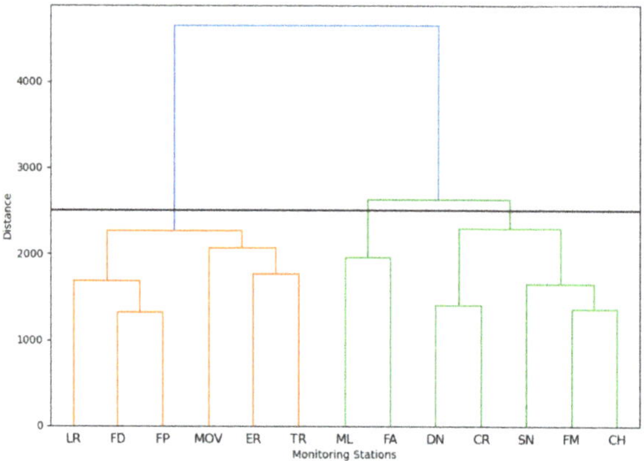

Figure 5. Dendrogram obtained from cluster analysis for 2012.

The daily average ozone concentration profiles (in $\mu m\ m^{-3}$) and the relative frequency (in percentage) of its maximum for the first four clusters are shown in Figure 6. The first cluster shows stations with a higher variation in the O_3 concentration during the day, presenting a decrease in the early morning followed by an increase in the afternoon, with most having their peak at 15 h, as shown in the maximum concentrations profile. In cluster 2, O_3 concentrations have similar behaviours, increasing between 15 h and 16 h. The O_3 concentrations show higher values (and reduced amplitude of values) than those observed in the previous cluster. The stations representing the Portuguese Islands are grouped in cluster 3. They show the lowest concentrations and the shortest variation between the daily minimum and maximum. In the final cluster, the ozone profile in both stations is closer to cluster 1; however, the daily maximum and minimum are lower. Except for stations in cluster 3, a similar daily profile of O_3 concentration was observed in all monitoring stations, the lowest occurring in the morning and the highest in the early afternoon. In FA and ST stations, the peak occurs at night. This event results from the horizontal and vertical transport of ozone and its precursors [13]. Studies of nocturnal ozone concentration increase were developed in the Portuguese continent to explain unexpected ozone levels [21,36]. The ozone enhancement events can be related to transport processes and the related usual meteorological conditions due to the absence of photochemical production. Additionally, vertical mixing was registered in the boundary layer in winter, contributing to surface nocturnal ozone peaks.

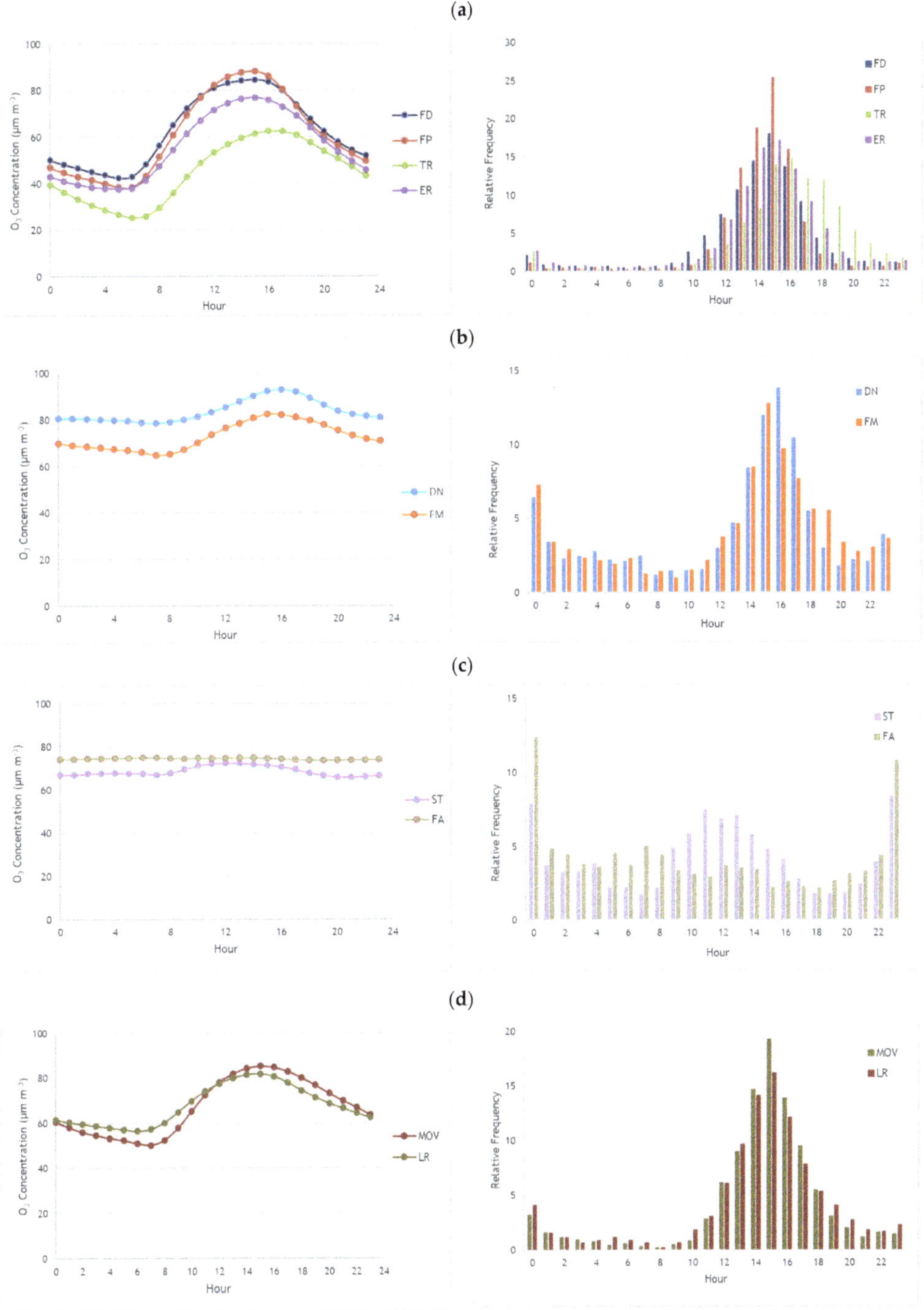

Figure 6. Daily average O_3 concentration profile (on the left) and daily evolution of the relative frequency (in percentage) of the daily maximum O_3 concentration for the analysed period (on the right), for cluster 1 (**a**), cluster 2 (**b**), cluster 3 (**c**) and cluster 4 (**d**).

3.2. Surface Ozone and NO$_x$ Relation

In suburban and urban areas, surface O$_3$ concentrations are lower than at rural sites. Figure 7 presents O$_3$, NO$_2$, and NO daily average concentrations for the EV, MD, and DN stations in different months. High O$_3$ concentrations in the DN station can be observed compared with the other two stations. For all stations, high concentrations were observed in the afternoon. As for NO$_2$ and NO, the concentrations were expectably higher in the suburban and urban areas due to more intensive anthropogenic emission rates. High values observed in the first hours of the day (especially in the NO$_2$ plot) can be explained by the lack of ozone formation during this period. As expected, the anthropogenic influence in rural stations is weaker, shown by lower levels of NO$_x$ in the DN station (Figure 7h,i). In general, an almost-symmetric daily profile is shown between O$_3$ and NO$_x$: in the first hours of the day, NO$_x$ levels are low due to low emissions; in the morning, with the beginning of anthropogenic activities (specifically traffic emissions), there is an increase in NO and NO$_2$ emissions followed by a decrease in ozone levels due to titration with NO; and during the afternoon, when there is a greater incidence of solar radiation, the NO$_2$ emitted and formed through the NO present in the air will lead to ozone formation. In the nighttime, ozone levels usually decrease due to the lack of solar radiation. In these conditions, NO$_2$ photolysis does not occur, and the existing NO can be oxidised back to NO$_2$ with the reaction with O$_3$.

Ferreira et al. [37] presented the same connection between O$_3$ and NO$_2$ levels in the Lisbon region, registering a higher ozone concentration in the periphery of the urban centre. In the southwest of the Iberian Peninsula, a study of O$_3$, NO, and NO$_2$ trends was developed at rural, urban, suburban, and industrial sites by Domínguez-López et al. [10]. According to their reports, most rural sites presented a low and constant NO$_x$ level. In suburban and urban stations, besides the morning NO$_x$ peak, there is a second in the evening due to a decrease in solar activity and an increase in traffic. In the same reference, an analysis of monthly variations showed similar results to the ones shown in Figure 7, with higher ozone levels in the spring and summer for all stations and NO$_x$ levels higher in the autumn and winter for urban and suburban sites, while NO$_2$ and NO at steady levels during the year were observed at rural sites. In Spain, urban/suburban sites also registered higher NO and NO$_2$ concentrations and O$_3$ higher levels in rural/remote regions [35], as shown in a study conducted in the UK too [38].

Considering the ozone concentrations at a chemical equilibrium state, that is, a relation to the NO$_2$/NO ratio (Equation (2), being K the equilibrium constant),

$$[O_3] = [NO_2] \cdot [O_2] / [NO] \cdot K, \tag{2}$$

Thus, the influence of the NO$_2$/NO ratio in the O$_3$ concentration was also included in the PC and MLR analysis. A correlation matrix was obtained (Tables 4–6) to demonstrate the relationship between ozone and its considered precursors in 2019.

Table 4. Pearson's correlation and Multiple Linear Regression results for EV station.

PC	O$_3$	NO$_2$	NO	NO$_2$/NO
O$_3$	1			
NO$_2$	−0.611	1		
NO	−0.565	0.846	1	
NO$_2$/NO	0.353	−0.237	−0.583	1
MLR	b_0	b_1	b_2	b_3
	46.5	−19.8	7.13	8.57

Table 5. Pearson's correlation and Multiple Linear Regression results for MD station.

PC	O_3	NO_2	NO	NO_2/NO
O_3	1			
NO_2	−0.120	1		
NO	−0.188	0.441	1	
NO_2/NO	0.054	0.444	−0.502	1
MLR	b_0	b_1	b_2	b_3
	82.2	-	−4.18	−1.05

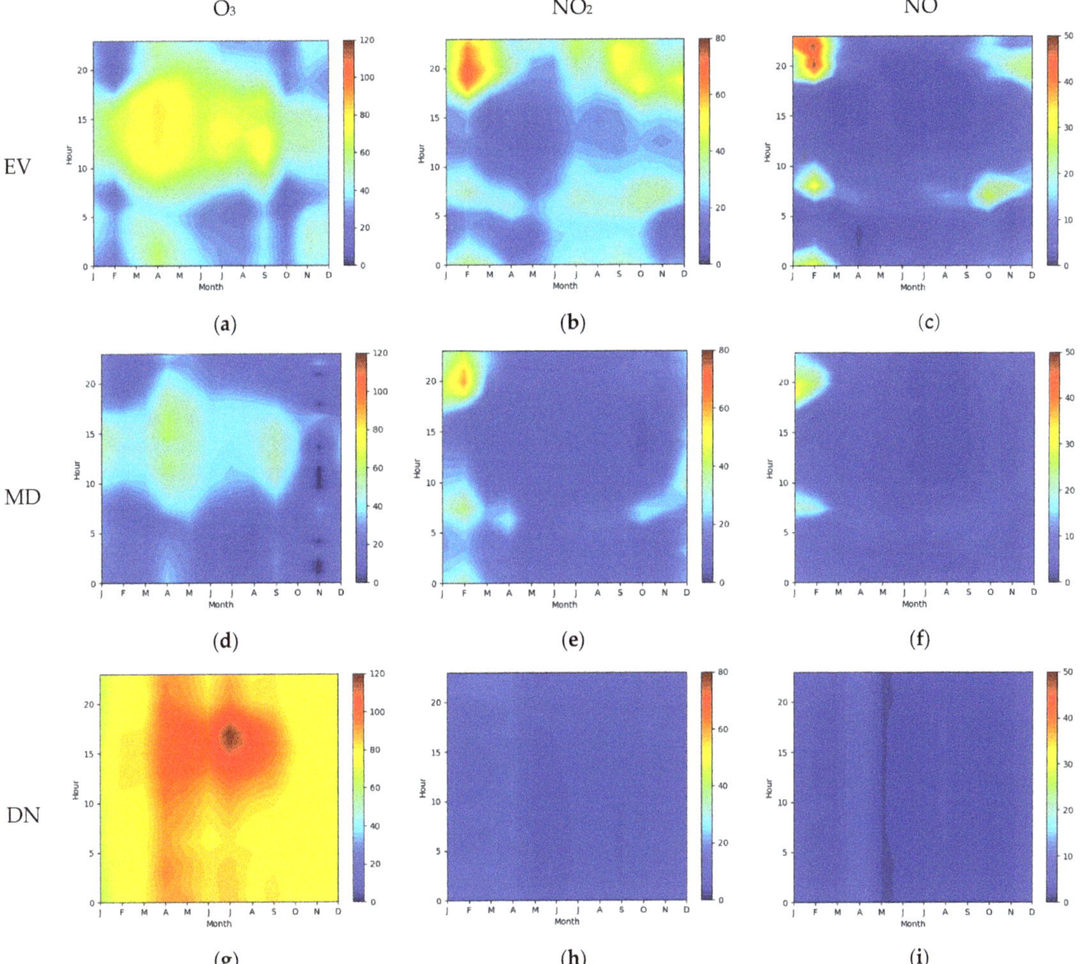

Figure 7. Contour plot of O_3 (**a**,**d**,**g**), NO_2 (**b**,**e**,**h**), and NO (**c**,**f**,**i**) concentrations (in µg m^{-3}) at EV (urban), MD (suburban), and DN (rural) stations.

Table 6. Pearson's correlation and Multiple Linear Regression results for DN station.

PC	O_3	NO_2	NO	NO_2/NO
O_3	1			
NO_2	−0.460	1		
NO	−0.286	0.609	1	
NO_2/NO	−0.075	0.206	−0.213	1
MLR	b_0	b_1	b_2	b_3
	28.8	−8.38	-	0.38

Considering the strength of variables association (see Table S2), Table 4 presents a significant negative correlation between O_3, NO_2, and NO and a medium positive association with NO_2/NO in the urban station (EV), which is what was expected for this type of environment. These results confirm the previous logic of the anti-correlation of ozone and its precursor's daily profile, particularly noticed in the urban sites. Regarding the combined effect of the selected environmental variables, MLR identified a negative relation between NO_2 and O_3 concentrations and a positive with NO, trends also known in other studies in the literature [39].

In the suburban sites (Table 5), a weaker relationship between O_3 and NO_x is expected than in the urban site (Figure 7b). The PC coefficients for O_3 and its precursors obtained are lower than those for the EV station. The NO shows a small correlation with O_3, and the NO_2/NO ratio shows no correlation (coefficient value close to 0). That result is also shown in the smaller b_3 value of MLR, compared to the previous value. The NO_2 shows a small negative association with O_3 and does not show statistical significance to its formation, according to the MLR analysis (b_1 was not considered statistically significant). Table 6 shows the PC and MLR results for the DN station. As is expected, the rural site shows a low (NO_2 and NO) or no relation (NO_2/NO ratio) between O_3 and its chemical precursors. According to the MLR results, NO_2 levels do not influence ozone formation at this site. Contrary to the suburban station, the NO concentration and NO_2/NO ratio have a small negative effect on O_3 levels.

The strength of the association varies with season [40]. The correlation analysis in each season showed negative correlations of O_3 with NO and NO_2 (higher correlation with NO_2 in spring and summer and with NO in the autumn and winter) [41]. This difference can be explained by the interference of meteorological parameters and distinct weather types that affect the pollutants' transport and mixing in the atmosphere.

3.3. Surface Ozone and Meteorological Parameters Relation

The permanence of pollutants in the atmosphere is not only affected by chemical reactions among themselves but also by meteorological conditions, whether at a micro or macro scale. Studying the weather conditions when peak pollutant levels are registered is important to prevent negative effects on humans and the ecosystem. The ozone formation is enhanced by solar radiation and other meteorological parameters. Since these have different levels throughout the year, a quarterly analysis was performed applying PC and MLR to O_3 and the meteorological parameters considered (Tables 7 and 8). The stations under analysis are suburban-type, located in sites with relative anthropogenic influence (e.g., the city airport-related activities). In the MP station, the first quarter presented a medium- (between 0.3 and 0.5) positive correlation with temperature (T) and solar radiation (SR) and a small negative association with relative humidity (RH), meaning that the increase in RH leads to a decrease in ozone (disregarding the results not statistically significant). The MLR, on the contrary, points to a small influence of pressure (P) in the ozone concentration, and relative humidity does not play a role in ozone formation/elimination. A small negative correlation of P and O_3 is pointed to in the second quarter, agreeing with the b_1 value for that period. T and SR show a continuous significant correlation (also seen in the 3rd and 4th quarters). The RH shows stronger negative correlations in the spring and summer seasons. According to MLR results, the SR has a greater impact in the April–May–June period than the October–November–December period. RH shows a significant negative relation with ozone in the 3rd quarter, and, in the same period, T does not influence O_3 levels.

Table 7. Pearson's correlation and Multiple Linear Regression results for MP station.

	PC	**P**	**T**		**RH**	**SR**
	1st quarter	0.029 *	0.414		−0.177	0.436
O_3	2nd quarter	−0.102	0.426		−0.402	0.472
	3rd quarter	0.012 *	0.441		−0.500	0.465
	4th quarter	−0.412	0.455		−0.288	0.395
	MLR	b_0	b_1	b_2	b_3	b_4
	1ST QUARTER	52.1	1.25	4.16	-	4.78
	2ND QUARTER	60.1	−3.54	2.12	−3.91	8.15
	3RD QUARTER	50.7	−2.11	-	−8.41	5.44
	4TH QUARTER	36.5	−8.61	2.57	−1.62	7.73

Note: the values with (*) are not statistically significant; $b_0 \rightarrow$ Y-intercept; $b_1 \rightarrow$ regression parameter for pressure; $b_2 \rightarrow$ regression parameter for temperature; $b_3 \rightarrow$ regression parameter for relative humidity; and $b_4 \rightarrow$ regression parameter for solar radiation.

Table 8. Pearson's correlation and Multiple Linear Regression results for VNT station.

	PC	**P**	**T**		**RH**	**SR**
	1st quarter	−0.193	0.435		−0.317	0.484
O_3	2nd quarter	−0.129	0.501		−0.469	0.523
	3rd quarter	0.037 *	0.513		−0.577	0.520
	4th quarter	−0.544	0.542		−0.306	0.407
	MLR	b_0	b_1	b_2	b_3	b_4
	1st quarter	52.5	−3.30	3.32	-	8.02
	2nd quarter	57.9	−4.21	3.04	−4.70	8.56
	3rd quarter	47.8	−1.48	1.62	−8.77	4.23
	4th quarter	50.9	−3.11	3.04	-	6.55

Note: the values with (*) are not statistically significant; $b_0 \rightarrow$ Y-intercept; $b_1 \rightarrow$ regression parameter for pressure; $b_2 \rightarrow$ regression parameter for temperature; $b_3 \rightarrow$ regression parameter for relative humidity; $b_4 \rightarrow$ regression parameter for solar radiation.

In the VNT station, P presents a higher correlation with ozone in the 4th quarter. The MLR results also show a negative influence of P on ozone levels. In this station, the correlation of T and O_3 presents stronger values, and, except in the 3rd quarter, there is a similar influence level of T in O_3 concentrations. The RH presents a high PC coefficient, but the MLR analysis shows only this parameter influence in the 2nd and 3rd quarters. As for SR, there is a positive correlation with high coefficient values. Ferreira et al. [37] studied the relationship between ozone and some parameters in Lisbon city. Accordingly, daily sea level medium pressure and medium relative humidity negatively influence ozone production in all seasons. As for temperature, the daily medium value only shows a positive (and weak) correlation in the winter months. Pires et al. [24] showed a weak positive correlation of T and a slight negative association of SR and RH with O_3 in a station located in Northern Portugal. The discrepancies between the two applied statistical methods expose the complex task of predicting ozone levels, although, in general, studies applying relationship methods show a positive correlation of O_3 with SR and T and a negative correlation with RH [39,41].

O_3 concentrations can also be influenced by transport and mixing phenomena. Thus, an analyse relating wind direction and speed with O_3 was implemented. Figure S4 presents a wind rose with the wind speed and direction distribution for the year 2015. The most prominent winds were from NW (stronger winds) and E (calmer winds) related to the air masses from the Atlantic Ocean and Spain that the northern Portuguese region is subjected to. On a macro scale (synoptic weather), it is essential to mention the influence of the Azores anticyclone in Europe [42–44], especially in the Mediterranean [25,35,45–48]. This type of weather system is related to a higher T and low RH and cloud cover (consequently contributing to higher SR activity), enhancing O_3 production. The relatively low wind activity provides the accumulation of this pollutant in the atmosphere.

A more significant contribution of air masses from the Atlantic Ocean for ozone higher concentrations is observed (median values represented in the violin plots, Figure 8). The peak O_3 values are registered for the N-NE and W-NW directions in the MP station. Located closer to shore, the higher concentrations are registered when the wind has lower speed values (more calm weather), which allows the accumulation of ozone. In the VNT station, the peak levels are reached under the North Atlantic air masses influence (NW-N). They are related to faster wind activity, possibly meaning that the high ozone concentrations in this site are associated with the circulation of polluted air masses. Santurtún et al. [48] related ozone trends to weather types in Spain. Accordingly, the stations studied were influenced by the anticyclone system and east and northeast flow (corresponding to the information in Figure S4). Knowing the comparable importance of both chemical and meteorological precursors to ozone formation, Li et al. [49] studied the increase of O_3 pollution in China, concluding that temperature is the meteorological parameter with more influence, but that it is related to anticyclonic conditions. Additionally, the authors pointed to the decrease in $PM_{2.5}$ and the unmitigated emissions of VOCs from anthropogenic sources.

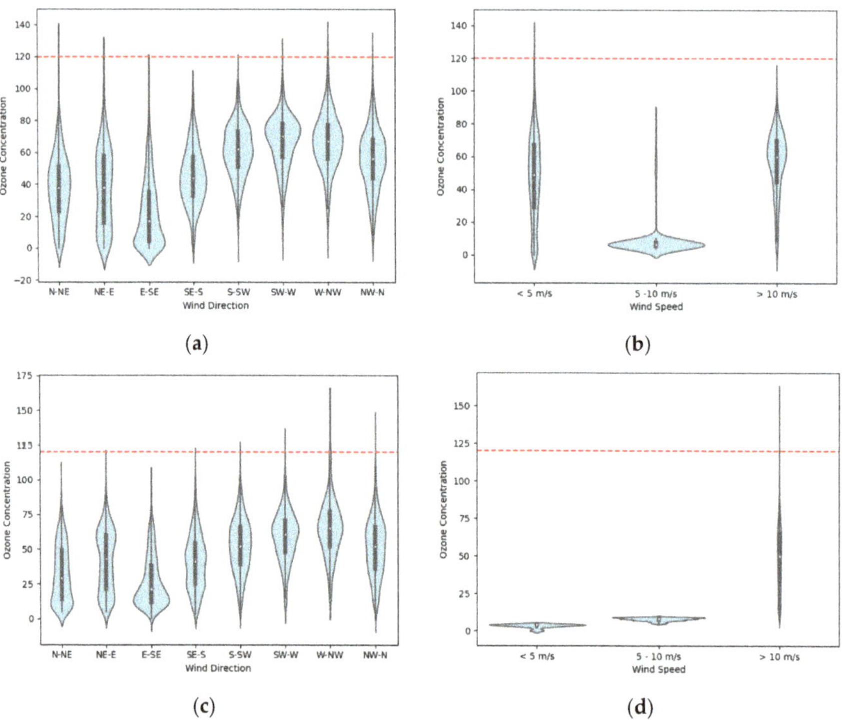

Figure 8. Relationship between wind direction and speed with ozone levels (in $\mu m\ m^{-3}$) at (**a**,**b**) MP and (**c**,**d**) VNT stations.

4. Conclusions

Annual average O_3 concentrations presented a decreasing trend during the analysed period. However, exceedances to EU legislated values for human health protection were still observed. CA identified several O_3 concentration patterns, showing the effectiveness of the current geographical distribution of the air quality monitoring stations. Negative correlations were determined between NO, NO_2, and O_3 concentrations in urban and suburban stations. The correlation with the NO_2/NO ratio only showed significance in the urban site. As for MLR results, in general, the stronger influence of NO_x levels was expected in the urban station. The O_3 concentrations in both suburban sites showed a strong positive correlation between O_3, T, and SR, a strong negative correlation with RH, and a weaker negative association with pressure. The meteorological parameter that shows a higher contribution in O_3 concentration was solar radiation, showing a stronger influence in the 2nd annual quarter, where the spring high average concentrations are registered. The wind direction distribution showed that the location was under the flow of NW and E-SE winds, related to the Azores anticyclone, and air masses from the Iberian Peninsula, inland.

Supplementary Materials: The following supporting information can be downloaded at: https://www.mdpi.com/article/10.3390/su14042383/s1, Figure S1: Map with the geographical distribution of Portugal's rural monitoring stations; Figure S2A: Dendrograms for years (a) 2009, (b) 2010, (c) 2011, (d) 2013, (e) 2014 and (f) 2015; Figure S2B: Dendrograms for years (a) 2016, (b) 2017, (c) 2018 and (d) 2019; Figure S3: Annual profile of O_3 concentration ($\mu g \ m^{-3}$) and monthly distribution for 2012 at different stations: (a) Douro-Norte; (b) Fundão; (c) Fernando Pó; (d) Faial; Figure S4: Wind rose with wind direction and speed distribution for the year 2015; Table S1: Matrix of relative frequencies depicting the relationship between the different stations; and Table S2: Guidelines for interpretation of PC coefficient values [50].

Author Contributions: Conceptualisation, J.C.M.P.; Methodology, R.C.V.S. and J.C.M.P.; Software, R.C.V.S. and J.C.M.P.; Validation, R.C.V.S. and J.C.M.P.; Formal Analysis, R.C.V.S. and J.C.M.P.; Investigation, R.C.V.S. and J.C.M.P.; Data Curation, R.C.V.S. and J.C.M.P.; Writing—Original Draft Preparation, R.C.V.S.; Writing—Review and Editing, R.C.V.S. and J.C.M.P.; Visualization, R.C.V.S. and J.C.M.P.; Supervision, J.C.M.P.; Funding Acquisition, J.C.M.P. All authors have read and agreed to the published version of the manuscript.

Funding: This work was financially supported by: LA/P/0045/2020 (ALiCE) and UIDB/00511/2020-UIDP/00511/2020 (LEPABE) funded by national funds through FCT/MCTES (PIDDAC). J.C.M.P. acknowledges the FCT Investigator 2015 Programme (IF/01341/2015).

Institutional Review Board Statement: Not applicable.

Informed Consent Statement: Not applicable.

Acknowledgments: The authors thank Instituto Português do Mar e da Atmosfera (IPMA) for providing the meteorological data.

Conflicts of Interest: The authors declare no conflict of interest.

References

1. Seinfeld, J.H.; Pandis, S.N. *Atmospheric Chemistry and Physics: From Air Pollution to Climate Change*; John Wiley & Sons, Inc.: Hoboken, NJ, USA, 1998.
2. Sousa, S.I.; Ferraz, M.A.; Pereira, M.C.; Martins, F.G. Avaliação das Concentrações Pré-industriais e Actuais de Ozono Superficial através de Séries Temporais. In Proceedings of the 8ª Conferência Nacional de Ambiente, Lisboa, Portugal, 27–29 October 2004.
3. Wallace, J.M.; Hobbs, P.V. *Atmospheric Science—An Introductory Survay*; Elsevier: Amsterdam, The Netherlands, 2006.
4. EEA. *Air quality in Europe—2020 Report*; EEA: Copenhagen, Denmark, 2020.
5. EEA. Exposure of Europe's Ecosystems to Ozone. Available online: https://www.eea.europa.eu/api/SITE/data-and-maps/indicators/exposure-of-ecosystems-to-acidification-15 (accessed on 4 June 2021).
6. Pleijel, H.; Broberg, M.C.; Uddling, J.; Mills, G. Current surface ozone concentrations significantly decrease wheat growth, yield and quality. *Sci. Total Environ.* **2018**, *613*, 687–692. [CrossRef] [PubMed]
7. Feng, Z.; De Marco, A.; Anav, A.; Gualtieri, M.; Sicard, P.; Tian, H.; Fornasier, F.; Tao, F.; Guo, A.; Paoletti, E. Economic losses due to ozone impacts on human health, forest productivity and crop yield across China. *Environ. Int.* **2019**, *131*, 104966. [CrossRef] [PubMed]

8. Finlayson-Pitts, B.J.; Pitts, J.N. Atmospheric chemistry of tropospheric ozone formation: Scientific and regulatory implications. *Air Waste* **1993**, *43*, 1091–1100. [CrossRef]
9. Li, K.; Jacob, D.; Liao, H.; Shen, L.; Zhang, Q.; Bates, K. Anthropogenic drivers of 2013–2017 trends in summer surface ozone in China. *Proc. Natl. Acad. Sci. USA* **2019**, *116*, 422–427. [CrossRef] [PubMed]
10. Domínguez-López, D.; Adame, J.A.; Hernández-Ceballos, M.A.; Vaca, F.; De La Morena, B.A.; Bolívar, J.P. Spatial and temporal variation of surface ozone, NO and NO_2 at urban, suburban, rural and industrial sites in the southwest of the Iberian Peninsula. *Environ. Monit. Assess.* **2014**, *186*, 5337–5351. [CrossRef] [PubMed]
11. Sun, H.; Shin, Y.; Xia, M.; Ke, S.; Yuan, L.; Guo, Y.; Archibald, A. Spatial Resolved Surface Ozone with Urban and Rural Differentiation during 1990–2019: A Space-Time Bayesian Neural Network Downscaler. *Environ. Sci. Technol.* **2021**, *5*, 167–174. [CrossRef] [PubMed]
12. Notario, A.; Díaz-de-Mera, Y.; Aranda, A.; Adame, J.A.; Parra, A.; Romero, E.; Parra, J.; Muñoz, F. Surface ozone comparison conducted in two rural areas in central-southern Spain. *Environ. Sci. Pollut. Res.* **2012**, *19*, 186–200. [CrossRef]
13. Pires, J.C.M.; Alvim-Ferraz, M.C.M.; Martins, F.G. Surface ozone behaviour at rural sites in Portugal. *Atmos. Res.* **2012**, *104–105*, 164–171. [CrossRef]
14. Charlson, R.; Schwartz, S.; Hales, J.; Cess, R.; Coakley, J.A., Jr.; Hansen, J.; Hofmann, D. Climate Forcing by Anthropogenic Aerosols. *Science* **1992**, *255*, 423–430. [CrossRef]
15. Lou, S.; Liao, H.; Zhu, B. Impacts of aerosols on surface-layer ozone concentrations in China through heterogeneous reactions and changes in photolysis rates. *Atmos. Environ.* **2014**, *85*, 123–138. [CrossRef]
16. Fang, C.; Wang, L.; Wang, J. Analysis of the spatial–temporal variation of the surface ozone concentration and its associated meteorological factors in Changchun. *Environments* **2019**, *6*, 46. [CrossRef]
17. Afonso, N.F.; Pires, J.C.M. Characterization of surface ozone behavior at different regimes. *Appl. Sci.* **2017**, *7*, 944. [CrossRef]
18. Barros, N.; Silva, M.P.; Fontes, T.; Manso, M.C.; Carvalho, A.C. Learning from 24 years of ozone data in Portugal. *WIT Trans. Ecol. Environ.* **2014**, *183*, 117–128. [CrossRef]
19. Fernández-Guisuraga, J.M.; Castro, A.; Alves, C.; Calvo, A.; Alonso-Blanco, E.; Blanco-Alegre, C.; Rocha, A.; Fraile, R. Nitrogen oxides and ozone in Portugal: Trends and ozone estimation in an urban and a rural site. *Environ. Sci. Pollut. Res.* **2016**, *23*, 17171–17182. [CrossRef] [PubMed]
20. Kulkarni, P.S.; Bortoli, D.; Domingues, A.; Silva, A.M. Surface ozone variability and trend over urban and suburban sites in Portugal. *Aerosol Air Qual. Res.* **2016**, *16*, 138–152. [CrossRef]
21. Kulkarni, P.S.; Bortoli, D.; Silva, A.M. Nocturnal surface ozone enhancement and trend over urban and suburban sites in Portugal. *Atmos. Environ.* **2013**, *71*, 251–259. [CrossRef]
22. Borrego, C.; Monteiro, A.; Martins, H.; Ferreira, J.; Fernandes, A.P.; Rafael, S.; Miranda, A.I.; Guevara, M.; Baldasano, J.M. Air quality plan for ozone: An urgent need for North Portugal. *Air Qual. Atmos. Health* **2016**, *9*, 447–460. [CrossRef]
23. Pires, J.C.M.; Gonçalves, B.; Azevedo, F.G.; Carneiro, A.P.; Rego, N.; Assembleia, A.J.B.; Lima, J.F.B.; Silva, P.A.; Alves, C.; Martins, F.G. Optimization of artificial neural network models through genetic algorithms for surface ozone concentration forecasting. *Environ. Sci. Pollut. Res.* **2012**, *19*, 3228–3234. [CrossRef]
24. Pires, J.C.M.; Martins, F.G.; Sousa, S.I.V.; Alvim-Ferraz, M.C.M.; Pereira, M.C. Selection and validation of parameters in multiple linear and principal components regression. *Environ. Model. Softw.* **2008**, *7*, 50–55. [CrossRef]
25. Carvalho, A.; Monteiro, A.; Ribeiro, I.; Tchepel, O.; Miranda, A.I.; Borrego, C.; Saavedra, S.; Souto, J.A.; Casares, J.J. High ozone levels in the northeast of Portugal: Analysis and characterization. *Atmos. Environ.* **2010**, *44*, 1020–1031. [CrossRef]
26. Monteiro, A.; Gouveia, S.; Scotto, M.G.; Lopes, J.; Gama, C.; Feliciano, M.; Miranda, A.I. Investigating ozone episodes in Portugal: A wavelet-based approach. *Air Qual. Atmos. Health* **2016**, *9*, 775–783. [CrossRef]
27. QualAR. Dados Estações. Available online: https://qualar.apambiente.pt/downloads (accessed on 8 January 2021).
28. Diário da República. *Decreto-Lei n° 102/2010*; Diário da República Electrónico: Lisbon, Portugal, 2010; pp. 4177–4205.
29. Afonso, P.A.F. *Concentrações de Ozono Superficial em Portugal: Avaliação dos Padrões Temporais e dos Contrastes Espaciais em Estações de Fundo*; Instituto Politécnico de Bragança: Bragança, Portugal, 2014.
30. Yan, Y.; Pozzer, A.; Ojha, N.; Lin, J.; Lelieveld, J. Analysis of European ozone trends in the period 1995. *Atmos. Chem. Phys.* **2018**, *18*, 5589–5605. [CrossRef]
31. Chang, K.L.; Petropavlovskikh, I.; Cooper, O.R.; Schultz, M.G.; Wang, T. Regional trend analysis of surface ozone observations from monitoring networks in eastern North America, Europe and East Asia. *Elementa* **2017**, *5*, 5. [CrossRef]
32. Proietti, C.; Fornasier, M.F.; Sicard, P.; Anav, A.; Paoletti, E.; De Marco, A. Trends in tropospheric ozone concentrations and forest impact metrics in Europe over the time period 2000. *J. For. Res.* **2021**, *32*, 543–551. [CrossRef]
33. Luo, J.; Liang, W.; Xu, P.; Xue, H.; Zhang, M.; Shang, L.; Tian, H. Seasonal Features and a Case Study of Tropopause Folds over the Tibetan Plateau. *Adv. Meteorol.* **2019**, *2019*, 4375123. [CrossRef]
34. Vingarzan, R. A review of surface ozone background levels and trends. *Atmos. Environ.* **2004**, *38*, 3431–3442. [CrossRef]
35. Massagué, J.; Contreras, J.; Campos, A.; Alastuey, A.; Querol, X. 2005–2018 trends in ozone peak concentrations and spatial contributions in the Guadalquivir Valley, southern Spain. *Atmos. Environ.* **2021**, *254*, 118385. [CrossRef]
36. Kulkarni, P.S.; Dasari, H.P.; Sharma, A.; Bortoli, D.; Salgado, R.; Silva, A.M. Nocturnal surface ozone enhancement over Portugal during winter: Influence of different atmospheric conditions. *Atmos. Environ.* **2016**, *147*, 109–120. [CrossRef]

37. Ferreira, F.C.; Torres, P.M.; Tente, H.S.; Neto, J.B. Ozone levels in Portugal: The Lisbon region assessment. In Proceedings of the Air and Waste Management Association's Annual Conference and Exhibition, AWMA, Indianapolis, IN, USA, 22–25 June 2004.
38. Finch, D.P.; Palmer, P.I. Increasing ambient surface ozone levels over the UK accompanied by fewer extreme events. *Atmos. Environ.* **2020**, *237*, 117627. [CrossRef]
39. Mazzuca, G.M.; Pickering, K.E.; New, D.A.; Dreessen, J.; Dickerson, R.R. Impact of bay breeze and thunderstorm circulations on surface ozone at a site along the Chesapeake Bay 2011. *Atmos. Environ.* **2019**, *198*, 351–365. [CrossRef]
40. Venkanna, R.; Nikhil, G.N.; Siva Rao, T.; Sinha, P.R.; Swamy, Y.V. Environmental monitoring of surface ozone and other trace gases over different time scales: Chemistry, transport and modeling. *Int. J. Environ. Sci. Technol.* **2015**, *12*, 1749–1758. [CrossRef]
41. Paraschiv, S.; Barbuta-Misu, N.; Paraschiv, S.L. Influence of NO_2, NO and meteorological conditions on the tropospheric O_3 concentration at an industrial station. *Energy Rep.* **2020**, *6*, 231–236. [CrossRef]
42. Katragkou, E.; Zanis, P.; Tegoulias, I.; Melas, D.; Kioutsioukis, I.; Küger, B.C.; Huszar, P.; Halenka, T.; Rauscher, S. Decadal regional air quality simulations over Europe in present climate: Near surface ozone sensitivity to external meteorological forcing. *Atmos. Chem. Phys.* **2010**, *10*, 11805–11821. [CrossRef]
43. Pope, R.J.; Butt, E.W.; Chipperfield, M.P.; Doherty, R.M.; Fenech, S.; Schmidt, A.; Arnold, S.R.; Savage, N.H. The impact of synoptic weather on UK surface ozone and implications for premature mortality. *Environ. Res. Lett.* **2016**, *11*, 124004. [CrossRef]
44. Solberg, S.; Derwent, R.G.; Hov, Ø.; Langner, J.; Lindskog, A. European abatement of surface ozone in a global perspective. *Ambio* **2005**, *34*, 47–53. [CrossRef]
45. Adame, J.A.; Lozano, A.; Bolívar, J.P.; De la Morena, B.A.; Contreras, J.; Godoy, F. Behavior, distribution and variability of surface ozone at an arid region in the south of Iberian Peninsula (Seville, Spain). *Chemosphere* **2008**, *70*, 841–849. [CrossRef]
46. Carnero, J.A.A.; Bolívar, J.P.; de la Morena, B.A. Surface ozone measurements in the southwest of the Iberian Peninsula (Huelva, Spain). *Environ. Sci. Pollut. Res.* **2010**, *17*, 355–368. [CrossRef]
47. Kalabokas, P.D.; Mihalopoulos, N.; Ellul, R.; Kleanthous, S.; Repapis, C.C. An investigation of the meteorological and photochemical factors influencing the background rural and marine surface ozone levels in the Central and Eastern Mediterranean. *Atmos. Environ.* **2008**, *42*, 7894–7906. [CrossRef]
48. Santurtún, A.; González-Hidalgo, J.C.; Sanchez-Lorenzo, A.; Zarrabeitia, M.T. Surface ozone concentration trends and its relationship with weather types in Spain (2001–2010). *Atmos. Environ.* **2015**, *101*, 10–22. [CrossRef]
49. Li, K.; Jacob, D.J.; Shen, L.; Lu, X.; De Smedt, I.; Liao, H. Increases in surface ozone pollution in China from 2013 to 2019: Anthropogenic and meteorological influences. *Atmos. Chem. Phys.* **2020**, *20*, 11423–11433. [CrossRef]
50. Statistics, L. Pearson's Product Moment Correlation. Statistical Tutorials and Software Guides 2020. Available online: https://statistics.laerd.com/statistical-guides/pearson-correlation-coefficient-statistical-guide.php (accessed on 12 June 2021).

Article

Mitigation of Suspendable Road Dust in a Subpolar, Oceanic Climate

Brian Charles Barr [1], Hrund Ólöf Andradóttir [2,*], Throstur Thorsteinsson [1,3] and Sigurður Erlingsson [2]

[1] Faculty of Earth Sciences, School of Engineering and Natural Sciences, University of Iceland, Sturlugata 7, 102 Reykjavík, Iceland; bcb1@hi.is (B.C.B.); thorstur@hi.is (T.T.)
[2] Faculty of Civil and Environmental Engineering, School of Engineering and Natural Sciences, University of Iceland, Hjardarhagi 2-6, 107 Reykjavík, Iceland; sigger@hi.is
[3] Faculty of Environment and Natural Resources, School of Engineering and Natural Sciences, University of Iceland, Sturlugata 7, 102 Reykjavík, Iceland
* Correspondence: hrund@hi.is

Abstract: Tire and road wear particles (TRWP) are a significant source of atmospheric particulate matter and microplastic loading to waterways. Road wear is exacerbated in cold climate by the widespread use of studded tires. The goal of this research was to assess the anthropogenic levers for suspendable road dust generation and climatic conditions governing the environmental fate of non-exhaust particles in a wet maritime winter climate. Sensitivity analyses were performed using the NORTRIP model for the Capital region of Reykjavík, Iceland (64.1° N). Precipitation frequency (secondarily atmospheric relative humidity) governed the partitioning between atmospheric and waterborne PM_{10} particles (55% and 45%, respectively). Precipitation intensity, however, increased proportionally most the drainage to waterways via stormwater collection systems, albeit it only represented 5% of the total mass of dust generated in winter. A drastic reduction in the use of studded tires, from 46% to 15% during peak season, would be required to alleviate the number of ambient air quality exceedances. In order to achieve multifaceted goals of a climate resilient, resource efficient city, the most important mitigation action is to reduce overall traffic volume. Reducing traffic speed may help speed environmental outcomes.

Keywords: particulate matter; microplastics; non-exhaust emissions; NORTRIP

Citation: Barr, B.C.; Andradóttir, H.Ó.; Thorsteinsson, T.; Erlingsson, S. Mitigation of Suspendable Road Dust in a Subpolar, Oceanic Climate. *Sustainability* **2021**, *13*, 9607. https://doi.org/10.3390/su13179607

Academic Editors: José Carlos Magalhães Pires and Álvaro Gómez-Losada

Received: 26 July 2021
Accepted: 22 August 2021
Published: 26 August 2021

Publisher's Note: MDPI stays neutral with regard to jurisdictional claims in published maps and institutional affiliations.

1. Introduction

Road infrastructure plays a significant role in sustainable cities. Besides enabling the transport of people and goods, roads must ideally be safe, economical, climate resilient, and non-compromising of the urban environment [1]. A key challenge to attaining these joint environmental, economic, and societal goals is the multifaceted pollution associated with the frictional contact between a moving vehicle and the road surface (Figure 1). Non-exhaust particles from tire-, road-, and break wear have been recognized as an important source of particulate matter pollution in the 2.5 to 10 μm diameter range [2–8]. More recently, tire and bitumen asphalt road wear particles have emerged as a major microplastics (MP) source to the environment [9–11], and the largest MP contributor to aquatic environments [12], accounting for 5–10% of all plastics in the oceans, with country estimates ranging from 0.9% in The Netherlands to 32% in Norway [13]. Over 30% of the coarse airborne tire and break wear particles ($\leq$10 μm diameter, PM_{10}) is ultimately deposited in the world's oceans; a similar order of magnitude as direct and riverine transport [14]. The frictional contact between tire and road also generates noise pollution and pavement deterioration, which is influenced by climatic factors such as air temperature, solar radiation, and precipitation, each of which governs road wetness [15,16]. Fine particulate matter and traffic noise have been found as the first and second most important environmental causes of ill health in Western Europe [17]. A growing concern is that the smallest waterborne MP particles can accumulate in the cells and tissues of aquatic organisms and enter the food chain [18].

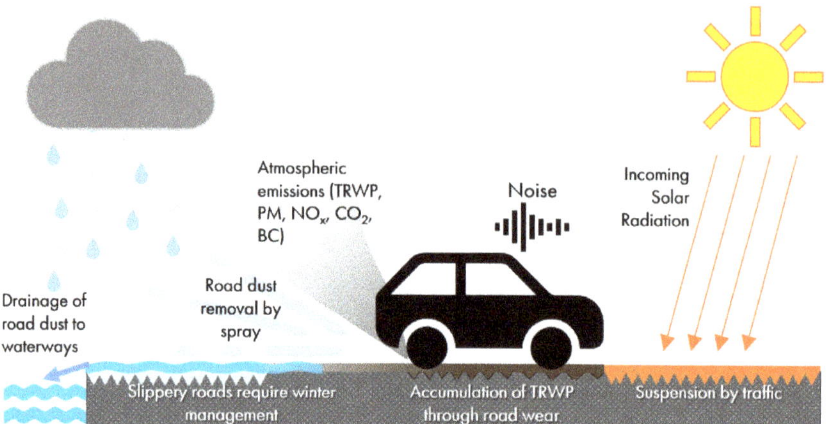

Figure 1. Road wear compromises the multifaceted goals of sustainable urban living, including safety, resource efficiency, amenities, and climate resilience.

Climatic conditions strongly influence the physical characteristics of pavements. Road safety and environmental problems are greatest near freezing air temperatures (± 2 °C) [19], which exacerbate the use of studded tires and deicing agents for anti-skid protection [5,20–22]. Studded tires paired with high vehicle speed further increases road abrasion rates [8,20,23,24]. Indirect emissions from winter road maintenance activities such as the application of road sand (and to a lesser extent road salt) for traction control can be important [8,25]. While only 0.5% of road salt became suspended, salting was attributed to 1–10% of total PM_{10} emissions [21]. Through its control on both road dust generation and emissions, local meteorology can cause to up to a 60% variation in mean winter PM_{10} concentrations [21,24]. Surface wetness is an effective particle binder, contributing to the buildup of dust to be released episodically during dry periods [24,26]. To the authors' best knowledge, no study has systematically assessed the contribution of individual weather parameters on the fate of suspendable road dust.

Even if many cities have readily switched to newer, lower-polluting vehicles (e.g., Euro 5 and Euro 6) and made major investments in sustainable and shared mobility, their ambient air still often exceeds the European standard for PM_{10} [27]. Many of these cities have applied a host of short- and long-term mitigation measures to limit road dust emissions, such as speed reductions, studded tire bans, stronger bitumen asphalt, and the application of dust-binding and road cleaning chemical agents with some success [21,26]. As winter temperature, atmospheric humidity, and precipitation increase in many parts of the world due to climate warming, maritime winter conditions with frequent snow and frost cycles may become more prevalent in high-latitude regions [28]. The degree to which these wet, oscillating weather conditions affect the magnitude and fate of road dust is not fully understood.

Attaining joint environmental, economic, and societal benefits of a sustainable city requires a detailed knowledge of the interplay between the road surface, traffic, and meteorological conditions. The goal of this research was, therefore, to assess the anthropogenic levers of road dust generation and meteorological conditions governing the environmental fate of non-exhaust particles, as a foundation for mitigation policy-making. To compliment previous research undertaken in the northern parts of continental Europe and North America, this study focuses on the subpolar, oceanic climate of Reykjavík, the capital and largest city of Iceland. Reykjavik undergoes frequent freeze–thaw cycles and high amounts of winter precipitation. The study addresses the following questions: (1) Which meteorological driver most strongly governs the partitioning between atmospheric and water bound traffic related PM_{10} particles? (2) Which actions are required to achieve the goals of no exceedances to air quality standards of particulate matter? (3) Which air quality mitigation method(s) can achieve the broadest benefits of sustainable cities? To answer these questions, a numerical model was used to incorporate the complex interplay and feedback mechanisms amongst anthropogenic processes and the hydro- and atmospheres. Simulations were compared to other sites with different climatic conditions.

2. Materials and Methods

2.1. Site

The Icelandic Capital Region (ICR), which includes the City of Reykjavík and five surrounding municipalities, has a subpolar, oceanic climate characterized by frequent precipitation, freeze–thaw cycles, and a narrow annual temperature range [29]. Despite a relatively small population (~220,000 inhabitants in 2020), particulate matter exceeds the 24-h European Health Safety Standards (EHSS) of 50 $\mu g/m^3$ of PM_{10} from 7 to 30 times per year on average (Figure 2). Asphalt wear was attributed to approximately half of the measured PM_{10} in a series of source apportionment studies [30,31]. Tire wear, and to a much lesser extent road markings, was estimated to contribute to approximately 80% of microplastic generation in Iceland [32]. Given the range of episodic, but large, sources of PM_{10} causing exceedances of the EHSS, such as ash and dust storms, local resuspension, long-range transport, and fireworks [33–36], there is a dire need to reduce the exceedances associated with traffic.

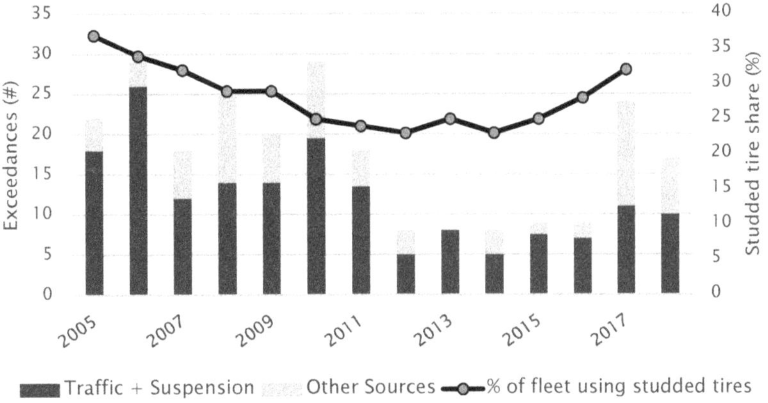

Figure 2. Annual exceedances of 24-h health safety limits and average share of Light Duty Vehicles on studded tires (data taken from in [37,38]).

2.2. Approach

The NORTRIP (Non-Exhaust Road Traffic Induced Particle) model is a comprehensive non-exhaust emissions model developed in collaboration between various Nordic governmental agencies and academics [39]. The model was selected because of its proven ability to (1) reproduce measured concentrations of particulate matter with satisfactory accuracy for sites in Fennoscandia [23,24,39,40], (2) assess the climatic and anthropogenic influences on the sources and sinks of non-exhaust particles, and (3) replicate the successes of mitigation strategies [21]. The model sensitivity to varying mitigation actions and different local meteorology, was tested. The results were interpreted in relation to previous studies.

2.3. Data

The Kauptún site is a four-lane urban traffic artery situated in an open lava field in the municipality of Garðabær (Figure 3a). The road experienced a steady increase in traffic volume in the aftermath of the 2007–2008 financial crisis (Figure 3b). The Kauptún site was equipped with traffic data counters and weight sensors to distinguish light- (LDV) or heavy-duty (HDV) vehicles, as well as a meteorological station recording wind, air temperature, road surface temperature, and conductivity. Moreover, information about road and winter management was available from the Icelandic Road & Coastal Administration [41]. Studded tire counts were conducted every five weeks based on a sample of 250 parked passenger vehicles in parking lots at two shopping centers, a university, and in the city center [38]. The chosen locations targeted vehicles owned by the local inhabitants. Rental vehicles, all of which were on studded tires in winter, may be underrepresented.

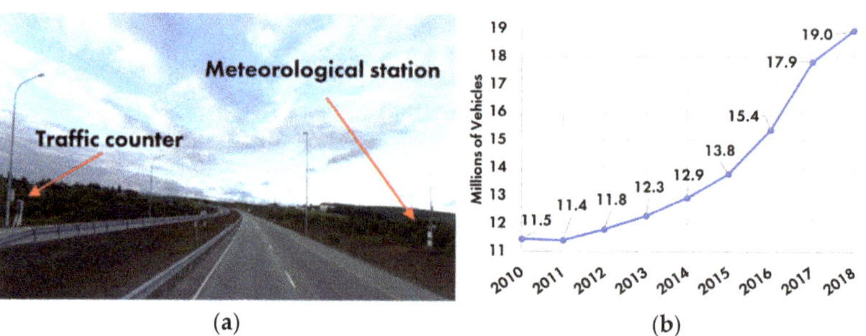

(a)   (b)

Figure 3. Kauptún study site. (**a**) Overview. (**b**) Traffic volume trends (data taken from in [41]).

Precipitation and net short-wave radiation data were obtained from the Icelandic Meteorological Office in Reykjavik [42]. The hourly precipitation was partitioned between rain (total 425 mm; max 6.3 mm/hr) and snow (total 81 mm; max 2.9 mm/hr), using a temperature threshold based on relative humidity (*RH*),

$$TRH = 0.75 + 0.85\,(100\% - RH). \tag{1}$$

If the air temperature exceeded *TRH*, then precipitation was treated as rainfall; otherwise, it was treated as snowfall [43].

Air quality data from the fixed, urban traffic station at Grensás, Reykjavík, were used for comparison [37]. The data sites had similar topographical features. Grensás, however, had a higher traffic (38 million cars annually) than the study site Kauptún (19 million cars annually).

2.4. Baseline Model Setup

The simulation period was defined as 15 October 2017 to 1 May 2018 representing 15 days before and after the legal period for use of studded tires in Iceland. Comparing the key model inputs to previous NORTRIP model sites (Table 1), Kauptún tends to be more humid, windy, and wet, with lower solar radiation to warm up and dry the bitumen road surface, and more winter road management activities such as salting and snow ploughing. The atmospheric PM_{10} concentration was simulated using the Operational Street Pollution Model (OSPM). Although designed to calculate dispersion within street canyons [44], a characteristic that does not apply to the study site, it generated significantly more stable and realistic concentrations than the primary NORTRIP option of utilizing the closest available NO_x data to calculate concentrations. Most of the default NORTRIP settings were used in accordance with previous applications of the model. However, some parameters such as asphalt characteristics were altered to better represent the specific conditions of the investigated site.

Table 1. Key model input and output parameters for baseline simulations in three Nordic cities.

		Hornsgatan [a]	RV4 [a]	Kauptún
		Stockholm	Oslo	Garðabær
Simulation	Latitude	59.3° N	59.9° N	64.1° N
	Period (winter; nr. of days)	2006–2007; 243	2004; 121	2017–2018; 196
Traffic & Pavement	Annual average daily traffic ($\times 10^3$ vehicles)	29.1	42.6	51
	Heavy duty vehicles share (%)	3	4.9	9
	Mean/Max studded tires (% LDV)	47/75	26/27	23/46
	Mean speed (km/hr)	43	75	80
	Pavement type factor (h_{pave})	0.83	1.6	1.62
Meteorology	Mean short wave radiation (W/m^2)	66 [b]	57 [b]	37
	Mean air temperature (°C)	5.8	1.0	1.0
	Relative humidity (%)	75	76	82
	Total precipitation (mm)	197	178	507
	Precipitation frequency (%)	5.8	13	17
	Mean wind speed at 10 m (m/s)	4.0	2.5	6.0 [c]
Winter Management	Total salt (ton/km)	6.3	39	14
	Salting events (NaCl)	45	113	526 [d]
	Ploughing events	2	9	115 [d]
Model Outcomes	Wet roads frequency (%)	43	48	53
	Net mean/90th percentile PM$_{10}$ (µg/m^3)	39/90	30/80	20/49

Notes: [a] Norman et al. (2016); [b] Global radiance, *I*, converted to incoming short-wave radiation using mean cloud cover, *n*, as $SW_{in} = I \times (1 - 0.75 \times n^{3.4})$; [c] Wind speed at 2 m (4.0 m/s) upscaled to 10 m elevation using the logarithmic law for neutral conditions $(10/2)^{0.25}$; [d] Multiple salting and ploughing events per day. Abbreviation: LDV = Light-Duty Vehicle.

Baseline simulations assumed a typical Icelandic asphalt, with local aggregate hardness as 7.9 Nordic ball mill (*NBM*). The maximum stone size (*MS*) was 16 mm, and the percentage of stone size greater than 4 mm was 65%. The pavement type factor (h_{pave}) was characterized using the Swedish Road Wear Model [45] as

$$h_{pave} = 2.49 + 0.144 \cdot NBM - 0.069 \cdot MS - 0.017 \cdot S_{>4mm}. \tag{2}$$

The resulting value was 1.62 g/km/veh corresponds to a higher wear rate than the reference in NORTRIP (4.68 vs. 2.88 g/km/veh, respectively). A more thorough description of the inputs and model setup is presented in the master's thesis by Brian C. Barr [46].

2.5. Sensitivity Analysis

A sensitivity analysis was conducted to understand which lever, either anthropogenic or meteorological, was most significant for PM generation and release of PM, to the atmosphere and waterways. Traffic-related parameters, such as studded tire share, traffic volume and speed, and the fraction of HDVs, were increased or decreased on percentage bases.

Incremental changes in meteorological conditions were simulated without changing road management practices. This approach gives an indication of lever strength but is not meant for forecasting purposes. Temperatures and relative humidity were raised or lowered by a uniform amount, and the amount of rain and snow values adjusted according to Equation (1). Precipitation frequency was reduced by eliminating the mildest precipitation (0.1–0.4 mm/hr) and increased by adding the minimum threshold precipitation (0.1 mm/hr) during dry periods with relative humidity exceeding different thresholds (>97, 98, and 99%). Precipitation intensity was investigated by proportionally increasing or reducing the amount of precipitation that occurred for each hourly data point. The interdependence of different meteorological variables, e.g., climate warming and relative humidity [29], and to which degree such parameters reinforce or balance each other was not a focus of this study.

Three non-realistic ("extreme") precipitation scenarios were applied to better understand the control of road wetness and precipitation phase. Furthermore, simulations were conducted with proportionally changing the amount of road salt applied and incorporating cleaning and wetting. No simulations were performed using sanding, because sanding is not applied at this site, and sanding is not yet fully developed in the NORTRIP model [24,40]. Last, the sensitivity of the model was tested by increasing the hardness of the aggregate (*NBM*).

2.6. Model Outputs

Air quality was primarily assessed based on three metrics: the average PM_{10} concentrations during simulation period as related to annual guidelines (40 and 20 $\mu g/m^3$ according to European Air Quality Directive and WHO guidelines, respectively [47,48]), the number of exceedances to the European Standard of 24 h PM_{10} of 50 $\mu g/m^3$, and the maximum daily PM_{10} concentrations.

Road surface conditions were primarily assessed by surface wetness, as the percentage of total time that roads were wet. The key metric used to gauge the need for winter management was the number of days when road ice was above the 0.1 mm threshold.

Road wear is mainly represented as a function of traffic (i.e., vehicle count, *LDV* vs. *HDV*, and speed), and road surface strength in the NORTRIP model [39]. Sand sources (road abrasion and crushing) were not relevant in this study. The generated road dust is either directly emitted to the atmosphere or retained on the road surface (wear retention), which is subsequently removed from the road surface via four sink mechanisms: (1) Atmospheric suspension due to the contact of tires with a dry road surface; (2) spray, because of tire contact with a wet road; (3) drainage into the stormwater collection system via street inlets; and (4) windblown material, negligible in all simulations and not discussed further. The sink terms are either presented as rates ($g/m^2/hr$) or as cumulative mass over the simulation period.

Cumulative dust generation, including both the instantaneous losses to atmosphere and wear retention, was estimated by summing up all sink terms during a simulation with continuous mild (0.1 mm/hr) rainfall as 2219 g/m^2.

Mass loading (g/m^2) represents the mass of road dust and sand that accumulates on the road surface.

3. Results

3.1. Baseline Simulation

The model was effective at representing the timing of observed PM_{10} episodes at the urban traffic air quality station in Reykjavík (Figure 4a), but tended to overpredict concentrations, particularly during the nearly sunless months of December and January. Concentrations of PM_{10} were overestimated by 22% on average during the simulation period, resulting in 20 modeled Health Safety Limit (HSL) exceedances, as opposed to 13 observed ones. The New Year's Eve exceedance was due to fireworks and as such, not represented in the model. Overall, the 0.57 correlation coefficient between modeled and observed particulate matter concentrations was comparable to other Scandinavian studies, and as such, is considered a satisfactory performance.

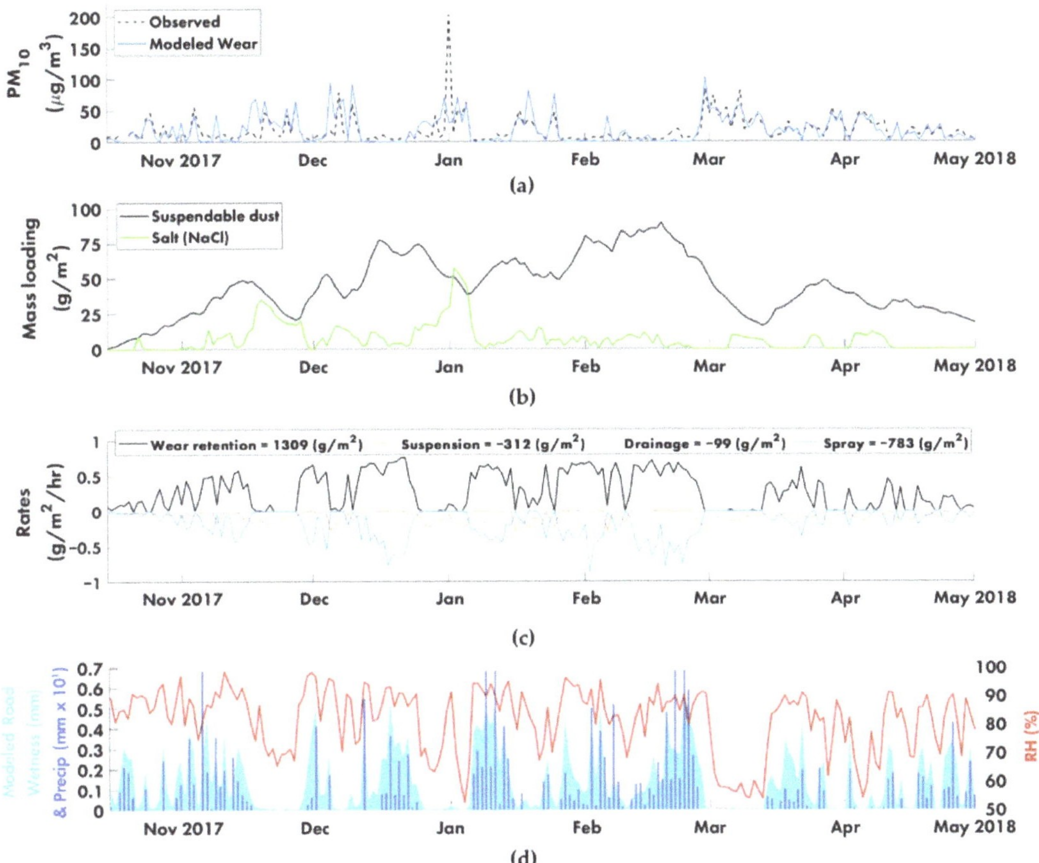

Figure 4. Daily baseline simulation outputs at Kauptún during winter 2017/2018. (**a**) Modeled and observed PM_{10} concentrations. (**b**) Amount of dust accumulated on the surface of the road that is available for resuspension. (**c**) Source and sink rates for accumulation of dust on the road surface. (**d**) Daily modeled road wetness (shade), precipitation (bars), and relative humidity (line).

A progressive accumulation of road dust was predicted as increasing number of vehicles using studded tires were included in the model (Figure 4b). More dust was retained on the roads during wet periods (Figure 4c), but at the same time, the removal mechanisms of spray and drainage became more effective, reducing the amount of dust available to be suspended in air. Once the road dried (reducing road wetness, Figure 4d), the source and water-related sink terms became negligible while suspension, due to tire contact with the road, begins, elevating the particulate matter concentrations in the air (Figure 4a) and reducing the mass loading on the road (Figure 4b). Of particular note are the few dry periods promoting atmospheric releases in the months of November through February. As a result, considerable amounts of dust accumulated on the road to be released during a series of closely spaced spring dust episodes.

Predicted road wetness correlated well with measured precipitation and relatively humidity (Figure 4d). Specifically, the road dried quickly after precipitation ended, usually within a few hours. Additionally, dry roads corresponded well with times when relative humidity dropped below 70%. Albeit infrequent, wet roads were predicted on days with no precipitation, but elevated relative humidity. This suggests that precipitation and relative humidity can be used as indicators of road wetness, and by extension, can be used to anticipate periods of elevated dust suspension.

3.2. Sensitivity to Traffic and Pavement Parameters

The model predicts a near linear deterioration in air quality in response to increasing traffic-related parameters (Table 2). The fraction of light-duty vehicles on studded tires is identified as the single most influential traffic-related parameter to air quality: A 10% reduction in the usage of studded tires resulted in an ~25% decrease in average PM_{10} concentrations, and a 10 $\mu g/m^3$ drop in maximum daily PM_{10} concentration. According to the model, the studded tire usage of the light-duty vehicle fleet needs to go down to 15% in order to achieve zero exceedances of the health safety limit.

Table 2. Model sensitivity to traffic and pavement.

Category	Alteration from Baseline	Δ Max PM_{10} ($\mu g/m^3$)	Δ Avg. PM_{10} ($\mu g/m^3$)	Δ HSL Exceedances (Days)
Baseline	46% Max ST	104	21	20
Studded Tires	35% Max	−21	−4	−7
	25% Max	−39	−7	−16
	15% Max	−58	−10	−20
	0% (full ST ban)	−80	−16	−20
Traffic Volume	−10%	−7	−2	−3
	−20%	−15	−4	−6
Traffic Speed	−10%	−4	−2	−2
	−20%	−9	−4	−6
Composition	HDVs Excluded	0	−2	−2
Wear resistant DGP	$h_{pave} = 0.93$	−40	−8	−16
	$h_{pave} = 1.3$	−19	−4	−7
	$h_{pave} = 1.5$	−9	−2	−3

Abbreviations: HSL = Health Safety Limit; ST = Studded tires; HDV = Heavy-Duty Vehicle; DGP = Dense Grade Pavement.

The model suggests that shortening the legal time to use studded tires does not reduce the number of exceedances to HSL (not shown), albeit it can moderate the concentrations early in the season. Average and maximum particulate matter concentrations dropped 2 and 7 $\mu g/m^3$, respectively, for every 10% reduction in traffic volume, which can be traced back to the reduction in studded tires vehicles; a 10% reduction in traffic volume is tantamount to an approximately 4.6% reduction in studded tires. A 10% decrease in traffic speed had a similar effect as 10% reduction in traffic volume, except that daily maximum PM_{10} was not reduced as much (Table 2). Removing heavy-duty vehicles (HDVs) from this particular road segment, which represented 9% of the total traffic count, was predicted to slightly reduce the average PM_{10} concentrations and cause two fewer HSL exceedances. Last, resurfacing the road with a more wear-resistant dense graded pavement (DGP), similar to that at Hornsgatan in Stockholm (Table 1), would strongly reduce road dust generation and exceedances, more so than reducing traffic volume by 20%.

3.3. Sensitivity to Meteorology and Winter Management

The model sensitivity to local meteorology was first tested by changing each parameter while maintaining others fixed (Table 3). The range tested was chosen to represent conditions at the two reference sites (Table 1). The results highlight that rainfall frequency, more so than rainfall intensity, controls the air quality (mean concentrations and exceedances) via the frequency of wet or icy roads. Relative humidity (RH) exerts a secondary control on road wetness, which translates to increasing atmospheric particulate matter mean concentrations and exceedances as RH is lower, but has limited effect on road ice and hence the need for road salting. Air temperature, however, highly influences road ice formation, and by extension, winter management practices, and it should be noted that the response was nonlinear. Lowering mean air temperature to or below freezing point affected average PM_{10} concentrations (and exceedances to ambient air quality standards) more than warming from the freezing point.

Table 3. Model sensitivity to meteorological parameters and road management practices.

Category	Alteration from Baseline [1]	Δ Avg. PM_{10} (µg/m³)	Δ HSL Exceedances (Days)	Δ% Wet Roads	Δ Road Ice > 0.1 mm (Days)
Baseline	507 mm; frequency 17.1%	20.5	20	53.3%	38
Precipitation Frequency	+7.7%; 544 mm total	−2.8	−4	5.7%	15
	+5.6%; 534 mm	−2.2	−3	4.2%	13
	+3.9%; 526 mm	−1.8	−3	3.9%	10
	−3.2%; 492 mm	0.5	0	−1.6%	0
	−6.3%; 463 mm	1.3	0	−3.8%	−2
	−8.2%; 436 mm	1.6	0	−4.5%	−4
	−9.8%; 406 mm	2.1	0	−6.5%	−11
Relative Humidity	−6%; mean: 77%	1.8	2	−4.8%	−1
	−4%; mean: 79%	1.2	2	−3.2%	−1
	−2%; mean: 81%	0.6	0	−1.6%	0
	+2%; mean: 84%; max: 100%	−1.0	−1	2.3%	0
Air Temperature	+5 °C; mean: 6 °C	0.5	0	−1.0%	−37
	+2 °C; mean: 3 °C	0.3	1	0.0%	−26
	+1 °C; mean: 2 °C	0.4	0	−1.0%	−13
	−1 °C; mean: 0 °C	−0.9	−1	2.0%	23
	−2 °C; mean: −1 °C	−1.6	−2	4.0%	40
Precipitation Intensity	×1.20; 608 mm total	−0.2	0	1.0%	2
	×0.80; 406 mm total	0.4	0	−1.0%	−1
"Extreme" Scenarios	No precip	6.1	7	−15%	−36
	No precip; No WM	7.1	3	−26%	−36
	Rain only ($T_{air} \geq 4$ °C); No WM	−2.3	−7	3.0%	−38
	Constant rain (0.1 mm/hr); 475 mm	−19.8	−20	47%	80
Salting	−50%	−0.8	1	1.0%	3
	50%	1.2	2	−1.0%	−3
Wetting	0.2 mm every four hours during long, dry periods	−4.2	−9	11%	16

Note: [1] Winter management (WM) practices same as in base line simulation, unless otherwise noted.

The "extreme" scenarios provide additional insights to the control of precipitation and road management on particulate matter pollution. In the absence of precipitation, the mean PM_{10} concentrations would be higher, and more exceedances of the air quality standards would occur, in accordance with wet removal processes being eliminated. Discontinuing winter salting and plowing results in fewer exceedances to air quality standards. If all the precipitation would fall as rainfall, so no winter management would be needed, the average PM_{10} concentration is lowered by 10 µg/m³, and the number of exceedances reduced by one-third. Therefore, winter management is a secondary, yet significant, contributor to particulate matter pollution.

A more aggressive salting scenario increased road dust on the same order of magnitude as a moderate change in meteorological conditions (1 µg/m³ mean PM_{10}). Salt has the potential to increase moisture, which itself has a mitigating effect on suspension (Denby et al., 2012). Road wetting on an as-needed basis (Table 3) almost reduces the exceedances to national health safety limits by half, in accordance with the strong dependence on precipitation frequency.

3.4. Fate of Road Dust

Of the 2219 g/m² of road dust generated over the 6.5-month simulation period, 41% was directly emitted to the atmosphere and 14% suspended during dry periods with traffic. The major wet removal process was spray, predicted to remove 35% of the generated dust off the road to the roadside curb. Direct drainage to waterways via the stormwater collection system constituted only 4.5%. Of the meteorological factors tested, rainfall frequency controlled most whether road dust was emitted to the atmosphere or became water bound (Figure 5). Most notably, more frequent rainfall, and to a lesser extent higher atmospheric humidity, shifted direct emissions to wet removal via spray (Figure 5a,b). In the unrealistic extreme condition of constant, mild rainfall, 89% of the generated dust would leave the road surface via spray. The response to air temperature was much less pronounced and nonlinear around freezing point, where atmospheric emissions were predicted to be at minimum (Figure 5c). Increasing rainfall intensity shifted only slightly the removal from the atmosphere to wet removal (Figure 5d). Of all the meteorological parameters tested though, rainfall intensity exerted one of the strongest controls on drainage to waterways. For example, a 10% increase in rainfall intensity resulted in a 6 g/m² increase equally to spray and drainage. At the end of the simulation period, all scenarios tested suggested that around 110 g/m² of the generated road dust was still present on the road to be removed at a later time (noted as mass loading; Figure 5).

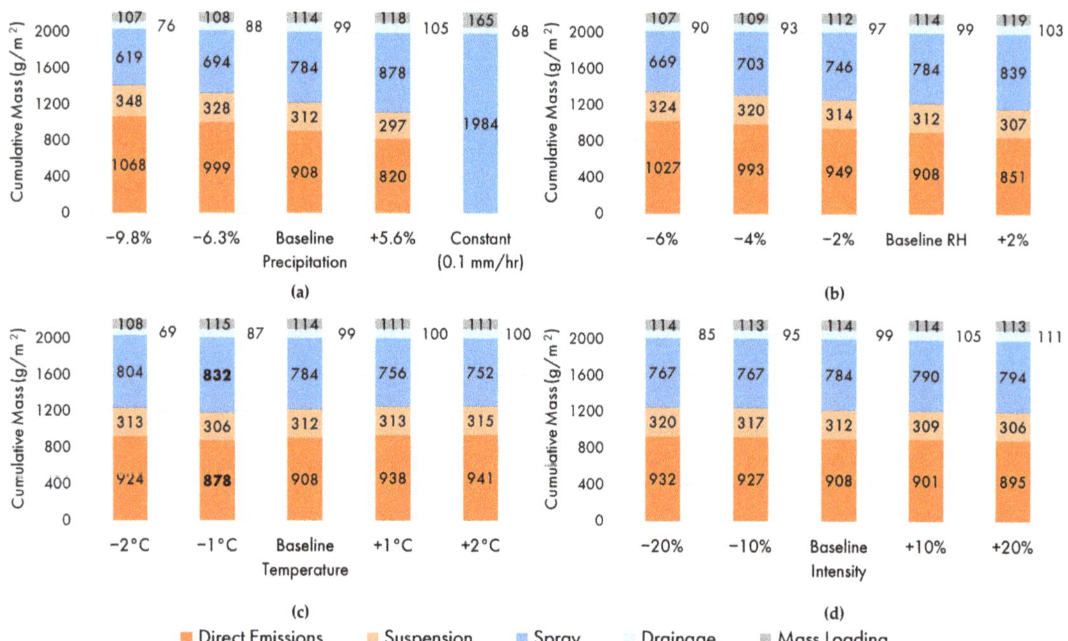

Figure 5. Sensitivity of wet (blue) and dry (orange) removal processes to individual meteorological parameters: (**a**) Precipitation frequency; (**b**) Relative humidity; (**c**) Air temperature and precipitation phase; (**d**) Precipitation intensity.

4. Discussion

4.1. Wet Maritime Climate

This study expands the current knowledge of non-exhaust particulate pollution by considering the wet, humid, windy, and sunless maritime climate in the Icelandic Capital Region compared to other studied capitals in Northern Europe (Table 1). NORTRIP model sensitivity analyses highlight these climatic attributes (frequent rain, high humidity) as efficient dust retention and wet removal processes, which are primarily spray and secondarily drainage (Table 3, Figure 5). Therefore, maritime, wet, and cold conditions promote microplastics loading to aquatic and terrestrial environments, estimated as representing 45% of the total mass of PM_{10} particles generated. The positive effect of this efficient wet removal of dust is the moderate average winter PM_{10} concentration that almost adheres to the WHO guidelines of 20 $\mu g/m^3$, and 50% to 100% lower when compared to Oslo and Stockholm (Table 1), despite the greater traffic volume. Unfortunate artifacts of prolonged wet roads (53% of time) are closely spaced, intense particulate matter episodes, both mid-winter and in springtime (Figure 4a). The potential effect of frequent freeze thaw cycles on road dust generation was not resolved in this study, as it was beyond the capability of the NORTRIP model. Previous research suggests a faster pavement deterioration in wet areas with frequent freeze–thaw cycles [16,49].

4.2. Mitigation Strategies for Atmospheric PM_{10}

The model identified the primary lever for road dust generation and abatement as reducing the number of light-duty vehicles on studded tires (Table 2). Traffic volume and asphalt type were secondary levers, followed by vehicle speed. Only a drastic reduction of studded tires usage in winter achieved the goals of adhering to ambient air quality standards. Yet, the share of light duty vehicles on studded tires has steadily increased since 2014 (Figure 2), partially as the public campaign *"Off with the studded tires"* (Icelandic: *"Burt með nagladekkin"*) was relaxed. In addition, episodic road ice conditions forming in October just before the studded tires legal interval (Nov. 15) prompt car owners to choose studded tires, despite that Reykjavík Municipality winter management services are frequent enough that studded tires are not needed within the city perimeter. Moreover, traffic volume has been drastically increasing over the past years (Figure 3b), both because of personal car ownership and tourism. Considering these historical trends in studded tire use and traffic volume, it is unlikely that focusing solely on the optimal levers will result in a timely improvement in air quality. Therefore, it is important to evaluate combined mitigation strategies; both short-term when episodes are expected, and long-term.

Short-term response to foreseeable PM episodes: The strong correlation between modeled road wetness, precipitation, and relative humidity (Figure 4d) suggests that the risk of PM_{10} episodes can be predicted based on periods of prolonged road dryness and high dust load. This allows authorities to implement preemptive, short-term mitigating actions to stifle the suspension of dust. One mitigation action that has been employed in cities facing severe pollution episodes, locally called "gray days", is to ban cars with an even or odd license plate from driving in the city. This aggressive, active, short-term measure alone, however, would not alleviate the problem with exceedances. However, coupling 50% reduction in traffic volume with a 10 km/h speed limit reduction brings the number of traffic related HSL exceedances to two (Table 4). An alternative, less intrusive approach is to wet the road system on gray days as a dust binding measure. This, combined with slight traffic and speed reductions, is predicted to lower the PM_{10} concentrations during episodes to one-third. While a significant reduction, it still does not achieve the Icelandic government's goal of zero traffic-related PM_{10} exceedances by 2029. It is important to keep in mind that short-term mitigation strategies are focused on reducing the suspension of PM_{10}, not the generation. The short timeframe of these measure does not influence the long-term dust generation, and as such, the loading of road dust to terrestrial and aquatic environments.

Table 4. Selected combined short and long-term mitigation strategies.

Mitigation Scenario	Targets				Air Quality Outcomes	
	Traffic Volume Reduction	Speed Reduction	Max. Studded Tire Usage	Road Re-Surfacing	Avg PM_{10} Reduction	Total HSL Exceedances
Short-Term—Aggressive	50%	15 km/h	Unchanged	No	44%	2
Short-Term—Moderate	10%	10 km/h	Unchanged	No	33%	5
Long-Term—Aggressive	15%	None	20%	No	50%	0
Long-Term—Moderate I	10%	10 km/h	25%	No	41%	2
Long-Term—Moderate II	10%	None	25%	Yes	63%	0

Long-term mitigation actions: The sensitivity analysis suggests that the long-term strategy needs to be focused on reducing the use of studded tires as much as possible; coupling this with a decrease in traffic volume would further decrease the overall number of studded tires. Historically, the lowest seasonal average percentage of studded tires was 23% in both 2012 and 2014 (Figure 2). From this perspective, an aggressive scenario is to half the number of studded tires (20% maximum) in 9 years. Yet, this provision does not suffice to achieve the goal of zero PM_{10} exceedances, so a 15% reduction in traffic volume must be implemented as well (Table 4). A less aggressive approach would be to reduce studded tire use to 25%, and traffic speed permanently by 10 km/h; this would lower the average PM_{10} concentrations by 41%. Finally, road resurfacing would offer significant long-term dust reduction, but its effectiveness increases substantially when coupled with reductions in studded tire usage. Road resurfacing with a similar asphalt strength as used at Hornsgatan resulted in a 40% reduction in emissions in the model; this reduction increases to 63% when coupled with aggressive studded tire reduction.

4.3. Mitigation Strategies for Sustainable Cities

As discussed in the previous section, a combination of different mitigation actions can help achieve the overall goal of zero road dust related particulate matter exceedances. However, in a broader context, it is important to recognize that some actions may provide auxiliary environmental benefits, such as greenhouse gas and noise abatement, while other may be more costly and take a long time to implement. Therefore, it is valuable to extend the criteria of beneficial outcomes to incorporate more multifaceted goals of a sustainable city. We will consider the anthropogenic levers tested in the NORTRIP model together with two new levers for comparison, the electrification of the car fleet and dust binding with agents such calcium magnesium acetate (CMA), which has effectively reduced peaks in PM_{10} concentrations [5]. Each lever was given a score based on its positive or negative contribution to goals of sustainability as stated in literature.

By considering more amenities than road dust generation, traffic reduction provided the most intense and diverse environmental benefits (Table 5). Moreover, fewer cars on the street drastically reduces collision rates and as such, this lever is ranked highest on safety. However, reducing personal cars requires a public transport alternative which may require substantial investment and may take a long time to implement [27]. Reducing studded tires is an effective noise and road dust abatement technique in cold climates. This measure, however, lacks climate resilience because of exhaust gas emissions and may raise winter safety concerns. Lowering speed limits contributes positively, albeit moderately, on all the diverse aspects of a sustainable city. Lower vehicle velocity generates less exhaust gases and noise [50] as well as road dust, but it should be kept in mind that more road dust is removed as drainage to local waterways due to lower suspension between tire and road surface. Its competitive advantage over the other traffic levers is, arguably, its ease and speed of implementation. Vehicle electrification provides only a reduction of harmful exhaust pollutants such as greenhouse gases [51], black carbon, and nitrogen oxides. While highly important, it cannot be the backbone of a strategy for sustainable cities because of concerns that electrical vehicles emit more road dust because of their greater weight compared to internal combustion vehicles [52], and of the natural resources depleted in converting the car fleet and time of implementation [53].

Table 5. Evaluation of multifaceted benefits of anthropogenic levers in a subpolar, oceanic climate.

Domain	Anthropogenic Lever	Environmental Benefits				Social Benefits	Logistics of Implementation	
		Non-Exhaust	Exhaust (e.g., GHG)	Noise	Resource Efficiency [1]	Safety	Cost	Time
Traffic	Reduce traffic volume	+	+	+	++	++	−−	12 years [2]
	Reduce studded tires	++	/	++	+	−	+	2+ years [2]
	Reduce speed	+	+	+	+	+	++	Days
	Increase electrical cars	−	++	/	−	/	/	12 years [2]
Pavement	Wear-resistant DGP	+	/	/	+	/	−−	6+ year [2]
	Open-graded OGP	−	/	+	−	+/	−	2 years [2]
Road Management	Dust binding	(+)	/	/	−	/	−	Hours
	Road wetting	(+)	/	/	−	−	−	Hours

Notes: Scale: ++ = Highly positive effect; + = Positive effect; / = Neutral/Unknown/Varies; − = Negative effect; −− = Highly negative effect; [1] Construction and rehabilitation needs for road infrastructure (pavements, bridges, parking lots). [2] Based on longevity of studded tires; personal vehicles; pavement. Abbreviations: GHG = Green House Gases; DGP/OGP = Dense/Open Graded Pavement.

The last four levers pertain to changing the properties of the road surface. The first lever is to increase the aggregate hardness (lower NBM) or the percentage of stones > 4 mm (Equation (2)) in dense pavement asphalt (DGP), as tested in the sensitivity analysis (Table 2). While this reduces road dust generation, and improves resource efficiency through less pavement wear, it is not expected to create additional benefits such as improved safety or reduced noise levels. Moreover, road resurfacing is only done at 6+ year intervals, and a stronger aggregate may not be found locally, as is the case in Iceland. An importation of stronger aggregates would thus increase greenhouse gas emissions from shipping. For comparison purposes, open-graded (OGP), permeable friction course pavements offer a host of safety and environmental benefits, including improved wet weather skid resistance, reduced splash and spray, reduced light reflection, reduced tire and pavement noise, improved pavement smoothness, reduced contribution to urban heat island effect, and reduced pollutant loadings in stormwater runoff [19]. However, they perform worse than DGP in winter, as they freeze faster and longer, need more deicing agents, their pores can store and retain snow and dust, and their aggregate structure makes them particularly susceptible to degradation, especially by studded tires [19,26,54]. Last, the short-term road management actions of dust binding provide no auxiliary benefits other than reducing particulate matter episodes. The amount of road dust generated will be the same, and it will ultimately be released to the atmosphere (at lower concentration) or hydrosphere. To conclude, our assessment suggests that increasing aggregate hardness is a good secondary option in a cold climate, after the traffic levers are applied, especially if such materials can be supplied locally.

5. Conclusions

With climate change, many regions of the world are experiencing warmer winter temperatures and more precipitation in the form of rainfall or rain on snow. Winter climates may transition from subpolar continental, characterized by little precipitation (that nearly always falls as snow) and low relative humidity, to subpolar oceanic with frequent, light, year-round precipitation, and a narrow temperature band. This study highlights that such changes, specifically more frequent rainfall, higher relative humidity, and higher precipitation intensity in winter, may on one hand reduce winter average PM_{10} concentrations; on the other hand, a maritime winter climate may promote air pollution episodes and more loading of microplastics to terrestrial and aquatic systems. Moreover, the indirect effect of frequent precipitation and fluctuating air temperature around the freezing point is ice formation, which calls for more winter management that generates more dust. The Icelandic experience is that it also prompts the usage of studded tires, even though local authorities regularly remind the population that the majority of the trips are conducted within the urban area where slippery conditions can be managed with plowing and salting. Short-term mitigation on grey days can alleviate air pollution episodes, but not the pollutant loading to waterways and concern of microplastics pollution entering the food chain. The strongest mitigation levers are to reduce the studded tire share and traffic volume, both of which take years to reverse unless strong measures are employed. Pairing restrictions of studded tires use with traffic speed reductions may both help with source reduction, and eliminate the competitive advantage of the personal vehicle being the fastest mean of transport. Last, electrification of the car fleet is no silver bullet in achieving a sustainable road system, as they contribute to road wear.

Author Contributions: Conceptualization and methodology, H.Ó.A., T.T., and S.E.; software, B.C.B., and T.T.; formal analysis, B.C.B.; data curation, B.C.B. and H.Ó.A.; writing—original draft preparation, B.C.B.; writing—review and editing, H.Ó.A., T.T., and S.E.; visualization, B.C.B.; supervision, project administration, funding acquisition, H.Ó.A. All authors have read and agreed to the published version of the manuscript.

Funding: This research was funded by the Icelandic Road and Coastal Administration Research Fund (Icelandic: Rannsóknasjóður Vegagerðarinnar; Grant nr. 1800-678).

Institutional Review Board Statement: Not applicable.

Informed Consent Statement: Not applicable.

Data Availability Statement: Model parameters and input data are available at https://tinyurl.com/3stjdnpe and https://tinyurl.com/22t5xshf.

Acknowledgments: Funding: data and expert knowledge from the Icelandic Road and Coastal Administration is greatly appreciated. Efla Consulting Engineers, Samrás ehf, the Environment Agency of Iceland and the Icelandic Meteorological Office are thanked for their help with data acquisition and interpretation. Assistance with the NORTRIP model is acknowledged.

Conflicts of Interest: The authors declare no conflict of interest. The funders had no role in the design of the study; in the collection, analyses, or interpretation of data; in the writing of the manuscript; or in the decision to publish the results.

References

1. United Nations Department of Economic and Social Affairs Sustainable Development. Available online: https://sdgs.un.org/goals/goal11 (accessed on 1 February 2021).
2. Segersson, D.; Eneroth, K.; Gidhagen, L.; Johansson, C.; Omstedt, G.; Nylén, A.E.; Forsberg, B. Health impact of PM10, PM2.5 and black carbon exposure due to different source sectors in Stockholm, Gothenburg and Umea, Sweden. *Int. J. Environ. Res. Public Health* **2017**, *14*, 742. [CrossRef] [PubMed]
3. Thorpe, A.J.; Harrison, R.M.; Boulter, P.G.; McCrae, I.S. Estimation of particle resuspension source strength on a major London Road. *Atmos. Environ.* **2007**, *41*, 8007–8020. [CrossRef]
4. Johansson, C.; Norman, M.; Gidhagen, L. Spatial & temporal variations of PM10 and particle number concentrations in urban air. *Environ. Monit. Assess.* **2007**, *127*, 477–487. [CrossRef] [PubMed]
5. Norman, M.; Johansson, C. Studies of some measures to reduce road dust emissions from paved roads in Scandinavia. *Atmos. Environ.* **2006**, *40*, 6154–6164. [CrossRef]
6. Omstedt, G.; Bringfelt, B.; Johansson, C. A model for vehicle-induced non-tailpipe emissions of particles along Swedish roads. *Atmos. Environ.* **2005**, *39*, 6088–6097. [CrossRef]
7. Laakso, L.; Hussein, T.; Aarnio, P.; Komppula, M.; Hiltunen, V.; Viisanen, Y.; Kulmala, M. Diurnal and annual characteristics of particle mass and number concentrations in urban, rural and Arctic environments in Finland. *Atmos. Environ.* **2003**, *37*, 2629–2641. [CrossRef]
8. Etyemezian, V.; Kuhns, H.; Gillies, J.; Chow, J.; Hendrickson, K.; McGown, M.; Pitchford, M. Vehicle-based road dust emission measurement (III): Effect of speed, traffic volume, location, and season on PM10 road dust emissions in the Treasure Valley, ID. *Atmos. Environ.* **2003**, *37*, 4583–4593. [CrossRef]

9. Järlskog, I.; Strömvall, A.M.; Magnusson, K.; Gustafsson, M.; Polukarova, M.; Galfi, H.; Aronsson, M.; Andersson-Sköld, Y. Occurrence of tire and bitumen wear microplastics on urban streets and in sweepsand and washwater. *Sci. Total Environ.* **2020**, *729*, 138950. [CrossRef]

10. Vogelsang, C.; Lusher, A.L.; Dadkhah, M.E.; Sundvor, I.; Umar, M.; Ranneklev, S.B.; Eidsvoll, D.; Meland, S. *Microplastics in Road Dust—Characteristics, Pathways and Measures*; Norwegian Institute for Water Research (NIVA): Oslo, Norway, 2019.

11. Sommer, F.; Dietze, V.; Baum, A.; Sauer, J.; Gilge, S.; Maschowski, C.; Gieré, R. Tire abrasion as a major source of microplastics in the environment. *Aerosol Air Qual. Res.* **2018**, *18*, 2014–2028. [CrossRef]

12. Hann, S.; Sherrington, C.; Jamieson, O.; Hickman, M.; Kershaw, P.; Bapasola, A.; Cole, G. *Investigating Options for Reducing Releases in the Aquatic Environment of Microplastics Emitted by (but Not Intentionally Added in) Products*; Eunomia Research and Consluting: Bristol, UK, 2018.

13. Kole, J.P.; Löhr, A.J.; van Belleghem, F.G.A.J.; Ragas, A.M.J. Wear and tear of tyres: A stealthy source of microplastics in the environment. *Int. J. Environ. Res. Public Health* **2017**, *14*, 1265. [CrossRef]

14. Evangeliou, N.; Grythe, H.; Klimont, Z.; Heyes, C.; Eckhardt, S.; Lopez-Aparicio, S.; Stohl, A. Atmospheric transport is a major pathway of microplastics to remote regions. *Nat. Commun.* **2020**, *11*, 3381. [CrossRef]

15. Freitas, E.; Pereira, P.; de Picado-Santos, L.; Santos, A. Traffic Noise Changes due to Water on Porous and Dense Asphalt Surfaces. *Road Mater. Pavement Des.* **2009**, *10*, 587–607. [CrossRef]

16. Llopis-Castelló, D.; García-Segura, T.; Montalbán-Domingo, L.; Sanz-Benlloch, A.; Pellicer, E. Influence of pavement structure, traffic, and weather on urban flexible pavement deterioration. *Sustainability* **2020**, *12*, 9717. [CrossRef]

17. Hänninen, O.; Knol, A.B.; Jantunen, M.; Lim, T.A.; Conrad, A.; Rappolder, M.; Carrer, P.; Fanetti, A.C.; Kim, R.; Buekers, J.; et al. Environmental burden of disease in Europe: Assessing nine risk factors in six countries. *Environ. Health Perspect.* **2014**, *122*, 439–446. [CrossRef] [PubMed]

18. Sharma, S.; Chatterjee, S. Microplastic pollution, a threat to marine ecosystem and human health: A short review. *Environ. Sci. Pollut. Res.* **2017**, *24*, 21530–21547. [CrossRef] [PubMed]

19. Akin, M.; Fay, L.; Shi, X. Friction and Snow—Pavement Bond after Salting and Plowing Permeable Friction Surfaces. *Transp. Res. Rec.* **2020**, *2674*, 794–805. [CrossRef]

20. Gustafsson, M.; Blomqvist, G.; Gudmundsson, A.; Dahl, A.; Swietlicki, E.; Bohgard, M.; Lindbom, J.; Ljungman, A. Properties and toxicological effects of particles from the interaction between tyres, road pavement and winter traction material. *Sci. Total Environ.* **2008**, *393*, 226–240. [CrossRef] [PubMed]

21. Norman, M.; Sundvor, I.; Denby, B.R.; Johansson, C.; Gustafsson, M.; Blomqvist, G.; Janhäll, S. Modelling road dust emission abatement measures using the NORTRIP model: Vehicle speed and studded tyre reduction. *Atmos. Environ.* **2016**, *134*, 96–108. [CrossRef]

22. Lundberg, J.; Janhäll, S.; Gustafsson, M.; Erlingsson, S. Calibration of the Swedish studded tyre abrasion wear prediction model with implication for the NORTRIP road dust emission model. *Int. J. Pavement Eng.* **2019**, *22*, 432–446. [CrossRef]

23. Denby, B.R.; Sundvor, I.; Johansson, C.; Pirjola, L.; Ketzel, M.; Norman, M.; Kupiainen, K.; Gustafsson, M.; Blomqvist, G.; Omstedt, G. A coupled road dust and surface moisture model to predict non-exhaust road traffic induced particle emissions (NORTRIP). Part 1: Road dust loading and suspension modelling. *Atmos. Environ.* **2013**, *77*, 283–300. [CrossRef]

24. Denby, B.R.; Sundvor, I.; Johansson, C.; Pirjola, L.; Ketzel, M.; Norman, M.; Kupiainen, K.; Gustafsson, M.; Blomqvist, G.; Kauhaniemi, M.; et al. A coupled road dust and surface moisture model to predict non-exhaust road traffic induced particle emissions (NORTRIP). Part 2: Surface moisture and salt impact modelling. *Atmos. Environ.* **2013**, *81*, 485–503. [CrossRef]

25. Ketzel, M.; Omstedt, G.; Johansson, C.; Düring, I.; Pohjola, M.; Oettl, D.; Gidhagen, L.; Wåhlin, P.; Lohmeyer, A.; Haakana, M.; et al. Estimation and validation of PM2.5/PM10 exhaust and non-exhaust emission factors for practical street pollution modelling. *Atmos. Environ.* **2007**, *41*, 9370–9385. [CrossRef]

26. Gustafsson, M.; Blomqvist, G.; Järlskog, I.; Lundberg, J.; Janhäll, S.; Elmgren, M.; Johansson, C.; Norman, M.; Silvergren, S. Road dust load dynamics and influencing factors for six winter seasons in Stockholm, Sweden. *Atmos. Environ. X* **2019**, *2*, 100014. [CrossRef]

27. Tomassetti, L.; Torre, M.; Tratzi, P.; Paolini, V.; Rizza, V.; Segreto, M.; Petracchini, F. Evaluation of air quality and mobility policies in 14 large Italian cities from 2006 to 2016. *J. Environ. Sci. Health Part A Toxic/Hazard. Subst. Environ. Eng.* **2020**, *55*, 886–902. [CrossRef]

28. Aygün, O.; Kinnard, C.; Campeau, S. Impacts of climate change on the hydrology of northern midlatitude cold regions. *Prog. Phys. Geogr. Earth Environ.* **2020**, *44*, 338–375. [CrossRef]

29. Andradóttir, H.Ó.; Arnardóttir, A.R.; Zaqout, T. Rain on snow induced urban floods in cold maritime climate: Risk, indicators and trends. *Hydrol. Process.* **2021**. [CrossRef]

30. Skúladóttir, B.; Thorlacius, A.; Larssen, S.; Bjarnason, G.G.; Þórdarson, H. *Method for Determining the Composition of Airborne Particle Pollution*; Nordtest: Espoo, Finland, 2003; p. 30.

31. Höskuldsson, P.; Thorlacius, A. *Uppruni Svifryks í Reykjavík*; EFLA Consulting Engineers: Reykjavík, Iceland, 2017; p. 17.

32. Sigurðsson, V.; Halldórsson, P. *Örplast í Hafinu við Ísland: Helstu Uppsprettur, Magn og Farvegir í Umhverfinu*; Biopol and Náttúrufræðistofa Norðurlands Vestra: Sauðárkróki, Iceland, 2019.

33. Butwin, M.K.; von Löwis, S.; Pfeffer, M.A.; Thorsteinsson, T. The effects of volcanic eruptions on the frequency of particulate matter suspension events in Iceland. *J. Aerosol Sci.* **2019**, *128*, 99–113. [CrossRef]

34. Thorsteinsson, T.; Jóhannsson, T.; Stohl, A.; Kristiansen, N.I. High levels of particulate matter in Iceland due to direct ash emissions by the Eyjafjallajökull eruption and resuspension of deposited ash. *J. Geophys. Res. Solid Earth* **2012**, *117*, 1–9. [CrossRef]
35. Thorsteinsson, T.; Gísladóttir, G.; Bullard, J.; McTainsh, G. Dust storm contributions to airborne particulate matter in Reykjavík, Iceland. *Atmos. Environ.* **2011**, *45*, 5924–5933. [CrossRef]
36. Andradottir, H.O.; Thorsteinsson, T. Repeated extreme particulate matter episodes due to fireworks in Iceland and stakeholders' response. *J. Clean. Prod.* **2019**, *236*, 117511. [CrossRef]
37. Environment Agency of Iceland. Hourly Air Quality Data at Grensás Urban Traffic Station 2017–2018. Unpublished Work. 2018.
38. EFLA Consulting Engineers. Studded Tires Counts in Reykjavík City 2000 to 2018. Unpublished Work.
39. Denby, B.R.; Sundvor, I. *NORTRIP Model Development and Documentation: NOn-Exhaust Road TRaffic Induced Particle Emission Modelling*; NILU: Oslo, Norway, 2012.
40. Denby, B.R.; Ketzel, M.; Ellermann, T.; Stojiljkovic, A.; Kupiainen, K.; Niemi, J.V.; Norman, M.; Johansson, C.; Gustafsson, M.; Blomqvist, G.; et al. Road salt emissions: A comparison of measurements and modelling using the NORTRIP road dust emission model. *Atmos. Environ.* **2016**, *141*, 508–522. [CrossRef]
41. Icelandic Road & Coastal Administration. Traffic Counts, Meteorology and Winter Management at Kauptún 2017—2019. Unpublished Work. 2019.
42. Icelandic Meteorological Office. Radiation and Precipitation at Reykjavík Station nr. 1. Unpublished Work. 2019.
43. Feiccabrino, J.; Graff, W.; Lundberg, A.; Sandström, N.; Gustafsson, D. Meteorological knowledge useful for the improvement of snow rain separation in surface based models. *Hydrology* **2015**, *2*, 266–288. [CrossRef]
44. Jensen, S.S.; Ketzel, M.; Brandt, J.; Becker, T.; Plejdrup, M.; Winther, M.; Ellermann, T.; Christensen, J.H.; Nielsen, O.-K.; Hertel, O. Air Quality at Your Street—Public Digital Map of Air Quality in Denmark. In Proceedings of the Sixth Scientific Meeting EuNetAir; Academy of Sciences: Prague, Czech Republic, 2016; pp. 14–17. [CrossRef]
45. Jacobson, T.; Wågberg, L.-G. *Utveckling och Uppgradering av Prognosmodell för Beläggningsslitage från Dubbade däck samt en Kunskapsöversikt över Inverkande Faktorer*; VTI: Linköping, Sweden, 2007.
46. Barr, B.C. Processes and Modeling of Non-Exhaust Vehicular Emissions in the Icelandic Capital Region. Master's Thesis, University of Iceland, Reykjavík, Iceland, 2020.
47. European Environmental Agency. *Air Quality in Europe—2020 Report*; EEA Report No 9/2020; European Environmental Agency: København, Denmark, 2020. [CrossRef]
48. World Health Organization. *Air Quality Guidelines Global Update 2005. Particulate Matter, Ozone, Nitrogen Dioxide and Sulfur Dioxide*; WHO: Geneva, Switzerland, 2005; ISBN 92 890 2192 6.
49. Ud Din, I.M.; Mir, M.S.; Farooq, M.A. Effect of Freeze-Thaw Cycles on the Properties of Asphalt Pavements in Cold Regions: A Review. *Transp. Res. Procedia* **2020**, *48*, 3634–3641. [CrossRef]
50. Pascale, A.; Fernandes, P.; Guarnaccia, C.; Coelho, M.C. A study on vehicle Noise Emission Modelling: Correlation with air pollutant emissions, impact of kinematic variables and critical hotspots. *Sci. Total Environ.* **2021**, *787*, 147647. [CrossRef]
51. Temporelli, A.; Carvalho, M.L.; Girardi, P. Life Cycle Assessment of Electric Vehicle Batteries: An Overview of Recent Literature. *Energies* **2020**, *13*, 2864. [CrossRef]
52. Timmers, V.R.J.H.; Achten, P.A.J. Non-exhaust PM emissions from electric vehicles. *Atmos. Environ.* **2016**, *134*, 10–17. [CrossRef]
53. Milovanoff, A.; Posen, I.D.; MacLean, H.L. Electrification of light-duty vehicle fleet alone will not meet mitigation targets. *Nat. Clim. Chang.* **2020**, *10*, 1102–1107. [CrossRef]
54. Lundberg, J.; Gustafsson, M.; Janhäll, S.; Eriksson, O.; Blomqvist, G.; Erlingsson, S. Temporal Variation of Road Dust Load and Its Size Distribution—A Comparative Study of a Porous and a Dense Pavement. *Water. Air. Soil Pollut.* **2020**, *231*, 561. [CrossRef]

Article

Considering Condensable Particulate Matter Emissions Improves the Accuracy of Air Quality Modeling for Environmental Impact Assessment

Doo Sung Choi [1,†], Jong-Sang Youn [2,†], Im Hack Lee [3], Byung Jin Choi [4] and Ki-Joon Jeon [5,*]

[1] Department of Building Equipment System & Fire Protection Engineering, University of Chungwoon, Incheon 22100, Korea; trebelle@chungwoon.ac.kr
[2] Department of Energy and Environmental Engineering, The Catholic University of Korea, Bucheon 14662, Korea; jsyoun@catholic.ac.kr
[3] Department of Environmental Engineering, University of Seoul, Seoul 02504, Korea; imhack@empas.com
[4] Jubix, Gyeonggi-do 16419, Korea; cbjin@jubix.co.kr
[5] Department of Environmental Engineering, Inha University, Incheon 22212, Korea
* Correspondence: kjjeon@inha.ac.kr
† Both authors contributed equally to this work.

Abstract: This study examines environmental impact assessment considering filterable particulate matter (FPM) and condensable particulate matter (CPM) to improve the accuracy of the air quality model. Air pollutants and meteorological data were acquired from Korea's national monitoring station near a residential development area in the target district and background site. Seasonal emissions of $PM_{2.5}$, including CPM, were estimated using the California puff (CALPUFF) model, based on Korea's national emissions inventory. These results were compared with the traditional environmental impact assessment results. For the residential development area, the seasonal $PM_{2.5}$ concentration was predicted by considering FPM and CPM emissions in the target area as well as the surrounding areas. In winter and spring, air quality standards were not breached because only FPM was considered. However, when CPM was included in the analysis, the results exceeded the air quality standards. Furthermore, it was predicted that air quality standards would not be breached in summer and autumn, even when CPM is included. In other words, conducting an environmental impact assessment on air pollution including CPM affects the final environmental decision. Therefore, it is concluded that $PM_{2.5}$ should include CPM for greater accuracy of the CALPUFF model for environmental impact assessment.

Keywords: environmental impact assessment; condensable particulate matter; air quality model; $PM_{2.5}$; air pollution

Citation: Choi, D.S.; Youn, J.-S.; Lee, I.H.; Choi, B.J.; Jeon, K.-J. Considering Condensable Particulate Matter Emissions Improves the Accuracy of Air Quality Modeling for Environmental Impact Assessment. *Sustainability* **2021**, *13*, 4470. https://doi.org/10.3390/su13084470

Academic Editor: José Carlos Magalhães Pires

Received: 10 March 2021
Accepted: 15 April 2021
Published: 16 April 2021

1. Introduction

Particulate matter (PM) includes ammonium sulfate ($(NH_4)_2SO_4$), ammonium nitrate (NH_4NO_3), and trace amounts of soil components, including titanium, iron, zinc, and lead [1,2]. It is composed of various components, such as trace metals, organic matter, and sea salt particles [3–5]. Considering their small size, these particles can be easily inhaled and cause pulmonary and respiratory diseases [6,7], lung diseases [8], and cardiovascular diseases [9]. Furthermore, PM has been reported to influence the premature mortality rate [10,11].

PM can be classified into primary PM (directly emitted from the source) and secondary PM (formed by photochemical reactions in the atmosphere after being emitted in the gaseous phase) [12–14]. In Korea, the national emission inventory estimates the annual emissions of major air pollutants, including carbon monoxide (CO), nitric oxide (NO_x), sulfur oxide (SO_x), total suspended particles (TSP), particulate matter less than 10 μm (PM_{10}), particulate matter less than 2.5 μm ($PM_{2.5}$), black carbon, and volatile organic compounds (VOCs) based on emission sources and regions [15,16]. The emission inventory for PM (TSP, PM_{10}, and $PM_{2.5}$) considers filterable PM (FPM), which is collected through a filter. However, unlike ambient air, the primary PM emitted from emission sources can be classified into FPM and condensable PM (CPM), and the aggregate of FPM and CPM is considered as the total PM [17]. CPM is in the gas phase under high-temperature conditions and condenses into PM immediately after emitting from the emission source. Corio and Sherwell [18] estimated that CPM accounts for about 76% of total PM10 emission from large stationary emission sources. Although the US EPA recognized the CPM issue in the early 1980s and developed a measurement method for stationary sources, it was not considered as a severe issue [19,20]. Sulfur trioxide (SO_3) in flue gas can react with water vapor to form sulfuric acid mist, which has often been misunderstood as CPM formation [21]. In recent years, not only studies on CPM emitted from stationary sources [22–24] but also studies using CPM as modeling inputs have been conducted [25]. In particular, Morino et al. [25] revealed the contribution of CPM to ambient PM by measuring CPM at stationary sources and confirmed the improvement in the prediction model for ambient PM by including CPM as a model input. Thus, it is necessary to include CPM in the air quality modeling of environmental impact assessments.

In atmospheric studies, AERMOD (the AMS/EPA regulatory model) is one of the most widely used models for environmental impact assessment [26,27]. It is a steady-state plume model that calculates atmospheric diffusion, based on the concept of turbulent structure and scaling in the atmospheric boundary layer. Moreover, it can be used in simple (planar) district scenarios and complex terrain scenarios [28]. The CALPUFF (California puff) model is another model that is used in the evaluation of large atmospheric emission sources, such as industrial complexes, power plants, and incinerators [26,29]. CALPUFF is a multi-layered, multi-stage, unsteady puff diffusion model that simulates the effects of temporally and spatially varying weather conditions on the transport of pollutants [30]. Furthermore, it can be applied to rough and complex terrains. Although the CALPUFF model is useful, it does not consider photochemical reactions and chemical reactions of secondary pollutants. Therefore, chemical transport models, such as CMAQ [31], which consider atmospheric chemical reactions, have been recently used in atmospheric environmental impact assessments.

To improve the CALPUFF model, this study considered CPM in the emission inventory and applied it to the seasonal environmental impact assessment of the target district.

2. Materials and Methods

In this study, the 2013 emission inventory provided by the National Center for Fine Dust Information [32] was used as input data for the CALPUFF model. In addition, total suspended particles (TSPs) and $PM_{2.5}$ (particulate matter with an aerodynamic diameter less than 2.5 µm) were selected as the target air pollutants. Moreover, CPM emission factors for stationary sources were used, as published by the National Institute of Environmental Research [33] (Table 1). Liquified natural gas (LNG), diesel, and B-C oil were measured in a boiler without a control device. Furthermore, the bituminous coal emission factor was measured at the end of the control devices in a power plant facility. The concentration of CPM emissions was calculated by multiplying the FPM to CPM emission factor ratio with $PM_{2.5}$, obtained from Korea's emission inventory data.

Table 1. Condensable particulate matter (CPM) emission factors estimated by Gong et al., 2016 [33].

Fuel Type	TPM	FPM	CPM	Note
LNG boiler (mg/m^3)	206.67	3.79	202.88	uncontrolled
Light oil boiler (mg/L)	65.78	3.38	62.40	uncontrolled
B-C oil boiler (mg/L)	371.47	143.83	227.64	uncontrolled
Bituminous power plant (g/ton)	71.65	6.55	65.10	uncontrolled

Figure 1 shows a schematic of the CALPUFF model. The CALPUFF model consists of a CALMET module, a meteorological model, and a CALPUFF module, an air pollution model. The CALMET module uses a land cover map, meteorological data, and aerological data as input data to calculate meteorological model results and then inputs the emission inventory to derive the CALPUFF model results. The control file serves to input commands to control each module. To simulate the FPM and CPM emission behavior, a software was developed to calculate the amount of PM emissions from major emission sources in the target district and link this with the meteorological data acquired from the automatic weather system data from the four monitoring stations near the target area. Thus, this software was used to establish a methodology to verify the accuracy of the concentrations of FPM and CPM emissions and understand atmospheric behavior prediction through case studies.

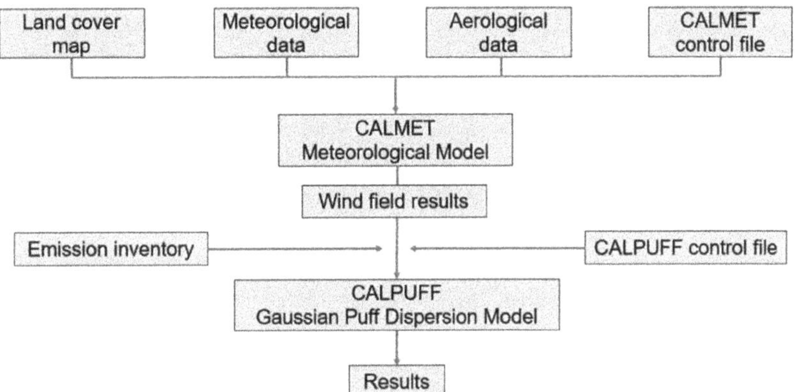

Figure 1. Schematic of the CALPUFF modeling process for environmental impact assessment.

The target area was the Bugok residential development district in Gunpo-si, Gyeonggi-do, with a project area of 470,000 m^2 (residential area of 200,795 m^2, commercial and business area of 6104 m^2, and public area of 263,101 m^2), accommodating approximately 9300 people. To compare and verify the accuracy of the PM concentration via model calculations, real-time data provided by the Korea Environment Corporation (KEC) monitoring station in and around the target district were used.

3. Results and Discussion

The CPM conversion factor management program developed in this study provided a function for searching and managing conversion factors by fuel type. Additionally, CPM emissions were calculated by applying the CPM conversion factor for each fuel to point, line, and surface emission sources. In the CPM emission factor management program of each fuel type, information on the FPM and CPM emission factors of PM$_{2.5}$ emission sources, including construction, buildings, and vehicles, was provided and managed during the development of the residential area. Thus, it was meticulously configured so that details on emission factors for each construction instrument, vehicle type, and building energy fuel could be searched. Seasonal variation is one of the most important parameters for environmental impact assessment, so we applied one month of each season [34,35].

3.1. Seasonal Results for Environmental Impact Assessment (Winter, January)

Figure 2a–d show the modeling results (5.5 × 5.5 km) by segregating the FPM and CPM emissions generated during the operational stage of the Bugok residential development district based on the area to be analyzed and on whether the emission sources outside the target district were considered. The FPM and CPM emissions were calculated by considering the emission sources in the Bugok residential development district. The FPM concentration was estimated to be 0.0038 $\mu g/m^3$, and the CPM concentration was predicted to be 0.27 $\mu g/m^3$. Meanwhile, $PM_{2.5}$ was analyzed by considering emission sources that were outside the target district and excluding the emission sources in the Bugok residential development district. The FPM concentration was predicted to be 10.63 $\mu g/m^3$, and the CPM concentration was estimated to be 14.82 $\mu g/m^3$. Therefore, the TPM (CPM + FPM) concentrations by emission sources inside and outside the target district were calculated to be 0.27 and 25.45 $\mu g/m^3$, respectively.

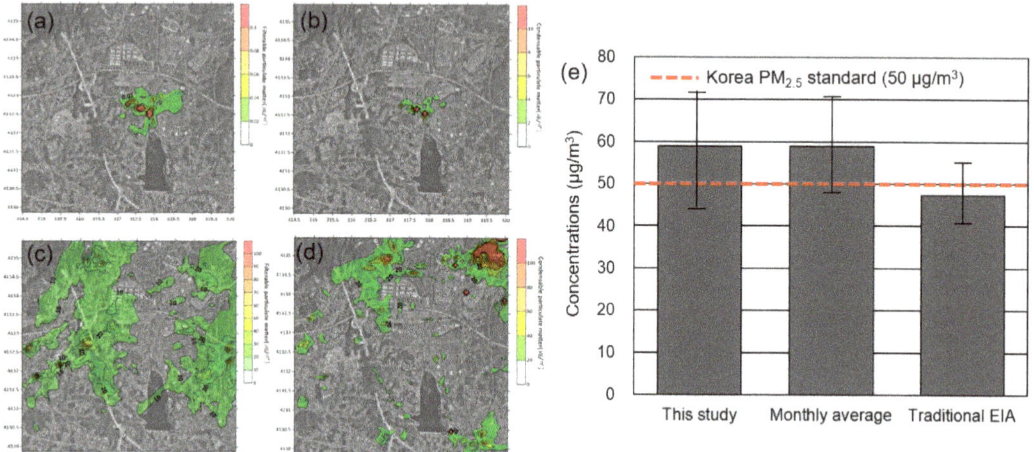

Figure 2. Winter modeling results of (**a**) FPM and (**b**) CPM concentrations in target district, and (**c**) FPM and (**d**) CPM concentrations outside target district. (**e**) Comparison of the results with the monthly average $PM_{2.5}$ concentration obtained from the monitoring station as well as the results of the traditional environmental impact assessment (EIA).

The modeling results were compared with $PM_{2.5}$ data acquired from the air pollutant monitoring stations near the target district. Figure 2e shows the results of this study, monthly averages of the monitoring stations, and traditional environmental impact assessment results. In terms of traditional environmental impact assessment, the $PM_{2.5}$ concentration at the study site was estimated to be 47.28 $\mu g/m^3$. This was the sum of the annual average PM concentration inside the target district and the $PM_{2.5}$ concentration data acquired from the monitoring station outside the target district. The modeling result of this study was 58.87 $\mu g/m^3$, including the long-range transboundary emissions (33.15 $\mu g/m^3$), the FPM concentration inside the target district (0.0038 $\mu g/m^3$), the FPM concentration outside the target district (10.63 $\mu g/m^3$), the CPM concentration inside the target district (0.27 $\mu g/m^3$), and the CPM concentration outside the target district (14.82 $\mu g/m^3$). In particular, the average concentration of TPM, the sum of FPM and CPM, in the target district was 25.72 $\mu g/m^3$, which was approximately two times lower than the concentration measured in winter at the monitoring station (58.87 $\mu g/m^3$).

Table 2 shows the $PM_{2.5}$ concentrations in winter in Deokjeok, Seogwipo, and Seosan, which are the national background concentration monitoring stations located in the far west and south side of Korea. As they are located in the far west and south side of Korea, data collected from monitoring stations are used for evaluation of air pollutants' long-range transportation from polluted areas [36,37]. The observed PM concentrations at the Deokjeok, Seogwipo, and Seosan stations were 31.55, 41.77, and 47.26 $\mu g/m^3$, respectively. The difference between the modeling results and the observed values estimated in this study was considered to be due to the effect of the long-range transboundary emissions from other countries.

Table 2. Seasonal PM$_{2.5}$ concentration collected from Korea's national background monitoring station.

(Unit: µg/m^3)	Deokjeok	Seogwipo	Seosan
Winter	47.26 ± 26.56	31.55 ± 19.48	41.77 ± 27.62
Spring	43.73 ± 22.65	58.67 ± 29.75	44.53 ± 18.12
Summer	23.45 ± 6.80	34.58 ± 6.88	23.79 ± 7.32
Autumn	30.45 ± 12.33	28.19 ± 9.63	29.53 ± 13.32

3.2. Seasonal Results for Environmental Impact Assessment (Spring, April)

Figure 3a–d show the results of the predicted spring PM concentration. The results of the prediction of FPM and CPM concentrations in the target district were 0.0026 µg/m^3 and 0.20 µg/m^3, respectively. An estimation of the PM concentration outside the target district revealed that the FPM concentration was 10.56 µg/m^3 and the CPM concentration was 9.59 µg/m^3. The concentration of TPM by emission source inside and outside the target district was 0.20 µg/m^3 and 20.16 µg/m^3, respectively. Thus, the TPM concentration around the target district was estimated to be 20.36 µg/m^3.

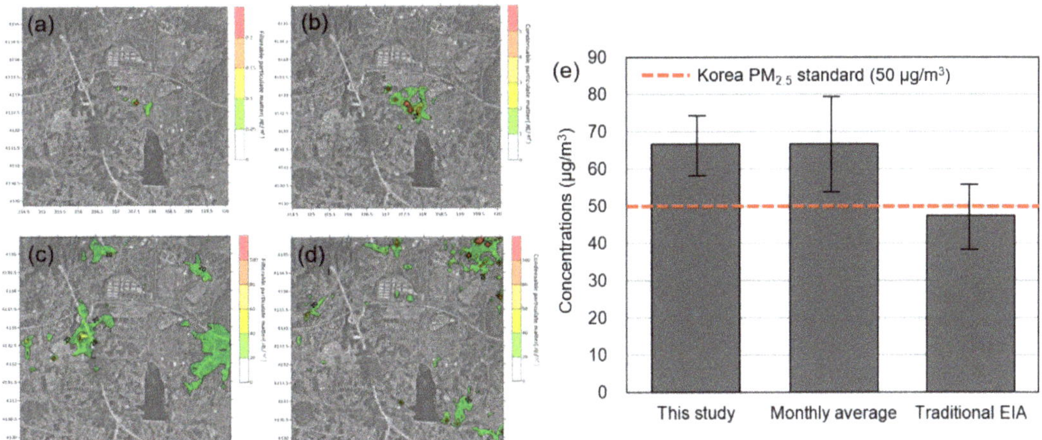

Figure 3. Spring modeling results of (**a**) FPM and (**b**) CPM concentrations in target district, and (**c**) FPM and (**d**) CPM concentrations outside target district. (**e**) Comparison of the results with the monthly average PM$_{2.5}$ concentration from the monitoring stations and the traditional EIA results.

Figure 3e shows the predicted results of this study, measurements of the monitoring station in spring, and the traditional environmental impact assessment results. The modeling result of this study, including the CPM concentration, was 66.70 µg/m^3, which included the long-range transboundary emissions (46.34 µg/m^3), the FPM concentration inside the target district (0.0026 µg/m^3), the FPM concentration outside the target district (10.56 µg/m^3), the CPM concentration inside the target district (0.20 µg/m^3), and the CPM concentration outside the target district (9.59 µg/m^3). In particular, the average concentration of TPM in the target district was 20.36 µg/m^3, which was approximately three times lower than the monthly average of 66.70 µg/m^3 at the monitoring station.

Table 2 shows the PM concentrations in winter in Deokjeok, Seogwipo, and Seosan. The observed PM concentrations at the Deokjeok, Seogwipo, and Seosan stations were 43.73 µg/m^3, 58.67 µg/m^3, and 44.53 µg/m^3, respectively. The difference between the modeling results and the observed values in this study was assumed to be the effect of the long-range transboundary emissions from other countries.

3.3. Seasonal Results for Environmental Impact Assessment (Summer, July)

Figure 4a–d show the results of the prediction of concentrations in summer. The results of the prediction of FPM and CPM concentrations inside the target district were evaluated as 0.0002 µg/m^3 and 0.04 µg/m^3, respectively. By estimating the PM concentration outside the target district, the FPM concentration was predicted to be 8.69 µg/m^3, and the CPM concentration was 7.67 µg/m^3. The concentration of TPM from emission sources inside and outside the target district was 0.04 µg/m^3 and 16.36 µg/m^3, respectively. The TPM concentration around the target district was 16.40 µg/m^3.

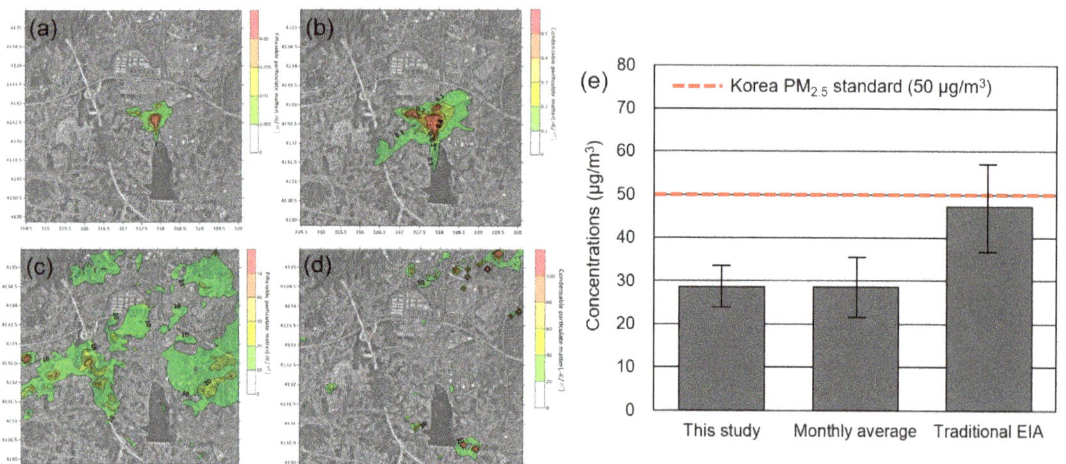

Figure 4. Summer modeling results of (**a**) FPM and (**b**) CPM concentrations in target district, and (**c**) FPM and (**d**) CPM concentrations outside target district. (**e**) Comparison of the results with the monthly average PM$_{2.5}$ concentration obtained from the monitoring stations and the results of traditional EIA.

Figure 3e shows the predicted results of this study, spring measurements of the monitoring station, and traditional environmental impact assessment results. The modeling result of this study, including the CPM concentration, was 28.65 µg/m^3, which included the long-range transboundary emissions (12.24 µg/m^3), the FPM concentration inside the target district (0.0002 µg/m^3), the FPM concentration outside the target district (8.69 µg/m^3), the CPM concentration inside the target district (0.04 µg/m^3), and the CPM concentration outside the target district (7.67 µg/m^3). The concentration of TPM in the target district was estimated to be 16.40 µg/m^3, which was 1.7 times lower than the average concentration measured at the monitoring station in summer (28.65 µg/m^3).

Differences in the predicted summer PM concentrations were considered to be due to the effect of the long-range transboundary emissions. Moreover, the observed PM concentrations at the three stations of Deokjeokdo, Seogwipo, and Seosan were 23.45 µg/m^3, 34.58 µg/m^3, and 23.79 µg/m^3, respectively (Table 2).

3.4. Seasonal Results for Environmental Impact Assessment (Autumn, October)

Figure 5a–d show the results of the autumn concentration prediction. The FPM and CPM concentrations inside the target district were estimated to be 0.0017 µg/m^3 and 0.20 µg/m^3, respectively. By evaluating the PM concentration outside the target district, the FPM concentration was predicted to be 11.29 µg/m^3, and the CPM concentration was predicted to be 9.63 µg/m^3. The concentration of TPM by emission source inside and outside the target district was 0.20 µg/m^3 and 20.92 µg/m^3, respectively. The TPM concentration around the target district was estimated to be 21.12 µg/m^3.

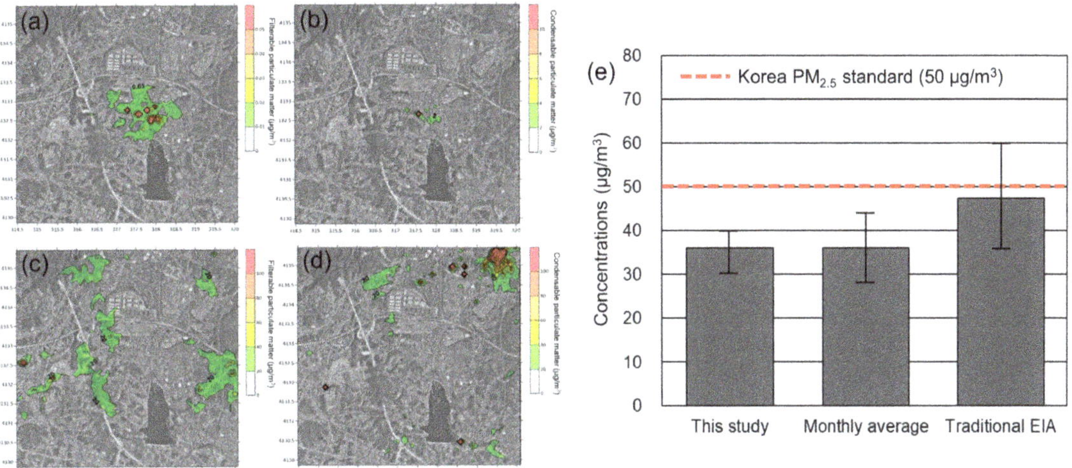

Figure 5. Autumn modeling results of (**a**) FPM and (**b**) CPM concentrations in target district, and (**c**) FPM and (**d**) CPM concentrations outside of target district. (**e**) The results of this study were compared with the monthly average PM$_{2.5}$ concentration obtained from the monitoring stations and the results of traditional EIA.

Figure 5e shows the predicted results of this study, spring measurements of the monitoring station, and traditional environmental impact assessment results. The predicted value of this study, including the CPM concentration was 36.06 µg/m^3, which included the long-distance inflow (14.95 µg/m^3), the in-pipe FPM concentration (0.0017 µg/m^3), the outside FPM concentration (11.29 µg/m^3), the in-pipe CPM concentration (0.20 µg/m^3), and the CPM concentration outside the building (9.63 µg/m^3). In particular, the average FPM and CPM concentrations in the study site were predicted to be 21.12 µg/m^3, which was 1.7 times lower than the average concentration measured in the autumn at the monitoring station (36.06 µg/m^3). The difference in the predicted autumn PM concentration was considered to be due to the long-distance migration. Moreover, it was confirmed that the PM$_{2.5}$ concentrations at Deokjeokdo, Seogwipo, and Seosan were 30.45 µg/m^3, 28.19 µg/m^3, and 29.53 µg/m^3, respectively.

Differences in the predicted autumn PM concentrations were considered to be due to the effect of the long-range transboundary emissions. The observed PM concentrations in the three stations recorded from Deokjeokdo, Seogwipo, and Seosan were 30.45 µg/m^3, 28.19 µg/m^3, and 29.53 µg/m^3, respectively (Table 2).

We included CPM and FPM in the PM prediction model which showed a less than 5% difference compared to the monitoring station data, while the results of traditional environmental impact assessment showed a difference of 20–40% compared to the monitoring station data. Ghim et al. [38] performed an evaluation of a PM2.5 prediction model including only FPM emissions, and the predicted value of PM2.5 was found to be 69% of the monitoring station data. Thus, including CPM emissions in the PM prediction model is one of the ways to increase the accuracy of the model for environmental impact assessment. Even with the results of measuring CPM and FPM at the large stationary emission source, the portion of CPM occupies more than 80% [39], which disproves that the PM concentration should be predicted using both CPM and FPM emissions. In addition, the US EPA recommends that the interim guidance for new source review permit programs should include the CPM in determining a new major stationary source permission [40]. Thus, environmental impact assessment should consider CPM as one of the factors of air quality analysis.

4. Conclusions

In this study, the PM concentration in the atmosphere was predicted by including CPM emissions in the environmental impact assessment. For the residential development area, the seasonal $PM_{2.5}$ concentration was predicted by considering the FPM and CPM emissions in the target area as well as the surrounding areas. In winter and spring, when only FPM was considered, the air quality standards were not breached. However, when CPM results were included in the analysis, air quality standards were exceeded. However, it was predicted that even if CPM is included, air quality standards would not be breached in summer and autumn. This means that air quality forecasting results, including CPM, may alter the results. In addition, the sum of the predicted values of seasonal CPM and FPM was 1.7 to 3 times lower than that of the actual measurement. Compared to the background concentration measurement, it was found to be a result of long-distance travel. Therefore, it is necessary to consider CPM in the emission inventory to carry out environmental impact assessment, air quality modeling, analysis and diagnosis of emissions according to the characteristics of each sector's emission source, and prediction of the $PM_{2.5}$ concentration in the surrounding areas. In this study, environmental impact assessment was performed by considering only primary PM using the CALPUFF model. What remains to be undertaken by future research is an environmental impact assessment including secondary PM.

Author Contributions: Conceptualization, J.-S.Y. and D.S.C.; methodology, I.H.L.; investigation, D.S.C.; writing—original draft preparation, J.-S.Y., D.S.C., and K.-J.J.; writing—review and editing, I.H.L. and K.-J.J.; visualization, B.J.C.; supervision, K.-J.J.; funding acquisition, D.S.C. All authors have read and agreed to the published version of the manuscript.

Funding: This research was funded by a grant (19CTAP-C130211-03) from the Technology Advancement Research Program (TARP) funded by the Ministry of Land, Infrastructure, and Transport of the Korean government.

Institutional Review Board Statement: Not applicable.

Informed Consent Statement: Not applicable.

Data Availability Statement: Not applicable.

Acknowledgments: The authors would like to thank the editors and anonymous reviewers for their constructive comments and valuable suggestions regarding this article.

Conflicts of Interest: The authors declare no conflict of interest.

References

1. Kelly, F.J.; Fussell, J.C. Size, source and chemical composition as determinants of toxicity attributable to ambient particulate matter. *Atmos. Environ.* **2012**, *60*, 504–526. [CrossRef]
2. Hand, J.L.; Schichtel, B.A.; Pitchford, M.; Malm, W.C.; Frank, N.H. Seasonal composition of remote and urban fine particulate matter in the United States. *J. Geophys. Res. Atmos. Phys.* **2012**, *117*. [CrossRef]
3. Wu, Y.-S.; Fang, G.-C.; Lee, W.-J.; Lee, J.-F.; Chang, C.-C.; Lee, C.-Z. A review of atmospheric fine particulate matter and its associated trace metal pollutants in Asian countries during the period 1995–2005. *J. Hazard. Mater.* **2007**, *143*, 511–515. [CrossRef]
4. Duplissy, J.; De Carlo, P.F.; Dommen, J.; Alfarra, M.R.; Metzger, A.; Barmpadimos, I.; Prevot, A.S.H.; Weingartner, E.; Tritscher, T.; Gysel, M.; et al. Relating hygroscopicity and composition of organic aerosol particulate matter. *Atmos. Chem. Phys. Discuss.* **2011**, *11*, 1155–1165. [CrossRef]
5. Van Donkelaar, A.; Martin, R.V.; Li, C.; Burnett, R.T. Regional Estimates of Chemical Composition of Fine Particulate Matter Using a Combined Geoscience-Statistical Method with Information from Satellites, Models, and Monitors. *Environ. Sci. Technol.* **2019**, *53*, 2595–2611. [CrossRef] [PubMed]
6. Mukherjee, A.; Agrawal, M. World air particulate matter: Sources, distribution and health effects. *Environ. Chem. Lett.* **2017**, *15*, 283–309. [CrossRef]
7. Kim, K.-H.; Kabir, E.; Kabir, S. A review on the human health impact of airborne particulate matter. *Environ. Int.* **2015**, *74*, 136–143. [CrossRef]
8. Paulin, L.; Hansel, N. Particulate air pollution and impaired lung function. *F1000Research* **2016**, *5*, 201. [CrossRef] [PubMed]
9. Du, Y.; Xu, X.; Chu, M.; Guo, Y.; Wang, J. Air particulate matter and cardiovascular disease: The epidemiological, biomedical and clinical evidence. *J. Thorac. Dis.* **2016**, *8*, E8–E19. [PubMed]

10. Hu, J.; Huang, L.; Chen, M.; Liao, H.; Zhang, H.; Wang, S.; Zhang, Q.; Ying, Q. Premature Mortality Attributable to Particulate Matter in China: Source Contributions and Responses to Reductions. *Environ. Sci. Technol.* **2017**, *51*, 9950–9959. [CrossRef]
11. Giani, P.; Anav, A.; De Marco, A.; Feng, Z.; Crippa, P. Exploring sources of uncertainty in premature mortality estimates from fine particulate matter: The case of China. *Environ. Res. Lett.* **2020**, *15*, 064027. [CrossRef]
12. Zhang, R.; Wang, G.; Guo, S.; Zamora, M.L.; Ying, Q.; Lin, Y.; Wang, W.; Hu, M.; Wang, Y. Formation of Urban Fine Particulate Matter. *Chem. Rev.* **2015**, *115*, 3803–3855. [CrossRef]
13. Jolliet, O.; Antón, A.; Boulay, A.-M.; Cherubini, F.; Fantke, P.; Levasseur, A.; McKone, T.E.; Michelsen, O.; Canals, L.M.I.; Motoshita, M.; et al. Global guidance on environmental life cycle impact assessment indicators: Impacts of climate change, fine particulate matter formation, water consumption and land use. *Int. J. Life Cycle Assess.* **2018**, *23*, 2189–2207. [CrossRef]
14. Faust, J.A.; Wong, J.P.S.; Lee, A.K.Y.; Abbatt, J.P.D. Role of Aerosol Liquid Water in Secondary Organic Aerosol Formation from Volatile Organic Compounds. *Environ. Sci. Technol.* **2017**, *51*, 1405–1413. [CrossRef] [PubMed]
15. Suh, H.H.; Bahadori, T.; Vallarino, J.; Spengler, J.D. Criteria air pollutants and toxic air pollutants. *Environ. Health Perspect.* **2000**, *108*, 625–633. [CrossRef]
16. Lee, H.-H.; Iraqui, O.; Gu, Y.; Yim, S.H.-L.; Chulakadabba, A.; Tonks, A.Y.-M.; Yang, Z.; Wang, C. Impacts of air pollutants from fire and non-fire emissions on the regional air quality in Southeast Asia. *Atmos. Chem. Phys. Discuss.* **2018**, *18*, 6141–6156. [CrossRef]
17. Feng, Y.; Li, Y.; Cui, L. Critical review of condensable particulate matter. *Fuel* **2018**, *224*, 801–813. [CrossRef]
18. Corio, L.A.; Sherwell, J. In-stack condensible particulate matter measurements and issues. *J. Air Waste Manag. Assoc.* **2000**, *50*, 207–218. [CrossRef]
19. *Estimation of the Importance of Condensed Particulate Matter to Ambient Particulate Levels*; PB84-102565; U.S. Environmental Protection Agency: Research Triangle Park, NC, USA, 1983.
20. Website of the U.S. Environmental Protection Agency, the United States of America. Method 202-Determination of Condensable Particulate Emissions from Stationary Sources. Available online: https://www.epa.gov/emc/method-202-condensable-particulate-matter (accessed on 25 March 2021).
21. Yao, Q.; Li, S.-Q.; Xu, H.-W.; Zhuo, J.-K.; Song, Q. Studies on formation and control of combustion particulate matter in China: A review. *Energy* **2009**, *34*, 1296–1309. [CrossRef]
22. Cano, M.; Vega, F.; Navarrete, B.; Plumed, A.; Camino, J.A. Characterization of Emissions of Condensable Particulate Matter in Clinker Kilns Using a Dilution Sampling System. *Energy Fuels* **2017**, *31*, 7831–7838. [CrossRef]
23. Li, X.; Zhou, C.; Li, J.; Lu, S.; Yan, J. Distribution and emission characteristics of filterable and condensable particulate matter before and after a low-low temperature electrostatic precipitator. *Environ. Sci. Pollut. Res.* **2019**, *26*, 12798–12806. [CrossRef]
24. Wu, B.; Bai, X.; Liu, W.; Lin, S.; Liu, S.; Luo, L.; Guo, Z.; Zhao, S.; Lv, Y.; Zhu, C.; et al. Non-Negligible Stack Emissions of Noncriteria Air Pollutants from Coal-Fired Power Plants in China: Condensable Particulate Matter and Sulfur Trioxide. *Environ. Sci. Technol.* **2020**, *54*, 6540–6550. [CrossRef] [PubMed]
25. Morino, Y.; Chatani, S.; Tanabe, K.; Fujitani, Y.; Morikawa, T.; Takahashi, K.; Sato, K.; Sugata, S. Contributions of Condensable Particulate Matter to Atmospheric Organic Aerosol over Japan. *Environ. Sci. Technol.* **2018**, *52*, 8456–8466. [CrossRef]
26. Tartakovsky, D.; Stern, E.; Broday, D.M. Comparison of dry deposition estimates of AERMOD and CALPUFF from area sources in flat terrain. *Atmos. Environ.* **2016**, *142*, 430–432. [CrossRef]
27. Barjoee, S.S.; Azimzadeh, H.; Kuchakzadeh, M.; MoslehArani, A.; Sodaiezadeh, H. Dispersion and Health Risk Assessment of PM10 Emitted from the Stacks of a Ceramic and Tile industry in Ardakan, Yazd, Iran, Using the AERMOD Model. *Iran. South Med. J.* **2019**, *22*, 317–332. [CrossRef]
28. Hadlocon, L.S.; Zhao, L.Y.; Bohrer, G.; Kenny, W.; Garrity, S.R.; Wang, J.; Wyslouzil, B.; Upadhyay, J. Modeling of particulate matter dispersion from a poultry facility using AERMOD. *J. Air Waste Manag. Assoc.* **2015**, *65*, 206–217. [CrossRef] [PubMed]
29. Jittra, N.; Pinthong, N.; Thepanondh, S. Performance Evaluation of AERMOD and CALPUFF Air Dispersion Models in Industrial Complex Area. *Air Soil Water Res.* **2015**, *8*. [CrossRef]
30. Holnicki, P.; Kałuszko, A.; Trapp, W. An urban scale application and validation of the CALPUFF model. *Atmos. Pollut. Res.* **2016**, *7*, 393–402. [CrossRef]
31. Kim, H.C.; Kim, E.; Bae, C.; Cho, J.H.; Kim, B.-U.; Kim, S. Regional contributions to particulate matter concentration in the Seoul metropolitan area, South Korea: Seasonal variation and sensitivity to meteorology and emissions inventory. *Atmos. Chem. Phys. Discuss.* **2017**, *17*, 10315–10332. [CrossRef]
32. Li, J.; Zhang, M.; Wu, F.; Sun, Y.; Tang, G. Assessment of the impacts of aromatic VOC emissions and yields of SOA on SOA concentrations with the air quality model RAMS-CMAQ. *Atmos. Environ.* **2017**, *158*, 105–115. [CrossRef]
33. Gong, B.; Kim, J.; Kim, H.; Lee, S.; Kim, H.; Jo, J.; Kim, J.; Gang, D.; Park, J.M.; Hong, J. A Study on the Characteristics of Condensable Fine Particles in Flue Gas. *J. Korean Soc. Atmos. Environ.* **2016**, *32*, 501–512. [CrossRef]
34. Norkko, J.; Thrush, S.F. Ecophysiology in environmental impact assessment: Implications of spatial differences in seasonal variability of bivalve condition. *Mar. Ecol. Prog. Ser.* **2006**, *326*, 175–186. [CrossRef]
35. Venugopal, T.; Giridharan, L.; Jayaprakash, M.; Periakali, P. Environmental impact assessment and seasonal variation study of the groundwater in the vicinity of River Adyar, Chennai, India. *Environ. Monit. Assess.* **2008**, *149*, 81–97. [CrossRef] [PubMed]

36. Jung, J.; Lee, K.; Cayetano, M.G.; Batmunkh, T.; Kim, Y.J. Optical and hygroscopic properties of long-range transported haze plumes observed at Deokjeok Island off the west coast of the Korean Peninsula under the Asian continental outflows. *J. Geophys. Res. Atmos.* **2015**, *120*, 8861–8877. [CrossRef]
37. Yu, Z.; Jang, M.; Kim, S.; Bae, C.; Koo, B.; Beardsley, R.; Park, J.; Chang, L.S.; Lee, H.C.; Lim, Y.-K.; et al. Simulating the Impact of Long-Range-Transported Asian Mineral Dust on the Formation of Sulfate and Nitrate during the KORUS-AQ Campaign. *ACS Earth Space Chem.* **2020**, *4*, 1039–1049. [CrossRef]
38. Ghim, Y.S.; Choi, Y.; Kim, S.; Bae, C.H.; Park, J.; Shin, H.J. Model Performance Evaluation and Bias Correction Effect Analysis for Forecasting PM2.5 Concentrations. *J. Korean Soc. Atmos. Environ.* **2017**, *33*, 11–18. [CrossRef]
39. Yang, H.-H.; Lee, K.-T.; Hsieh, Y.-S.; Luo, S.-W.; Huang, R.-J. Emission Characteristics and Chemical Compositions of both Filterable and Condensable Fine Particulate from Steel Plants. *Aerosol Air Qual. Res.* **2015**, *15*, 1672–1680. [CrossRef]
40. Interim Guidance on the Treatment of Condensable Particulate Matter Test Results in the Prevention of Significant Deterioration and Nonattainment New Source Review Permitting Programs. Available online: https://www.epa.gov/nsr/interim-guidance-treatment-condensable-particulate-matter-test-results-prevention-significant (accessed on 9 April 2021).

 *sustainability*

Article

Use of Simulated and Observed Meteorology for Air Quality Modeling and Source Ranking for an Industrial Region

Awkash Kumar [1,2,*], Anil Kumar Dikshit [1] and Rashmi S. Patil [1]

[1] Environmental Science and Engineering Department, Indian Institute of Technology, Bombay, Mumbai 400 076, Maharashtra, India; rspatil@iitb.ac.in (R.S.P.); dikshit@iitb.ac.in (A.K.D.)

[2] SAGE, Sustainable Approach for Green Environment, Powai, Mumbai 400 076, Maharashtra, India

* Correspondence: awkash.narayan@gmail.com; Tel.: +720-824-6617; Fax: 022-2576-4650

Abstract: The Gaussian-based dispersion model American Meteorological Society/Environmental Protection Agency Regulatory Model (AERMOD) is being used to predict concentration for air quality management in several countries. A study was conducted for an industrial area, Chembur of Mumbai city in India, to assess the agreement of observed surface meteorology and weather research and forecasting (WRF) output through AERMOD with ground-level NO_x and PM_{10} concentrations. The model was run with both meteorology and emission inventory. When results were compared, it was observed that the air quality predictions were better with the use of WRF output data for a model run than with the observed meteorological data. This study showed that the onsite meteorological data can be generated by WRF which saves resources and time, and it could be a good option in low-middle income countries (LIMC) where meteorological stations are not available. Also, this study quantifies the source contribution in the ambient air quality for the region. NO_x and PM_{10} emission loads were always observed to be high from the industries but NO_x concentration was high from vehicular sources and PM_{10} concentration was high from industrial sources in ambient concentration. This methodology can help the regulatory authorities to develop control strategies for air quality management in LIMC.

Keywords: meteorology; WRF; air quality; AERMOD; source apportionment

Citation: Kumar, A.; Dikshit, A.K.; Patil, R.S. Use of Simulated and Observed Meteorology for Air Quality Modeling and Source Ranking for an Industrial Region. *Sustainability* **2021**, *13*, 4276. https://doi.org/10.3390/su13084276

Academic Editor: Sara Egemose

Received: 3 February 2021
Accepted: 9 April 2021
Published: 12 April 2021

Publisher's Note: MDPI stays neutral with regard to jurisdictional claims in published maps and institutional affiliations.

1. Introduction

Urbanization-related issues have become very prominent across the world [1–3], especially in developing countries like India where cities have started facing an acute air pollution problem due to urbanization [4]. Many Indian megacities, such as Delhi, Mumbai, Bangalore, and Kolkata, are witnessing increasing health problems due to rapid increase in air pollution [5]. The total health cost due to air pollution for Mumbai was about USD 8 billion for the year 2012 [6]. This problem becomes particularly complex to resolve in urban areas because of diverse emission sources such as vehicles, industries, bakeries, hotels, diesel generating sets and combustion of solid fuels in the domestic sector.

Air quality monitoring networks have been installed at various locations in many cities. Also, installation and operation of a large number of air quality monitoring stations need considerable financial resources from government which may not be supported in low-middle income countries (LIMC). This monitoring data is increasingly used to communicate the existing status of air quality. However, it doesn't contribute to the understanding of sources and meteorological factors. Although the observed data represent air quality status for a particular location only, the use of dispersion models can provide information about much larger areas. Further, modeling helps in the determination of concentration plots on spatial and temporal scales and contributions from different types of source for air pollutants [7–11]. The dispersion model can also be used to identify pollution sources with the help of emission inventory [12,13]. This is very useful in making rational management strategies [7,14–20]. A dispersion model can also determine the contribution of various sources in a region whereas a receptor model determines source contribution at a particular location [21]. Data required for dispersion models include emission inventory, geographical data, and meteorological data of the region [22]. Data availability, especially meteorological data, is an important factor for the assessment of air quality in LMIC because running a meteorological monitoring station requires resources. The use of poor-quality meteorological data in air quality models may contain significant adverse effect model output quality [23,24]. Meteorological data is generally taken from a nearby meteorological station and is used for the study region. The results of air quality model may have significant error despite advanced computer technology, and various techniques like numerical modeling techniques, performance evaluations of state-of-the-art [23–27]. A survey has been done for the air quality and meteorological model [28].

The observed meteorological data from a monitoring station may not give good performance in air quality modeling for the urban industrial region where several emission sources are present at multiple heights and variation in topography. An alternative is to generate onsite meteorological data using a meteorological model which could be an effective option in LMIC. A study was conducted on the coupling of American Meteorological Society/ Environmental Protection Agency Regulatory Model (AERMOD) with Weather Research and Forecasting (WRF) model in Pune city (India) for a single pollutant PM_{10} [29]. However, their predicted concentration obtained by the WRF-generated meteorology and observed values have not been compared and contribution of various sources in the study region has not been estimated. Short term air quality forecasting also has been carried out using WRF forecasted meteorology and AERMOD for five days for Chembur region [30]. In these studies, the requirement of horizontal homogeneous hourly surface and upper meteorological data has been fulfilled from WRF model for AERMOD. The main objective of this study was to generate onsite meteorological data at mesoscale using WRF model and compare the results with observed meteorological data. Then, we proposed to use both the data in air quality modeling and to evaluate the option of making WRF coarse resolution output feasible in LMIC. This study was also continued to rank the contribution of emissions and ambient concentrations from sources for NO_x and PM_{10}. This will be useful for air quality management of the urban area for regulation purposes [31] in LMIC.

2. Study Area

The study area, Chembur, represents an industrial site of Mumbai city in India with global coordinates 19.05° N and 72.89° E. This area covers M East and M West wards of Municipal Corporation of Greater Mumbai, which is one of the financial centers of India as shown in Figure 1a. Chembur has a population of 1.2 million. It measures 6.5 km east-to-west and 8.45 km north-to-south, as shown in Figure 1c. This region has marine alluvium type of soil and North-South running basalt hills to its South [32]. The topographical features have been shown in Figure 1d in the Universal Transverse Mercator coordinate system. The elevation is maximum at the central part of the study area and minimum along the boundary of the Eastern study area. The elevation ranges from 1 to 200 m. The elevation just above the location of Rashtriya Chemicals and Fertilizers Limited (RCFL) is 100 to 200 m. Major industries in this area are Bharat Petrochemical Corporation Limited (BPCL), Hindustan Petroleum Refinery Corporation Limited (HPCL), Tata Thermal Power Corporation Limited (TPCL) and RCFL. Containers and heavy-duty vehicles from this area use the Port Trust Road, Mahul Road and Ramakrishna Chemburkar Marg (R C Marg). Road conditions are poor due to the continuous movement of heavy vehicles. The residential areas spread over the north boundary of the study area has a residential zone comprising Chheda Nagar (between point 2 and 3) and Shramjivi Nagar (Left side area of point 2), the south boundary is adjacent to the Tata Thermal Power Plant. The west boundary lies by RCFL and Mahul, and the east boundary is aligned with Shahyadri Nagar and Prayag Nagar.

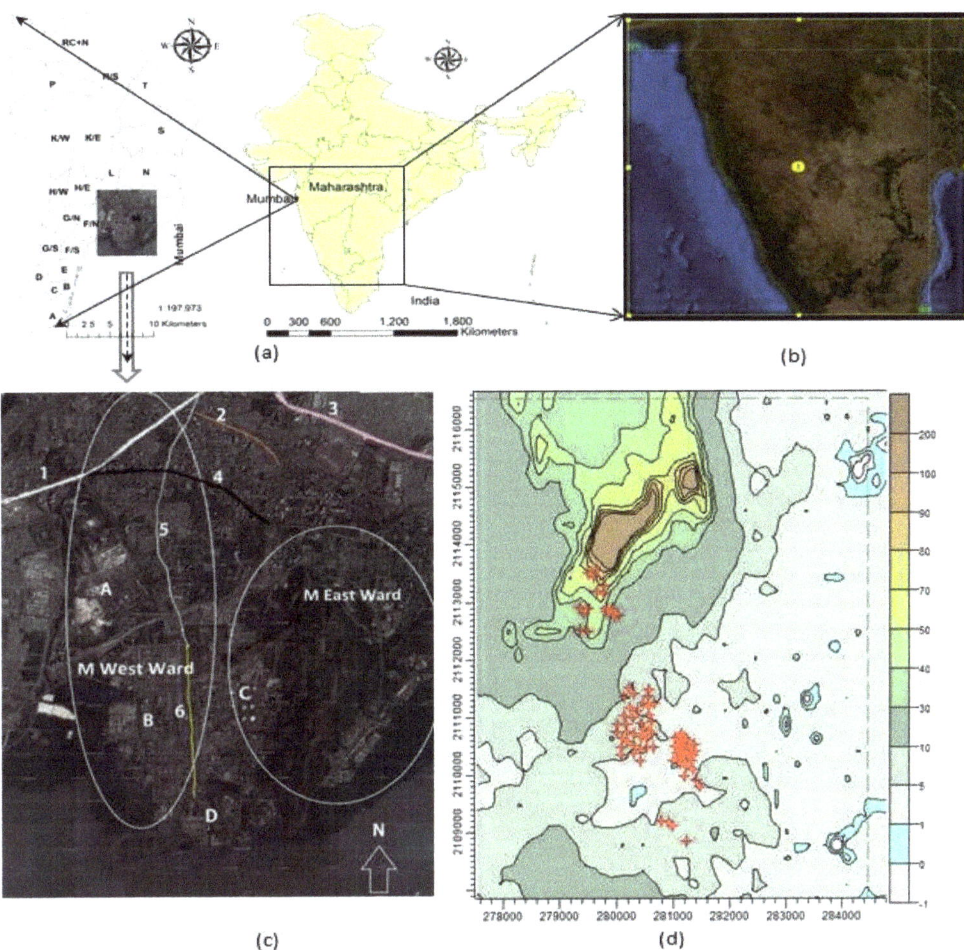

Figure 1. (a) Mumbai Area, (b) WRF Domain (c) Study Area Chembur (Downloaded from Google Earth) (d) Terrain Map. Note: Figure (d) is given in Universal Transverse Mercator coordinate system. 1-Eastern Express Highway Road, 2-NGA Marg, 3-Ghatkopar-Mankhurd Road, 4-V N Purav Road, 5-R C Marg, 6- B D Patil Marg, A-RCFL, B-BPCL, C-HPCL, D-TPCL.

Around twenty years ago, Chembur was one of the most polluted regions in Mumbai. With the sustained effort and pressure from authorities and industries for implementing a series of control measures, the region has witnessed an improvement in air quality. In the last two decades, the region characteristics have improved due to the closure of many industries, but residential development and vehicular density has increased [33]. Chembur still needs appropriate air quality studies for developing management strategies, as its ambient air quality is poor when compared with the National Ambient Air Quality Standard 2009, Central Pollution Control Board (CPCB) New Delhi (India). CPCB has published a document, giving a Comprehensive Environmental Pollution Index (CEPI) score for various industrial regions in the country. Chembur has a score of 69.19 CEPI in this report [34]. This score shows that this region should be rated as a severely polluted area. Hence, the region requires better understanding of air quality processes so that effectiveness of the action plan can be realized.

3. Methodology and Data

The schematic data flow of the study has been shown in Figure 2. AERMOD requires emission inventory and nine hourly meteorological parameters (wind speed, wind direction, rain fall, temperature, humidity, pressure, ceiling height, global horizontal radiation and cloud cover) as the input data. These meteorological parameters were generated from the WRF model for the year. Also, the meteorological parameters were observed at RCFL for the same time period and both data sets were compared. Prediction of concentrations using an air quality model (AERMOD) was carried out with the observed meteorological data and WRF generated data. Meteorological parameters were prepared in columns and temporal resolution was prepared in rows of a spreadsheet. This spreadsheet was processed in AERMET which is a pre-processor of AERMOD. The terrain data at 90 m resolution of Shuttle Radar Topography Mission (SRTM) was used in AERMAP which is also the pre-processor of AERMOD. Then, the AERMOD model was used to predict concentration of NO_x and PM_{10} as shown in Figure 2. Also, comparisons of both the models, WRF and AERMOD, were done. The metrological model, the setup of parameterization of variables, dispersion model, emission load, and observations have been presented section-wise.

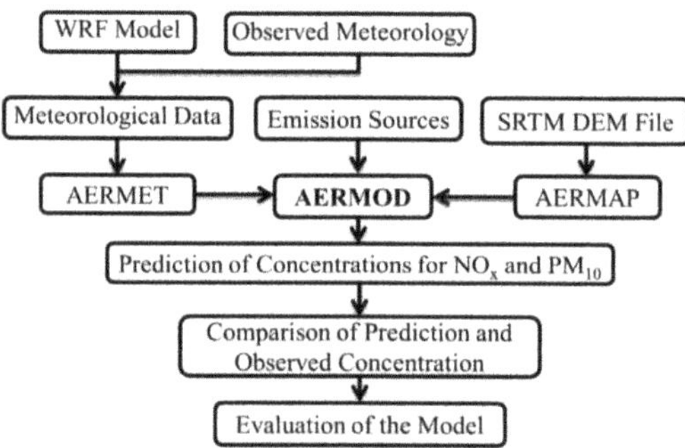

Figure 2. Schematic Flow for the Study.

4. Meteorological Model

The mesoscale model, Advanced Research WRF model version 3.2, has been used in this study [35]. This model is designed to assist both atmospheric research and operational forecasting needs [36]. NCEP FNL (Final) Operational Global Analysis data have been used as an input for WRF, which are on 1.0×1.0 degree grids prepared operationally every six hours. This product is from the Global Data Assimilation System (GDAS), which continuously collects observational data from the Global Telecommunications System (GTS) and other sources for analyses [37]. It is a limited area, non-hydrostatic primitive equation model with multiple options for various physical parameterization schemes. This version employs Arakawa C-grid staggering for the horizontal grid and a fully compressible system of equations. A terrain following hydrostatic pressure coordinates with vertical grid stretching is implemented vertically. The time split integration uses a third order Runge Kutta scheme with smaller time step for acoustic and gravity wave modes. The WRF model physical options used in this study consist of the WRF model Single Moment 6-class simple ice scheme for microphysics, the Kain–Fritsch scheme for the cumulus convection parameterization, and the Yonsei University planetary boundary layer scheme. The rapid radiative transfer model and the Dudhia scheme are used for longwave and shortwave radiation, respectively, while the Noah land surface model has been selected. All these parameterizations constitute a well-tested suite of schemes over the Indian region [38–40]. The model domain extends between 71° E to 81° E zonally and 11° N to 21° N meridionally, consisting of 100 by 100 grid points with 25 km grid spacing as shown in Figure 1b. The model was run from 1st January to 31st December of the year. The model has 28 vertical levels with the top of the model at 10 hPa. Topography as well as snow cover information have been obtained from the United States Geological Survey. The meteorological parameters have been extracted from the WRF model at ground level. The WRF model has been run at 25 km resolution which provides time series meteorological parameters for a specific period at a particular location. In this study, 9 hourly meteorological parameters (cloud cover, temperature, pressure, relative humidity, wind direction, wind speed, ceiling height, rainfall, and global horizontal radiation) have been simulated for the year using WRF. WRF gives output in network common data format, and GRADS v 2.2 is post processing software which reads the network common data format. The output from WRF was fed to Grads 2.2 to generate digital hourly meteorological data to arrange in an Excel spreadsheet. The AERMET required data in excel spreadsheet or other formats. The input in AERMET was given in excel spreadsheet which was prepared using the data obtained by Grads 2.2. Here, hourly data for each meteorological parameter are provided in different columns. The spreadsheet meteorological data were imported in AERMET which is pre-processor of AERMOD.

5. Dispersion Model

Dispersion model uses emission inventory, geographical and meteorological data to predict concentration at the receptor's point in the study region. The format of the input data varies with different models. There are many specific models for vehicular and industrial sources as well as for a variety of sources [19,41–43]. Industrial Source Complex (ISC3), developed by the United State Environmental Protection Agency, is a steady-state Gaussian plume model which can be used to assess pollutant concentrations from a wide variety of sources associated with an industrial source complex [44]. ISC3 operates in both long-term (ISCLT3) and short-term (ISCST3) models. ISCST3 model is the regulatory model in India and it has been used in many case studies [45]. Later on, it was updated to AERMOD whose performance was appreciable as they added some advanced algorithms to get more accurate results [46]. The air quality model that we use, AERMOD, has been applied to evaluate dispersion of several pollutants, including PM_{10}, HCN, SO_2, SF_6, and VOCs and is recommended widely by regulatory authorities [47–51].

The study area (as given Figure 1c) was given in AERMOD and emission locations were digitized according to real earth surface reference and quantities of emissions were put based on estimated emission inventory. Therefore, there is no resolution concept for emission inventory in this study. The meteorological data output from the WRF model was processed in AERMET and its output was fed into AERMOD. The pre-procedure AERMAP of AERMOD calculates representative terrain-influence height, also referred to as the terrain height scale, at a receptor in modeling of air quality. Cartesian uniform gridded receptors were given, apart from discrete receptors and all receptors were at 2 m height. Anemometer height was 10 m and surface roughness length was 1 m in this model run. Building downwash terminology was not considered. AERMOD calculates concentration for each hour using hourly meteorological data for each pollutant separately.

6. Emission and Concentration Data

Emission load has been computed for point sources (specifically 36 stacks of BPCL, 30 stacks of HPCL, 4 stacks of Tata Power and 17 stacks of RCFL), line sources (the 6 roads of Chembur), and area sources (e.g., bakeries, hotels and restaurants, crematorium and domestic sector). These area sources were taken from a previous study (CSIR-NEERI) [52] for M East and M West wards, where area sources emission load has been computed. These sources are for the region of Chembur (M East and West Ward) where domestic sectors are available. Industrial emission data are collected from industries and vehicular emission inventory are prepared based on field survey data for the study period. Vehicular emission rate is estimated using the actual number of vehicles in unit time, emission factors and vehicle kilometer travelled [53]. The percentage of contribution is estimated from type of the sources after making emission inventory of the region. PM_{10} and NO_x emissions are mainly caused by industries (94% and 64%, respectively). Remaining 36% of NO_x emission is contributed by vehicles (18%), domestic sources (17%) and others (1%). The emission factors of vehicles are available for particulate matter, and this has also been taken as PM_{10}. Figure 3 depicts emission loads of NO_x and PM_{10} from various types of sources in the study area. The observed concentration data were collected for NO_x and PM_{10} at industrial sites, i.e., HPCL and BPCL at a height of about 3.5–4.0 m. Continuous (hourly) ambient air quality monitoring was done at these sites using Telydene instrument for NO_x and PM_{10}.

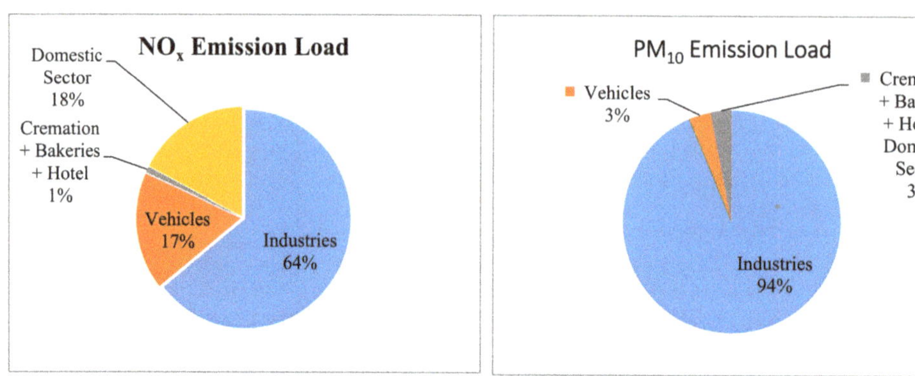

Figure 3. Emission load of NO_x and PM_{10} in Chembur (in kg/day).

7. Results and Discussion

The results of WRF were used as inputs in estimation of concentration by AERMOD. NCEP FNL (Final) Operational Global Analysis data was used to process as an input for WRF. It produced 30 meteorological parameters for the required time resolution and study period. AERMET (pre-processor of AERMOD) required nine meteorological parameters and these were extracted, out of 30 meteorological parameters for air quality modeling. After this, the contributions from various sources in ambient concentration were estimated in this study.

8. Validation of Wind and Temperature Time Series

WRF generated the nine meteorological parameters hourly which were used in air quality modeling. The validation of meteorological output from WRF was done. In the validation of WRF, output time series temperature and wind are significant for air pollution Therefore, the point of validation of temperature was conducted using the hourly temperature of the year of WRF with observed temperature data of RCFL industry, Chembur. A fair estimate of the dispersion of pollutants in the atmosphere is possible based on the frequency distribution of wind direction as well as wind speed. Wind transports pollutants from various sources, causing turbulent mixing and diluting pollution. Boundary layer cumulus clouds vent pollution into the free troposphere, and temperature and humidity levels in the boundary layer affect chemical reactions and the rates at which many dangerous compounds are formed.

8.1. Validation of Temperature

Figure 4 shows a comparison of surface temperature of the study area, between WRF simulation and observations, and it can be seen that they are in good agreement. The average percentage error between simulated and observed temperature is −5.3%. The standard deviation of percentage error is 10.4.

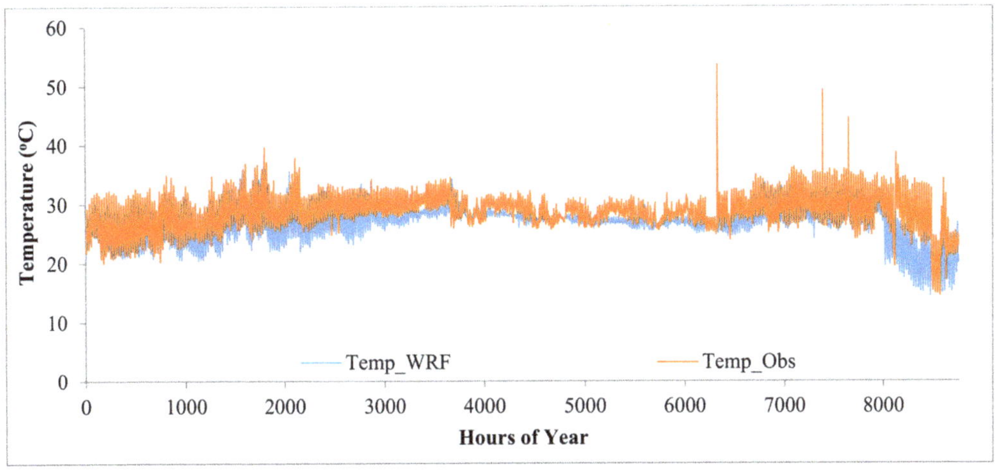

Figure 4. Comparison of Hourly Temperature of WRF with Observed Data.

8.2. Validation of Wind

Wind data was derived using WRF model at the height of 10 m from the ground for a whole year, and wind rose of Chembur, which was simulated by (a) WRF model and (b) RCFL observed data for the year, was plotted (Figure 5). Maximum wind persistence corresponds from the west and south-west direction in WRF, while in RCFL observed data, the wind was found to blow in the west and north-west direction for most of the time. Observed wind data had more periods of calmness than the data simulated by WRF. Observed wind rose had 62% calm condition whereas simulated wind rose had 4% calm condition over a period. The root mean square error and mean bias error between predicted and observed wind speed were 4.05 and 3.29, respectively. The root mean square error and mean bias error between predicted and observed temperature were 3.69 and -1.5, respectively. The observed data of wind was collected at RCFL industry at the height of 6 m, and it represents the microscale domain. Chembur has considerable variation in topography as shown Figure 1d. The topography changes after a few meters and this causes the wind to divert. Consequently, this variation of topography may cause the mismatch of wind rose of WRF and observed data.

Modeling was performed with both the meteorological parameters of WRF output and RCFL observed data. WRF performance of wind is poor due to the surface inhomogeneity. However, for the purpose of dispersion modeling, we consider WRF as a good representation of the mesoscale flow. The observed wind at RCFL is not representative of the entire Chembur region for dispersion modeling. This may be because the maximum emission is from industrial sources in this region, which are at an elevated height. Hence, for these sources, the mesoscale meteorology generated by the WRF model may be more appropriate than the observed microscale meteorology.

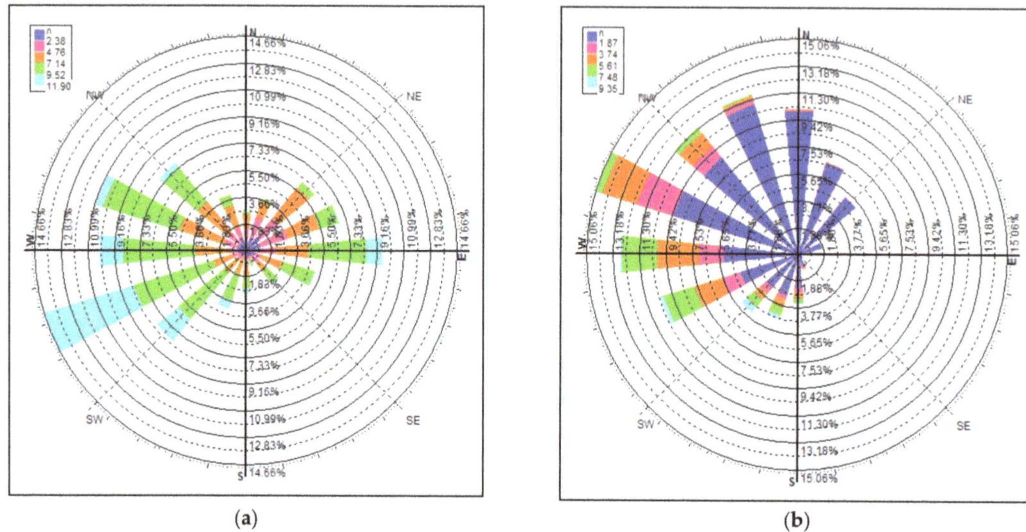

(a)   (b)

Figure 5. Annual Wind Rose Simulated by (**a**) WRF and (**b**) RCFL Observed Data.

9. Validation of NO_x and PM_{10}

The annual PM_{10} and NO_x concentration contour plots for all sources of the study domain are shown in Figure 6a,b respectively. A comparison of simulated concentration using WRF output with the observed concentration of PM_{10} and NO_x are in Table 1, and the model was well-compared for this study area. The root mean square error and mean bias error between predicted and observed concentrations for NO_x were 1.76 and 0.063, respectively, while they were 0.41 and 0.83 for PM_{10}, respectively. The standard deviations for NO_x at BPCL and HPCL were 33.6 and 30.2, respectively, and the standard deviations for PM_{10} at BPCL and HPCL were 16.4 and 12.4, respectively. It was seen that the values obtained through air quality modeling were closer to the observed concentrations with the mesoscale meteorology than the surface level meteorology. The model results using observed surface meteorology were high. Modeled values were in good agreement with the observed values at both locations for NO_x, but for PM_{10}, simulated concentrations were lesser than the observed concentration at HPCL. This can be due to the vehicular congestion and resuspended particles. The model performed well with mesoscale meteorology after all the sources and the entire region were considered. As Chembur has immense variation in topography, land use and geographical structure, as shown in Figure 1d, microscale meteorology varies with these factors. Further, mesoscale meteorology has been used for other analyses for air quality. The contours were plotted using the model for NO_x and PM_{10} concentrations based on one-hour average values for one year for all sources in the study area. The same analysis was repeated only for industrial sources and vehicular emission sources separately. It was observed that the maximum concentration of PM_{10} was 71.8 $\mu g/m^3$. This concentration was observed near the Chembur Gaothan area where the vehicular congestion was more intense. Around BPCL and HPCL area, PM_{10} concentration was around 50 $\mu g/m^3$. The minimum concentration of PM_{10} was less than 42 $\mu g/m^3$, and this concentration was observed near the southern boundary of the study region, which has been represented by deep violet color. The PM_{10} modeling was carried out using 30 $\mu g/m^3$ as the background concentration. This concentration was estimated when modeled concentrations matched with observed concentrations. This background concentration also includes resuspended particulate matter (RSPM), which is induced by vehicular congestion and other factors. In PM_{10} modeling, this background concentration can be taken as a correction factor. The maximum concentration of NO$_x$ was observed to be 53 µg/m^3 near the Ghatkopar–Mankhurd link road. This can be due to the vehicular congestion in Deonar Village, BPCL and HPCL area. Annual minimum concentration was less than 10 µg/m^3 along the eastern, western and southern boundaries of the study area, which has been shown in deep violet color.

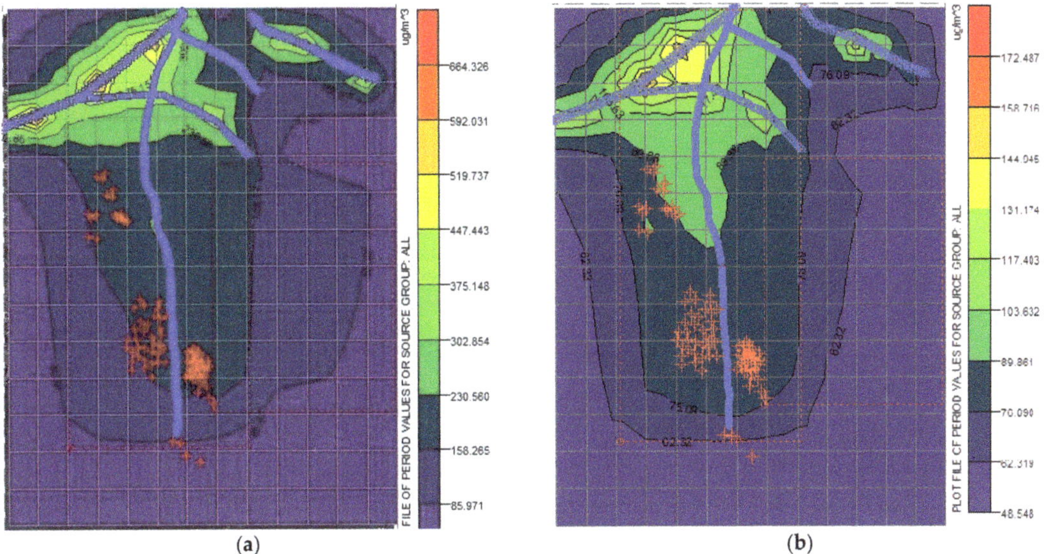

Figure 6. Annual Concentration of (**a**) NO$_x$ and (**b**) PM$_{10}$ (in µg/m^3) using observed meteorology for Chembur Area.

Table 1. Monthly and Annual Comparison of Simulated Concentration with Ambient Observed Concentration (µg/m^3).

Month-Location	NO$_x$			PM$_{10}$		
	Simulated	Observed	% Error	Simulated	Observed	% Error
Jan-BPCL	29	26	11.5	53	62	−14.5
Jan-HPCL	36	22	63.6	55	65	−15.4
Feb-BPCL	36	34	5.9	54	54	0.0
Feb-HPCL	43	28	53.6	57	68	−16.2
Mar-BPCL	28	27	3.7	53	56	−5.4
Mar-HPCL	39	25	56.0	56	64	−12.5
Apr-BPCL	31	34	−8.8	60	55	9.1
Apr-HPCL	27	28	−3.6	65	54	20.4
May-BPCL	21	32	−34.4	51	58	−12.1
May-HPCL	28	23	21.7	55	56	−1.8
Jun-BPCL	15	16	−6.3	51	35	45.7
Jun-HPCL	21	26	−19.2	55	58	−5.2
Jul-BPCL	8	20	−60.0	49	43	14.0
Jul-HPCL	15	18	−16.7	53	50	6.0
Aug-BPCL	7	20	−65.0	49	43	14.0
Aug-HPCL	18	17	5.9	55	48	14.6
Sep-BPCL	22	19	15.8	51	41	24.4

Table 1. *Cont.*

Month-Location	NO$_x$			PM$_{10}$		
	Simulated	Observed	% Error	Simulated	Observed	% Error
Sep-HPCL	26	21	23.8	54	54	0.0
Oct-BPCL	39	25	56.0	52	42	23.8
Oct-HPCL	40	32	25.0	55	55	0.0
Nov-BPCL	31	24	29.2	55	53	3.8
Nov-HPCL	23	26	−11.5	50	59	−15.3
Dec-BPCL	28	25	12.0	56	56	0.0
Dec-HPCL	15	21	−28.6	49	62	−21.0
Annual-BPCL	24.6	25.2	−3.4	52.8	49.8	8.6
Annual-HPCL	27.6	23.9	14.2	54.9	57.8	−3.9

Emission load does not represent the rank-wise contribution of sources to the ambient concentration of the region. Hence, modeling was carried out for industrial sources, vehicle sources, and low duty diesel vehicles (LDDVs) to observe the relative source-wise contribution to the ambient air quality for future scope of implementation of control strategies and environmental management.

9.1. Results of Industrial Sources

For industrial emission sources modeling, four industries (BPCL, HPCL, RCFL and TPCL) were considered in Chembur. In this study, NO$_x$ and PM$_{10}$ emissions were modeled for the year to find out the dominant source in the study domain.

9.2. Contribution of NO$_x$ and PM$_{10}$ Concentration by Industries

NO$_x$ and PM$_{10}$ emission load were 64% and 94%, respectively from industries in Chembur (Figure 3). However, it contributes less to ambient concentration in the study area. The southern part of the study area is dominated by industrial sources, and due to meteorology, the maximum concentration of NO$_x$ is 6.2 µg/m^3, seen at HPCL. Also, NO$_x$ concentration is 4.8 µg/m^3 in the southern part of BPCL. The maximum concentration of PM$_{10}$ is 35 µg/m^3 at BPCL and 33 to 34 µg/m^3 at HPCL and RCFL. Table 2 shows the comparison of the simulated concentration of NO$_x$ and PM$_{10}$ for industrial sources only and ambient simulated concentration of NO$_x$ and PM$_{10}$ for this study area, respectively.

Table 2. Comparison of simulated concentration from industries with ambient simulated concentration of NO$_x$ and PM$_{10}$.

Pollutant	Location	Simulated Concentration for Industries (µg/m^3)	Ambient Simulated Concentration (µg/m^3)	Contribution of Industrial Source to Ambient Air Quality
NO$_x$	BPCL	4.8	28.2	17%
	HPCL	6.1	21.7	28%
PM$_{10}$	BPCL	35.5	56.9	62%
	HPCL	33.2	52.1	64%

9.3. Results of Line Sources

In line source modeling, six roads in Chembur area have been considered. In the present study, NO$_x$ and PM$_{10}$ emissions have been modeled for the year to find out the dominant sources in the study domain. Vehicular emission varies with time such as morning peak, evening peak, off peak and the lean peak of the day.

9.4. NO_x and PM_{10} Concentration Contribution by Vehicles

NO_x and PM_{10} emission loads from vehicles in Chembur are 17% and 3%, respectively (Figure 3). Nevertheless, these are contributing considerably to ambient concentration because they are ground emission sources. The northern part of the study area is dominated by vehicular sources and high density of vehicles. The maximum concentration of NO_x was 43 $\mu g/m^3$ at Chedda Nagar and 40.1 $\mu g/m^3$ at Chembur Naka. The southern part of the study area has an inconsequential effect on vehicular pollution (only 2–5 $\mu g/m^3$). At Chheda Nagar, the concentration of ambient air quality from all the sources was 54 $\mu g/m^3$, while the concentration by vehicular sources was 41 $\mu g/m^3$. Chheda Nagar is in the northern region of the study area, and this part is affected less by industrial emissions. The northern part of the study domain is highly populated compared to the southern part. The southern part has lesser contribution from vehicles. At Chheda Nagar, the maximum concentration of PM_{10} from vehicles was 37.84 $\mu g/m^3$ and from the other sources was 70.8 $\mu g/m^3$.

9.5. NO_x and PM_{10} Concentration Modeling by Diesel Car and LDDV

Northern and western corners in this analysis are dominated by diesel cars and light duty diesel vehicles (LDDVs), while the other areas are almost free from NO_x pollutant. The maximum concentration of NO_x was 12.5 $\mu g/m^3$ and it was found at Chheda Nagar. The entire area of Shramjivi Nagar showed NO_x concentration of 7-9 $\mu g/m^3$ sourced from diesel cars and LDDVs. Concentration contribution in ambient air quality from vehicles was 43.1 $\mu g/m^3$. Thus, it can be concluded that diesel cars and LDDVs are contributing to one fourth of the line source emission. In PM_{10} concentration modeling, northern and western corners are affected by diesel cars and LDDVs. The maximum PM_{10} concentration of 32.8 $\mu g/m^3$ was observed at Everard Nagar (Point 1 area) and 32.5 $\mu g/m^3$ at Chembur Naka from diesel cars and LDDVs with 30 $\mu g/m^3$ background concentration. Thus, it can be concluded that diesel cars and LDDVs are contributing to around 17% to line sources and 25% to the total concentration.

10. Summary and Conclusions

The aim of the study was to generate onsite meteorological data for usage in air quality modeling to see the feasibility of coarse resolution of WRF output in LMIC. It generated onsite and real time meteorological data, which was fed in AERMET, the pre-processor of the dispersion model AERMOD. AERMOD calculated concentrations for NO_x and PM_{10} using compiled emission inventory for all the sources, namely industries and vehicles, of the study area. Air quality modeling results showed that in this particular case, the meteorological data from WRF output at mesoscale performed better than the observed meteorological data. WRF output could be a good option which may represent a better meteorology for the purpose of dispersion modeling. Also, this may be because industrial sources have a significant contribution in the region. The results were used to identify the critical areas and relative contribution of various sources to ambient air quality. The use of WRF model is very economical in resources and time.

The ambient concentration load of the study area is shown in Figure 7 for NO_x and PM_{10}. Here, for ambient concentration, the vehicular emissions are dominating in the region because these are ground level sources. Based on the study, following conclusions can be drawn:

- ➤ Amongst total emissions, PM_{10} emission load was 3%, and NO_x emission load was 17% from vehicles. Industr sources contributed 64% and 94% of NO_x and PM_{10} load, respectively. The domestic sector contributed significantly to NO_x emission as 18% of total emission load.
- ➤ NO_x emission load of industries was 64% of the total emission load, but it contributed only 25–30% of NO_x concentration in the ambient air.
- ➤ There was 94% contribution to total emission load of PM_{10} from industries in the study domain, but only 57% contribution to ambient air quality level.

➢ NO_x emission contribution from vehicles was 17% of total emission, but in ambient air quality it contributed only 26% of the total because it is a ground level source.

➢ Vehicular PM_{10} emission contribution was 3% of the total emission load, but in ambient air quality it contributed 25% of the total ambient PM_{10} concentration.

➢ At ambient concentration of NO_x, diesel cars and LDDVs contributed one fourth of the line sources for this study domain.

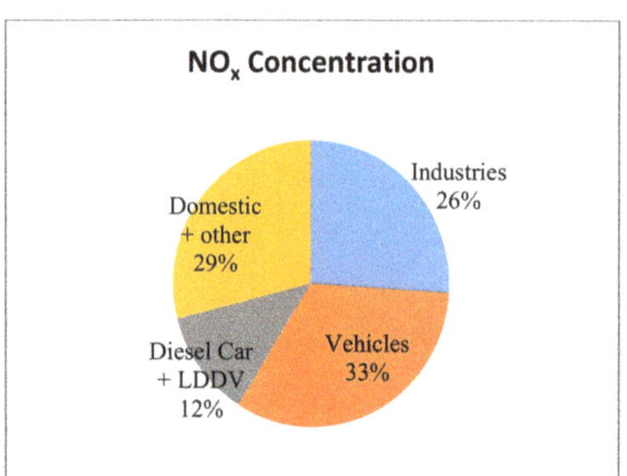

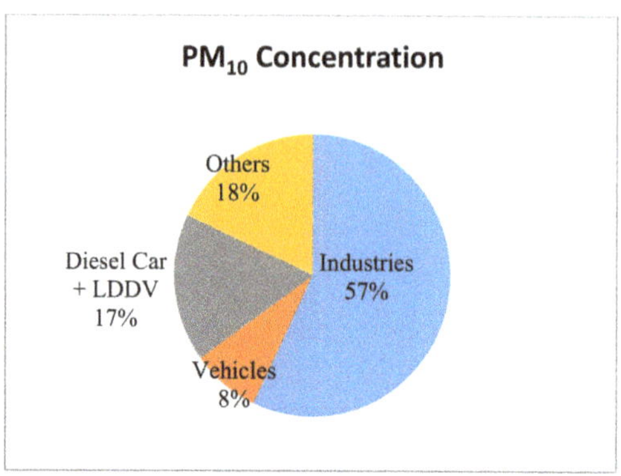

Figure 7. Comparison between emission load and ambient concentration load of NO_x and PM_{10} in Chembur area.

Micro-meteorology may vary a lot due to the topography of the region. Topographic features could be one of the limitations of the use of WRF output meteorology for air quality modeling because it may not capture the high-rise buildings in the region. In AERMOD, only one meteorological profile can be used but meteorology may not be uniform for the entire region. It has been found many times that the observed meteorological parameters carry some measurement errors [54–56].

This will be very useful in the forecast, implementation, control, and management strategies for improving air quality and also for performing a risk analysis of different types of sources in the region as future scope. Also, this work can be useful to regulatory authorities to develop a framework for air quality management in LMIC. The shortage of research was (a) the meteorological data were available at only one location which can be observed at other locations and comparison can be done for the same and (b) suspended dust can be estimated and incorporated in emission inventory. Various physics option parameters to set up the WRF model and simulation of the microscale meteorology with a comparison of observed meteorology could be possible future researches. Health benefit analysis can also be done by estimating population exposure with air quality [57].

Author Contributions: Conceptual, methodology development, software operation and writing, A.K., review, data procurement and guidance, A.K.D. and R.S.P. All authors have read and agreed to the published version of the manuscript.

Funding: This research received no external funding.

Institutional Review Board Statement: Ethical review and approval are not required for the study as the research does not involve humans.

Informed Consent Statement: Not applicable.

Data Availability Statement: Data available on request.

Conflicts of Interest: The authors declare no conflict of interest.

References

1. Huff, G.; Angeles, L. Globalization, industrialization and urbanization in Pre-World War II Southeast Asia. *Explor. Econ. Hist.* **2011**, *48*, 20–36. [CrossRef]
2. DeFries, R.; Pandey, D. Urbanization, the energy ladder and forest transitions in India's emerging economy. *Land Use Policy* **2011**, *27*, 130–138. [CrossRef]
3. Duh, J.-D.; Shandas, V.; Chang, H.; George, L.A. Rates of urbanisation and the resiliency of air and water quality. *Sci. Total. Environ.* **2008**, *400*, 238–256. [CrossRef]
4. Patankar, A.; Trivedi, P. Monetary burden of health impacts of air pollution in Mumbai, India: Implications for public health policy. *Public Health* **2011**, *125*, 157–164. [CrossRef] [PubMed]
5. World Health Organization (WHO). *World Health Statistics, Department of Public Health, Environmental and Social Determinants of Health*; WHO Press: Geneva, Switzerland, 2014.
6. Kumar, A.; Gupta, I.; Brandt, J.; Kumar, R.; Kumar, A.; Patil, R.S. Air quality mapping using GIS and economic evalu-ation of health impact for Mumbai city, India. *J. Air Waste Manag. Assoc.* **2016**, *66*, 470–481. [CrossRef] [PubMed]
7. Banerjee, T.; Barman, S.; Srivastava, R. Application of air pollution dispersion modeling for source-contribution assessment and model performance evaluation at integrated industrial estate-Pantnagar. *Environ. Pollut.* **2011**, *159*, 865–875. [CrossRef]
8. Kumar, A.; Dikshit, A.K.; Fatima, S.; Patil, R.S. Application of WRF Model for Vehicular Pollution Modelling Using AERMOD. *Atmos. Clim. Sci.* **2015**, *5*, 57–62. [CrossRef]
9. Kumar, A.; Patil, R.S.; Dikshit, A.K.; Islam, S.; Kumar, R. Evaluation of control strategies for industrial air pollution sources using American Meteorological Society/Environmental Protection Agency Regulatory Model with simulated meteorology by Weather Research and Forecasting Model. *J. Clean. Prod.* **2016**, *116*, 110–117. [CrossRef]
10. Kumar, A.; Patil, R.S.; Dikshit, A.K.; Kumar, R. Comparison of predicted vehicular pollution concentration with air quality standards for different time periods. *Clean Technol. Environ. Policy* **2016**, *18*, 2293–2303. [CrossRef]
11. Madala, S.; Satyanarayana, A.; Srinivas, C.; Kumar, M. Mesoscale atmospheric flow-field simulations for air quality modeling over complex terrain region of Ranchi in eastern India using WRF. *Atmos. Environ.* **2015**, *107*, 315–328. [CrossRef]
12. Karagulian, F.; Belis, C.A.; Dora, C.F.C.; Prüss-Ustün, A.M.; Bonjour, S.; Adair-Rohani, H.; Amann, M. Contributions to cities' ambient particulate matter (PM): A systematic review of local source contributions at global level. *Atmos. Environ.* **2015**, *120*, 475–483. [CrossRef]
13. Thunis, P.; Degraeuwe, B.; Pisoni, E.; Trombetti, M.; Peduzzi, E.; Belis, C.; Wilson, J.; Clappier, A.; Vignati, E. $PM_{2.5}$ source allocation in European cities: A SHERPA modelling study. *Atmos. Environ.* **2018**, *187*, 93–106. [CrossRef]
14. Abhijith, K.; Gokhale, S. Passive control potentials of trees and on-street parked cars in reduction of air pollution exposure in urban street canyons. *Environ. Pollut.* **2015**, *204*, 99–108. [CrossRef] [PubMed]
15. Gulia, S.; Shrivastava, A.; Nema, A.K.; Khare, M. Assessment of Urban Air Quality around a Heritage Site Using AERMOD: A Case Study of Amritsar City, India. *Environ. Model. Assess.* **2015**, *20*, 599–608. [CrossRef]
16. Jiménez-Guerrero, P.; Jorba, O.; Baldasano, J.M.; Gassó, S. The use of a modelling system as a tool for air quality management: Annual high-resolution simulations and evaluation. *Sci. Total. Environ.* **2008**, *390*, 323–340. [CrossRef]
17. Kumar, A.; Ketzel, M.; Patil, R.S.; Dikshit, A.K.; Hertel, O. Vehicular pollution modeling using the operational street pollution model (OSPM) for Chembur, Mumbai (India). *Environ. Monit. Assess.* **2016**, *188*, 349. [CrossRef] [PubMed]
18. Kumar, A.; Patil, R.S.; Dikshit, A.K.; Kumar, R.; Brandt, J.; Hertel, O. Assessment of impact of unaccounted emission on ambient concentration using DEHM and AERMOD in combination with WRF. *Atmos. Environ.* **2016**, *142*, 406–413. [CrossRef]
19. Mumovic, D.; Crowther, J.; Stevanovic, Z. Integrated air quality modelling for a designated air quality management area in Glasgow. *Build. Environ.* **2006**, *41*, 1703–1712. [CrossRef]
20. Kumar, A.; Patil, R.S.; Dikshit, A.K.; Kumar, R. Assessment of Spatial Ambient Concentration of NH_3 and its Health Impact for Mumbai City. *Asia J. Atmos. Environ.* **2019**, *13*, 11–19. [CrossRef]
21. Patil, R.S.; Kumar, R.; Menon, R.; Kumar, M.; Sethi, V. Development of particulate matter speciation profiles for major sources in six cities in India. *Atmos. Res.* **2013**, *132–133*, 1–11. [CrossRef]
22. Mohan, M.; Bhati, S.; Sreenivas, A.; Marrapu, P. Performance Evaluation of AERMOD and ADMS-Urban for Total Suspended Particulate Matter Concentrations in Megacity Delhi. *Aerosol Air Qual. Res.* **2011**, *11*, 883–894. [CrossRef]
23. Seaman, N.L. Meteorological modeling for air-quality assessments. *Atmos. Environ.* **2000**, *34*, 2231–2259. [CrossRef]
24. Sistla, G.; Zhou, N.; Hao, W.; Ku, J.-Y.; Rao, S.; Bornstein, R.; Freedman, F.; Thunis, P. Effects of uncertainties in meteorological inputs on urban airshed model predictions and ozone control strategies. *Atmos. Environ.* **1996**, *30*, 2011–2025. [CrossRef]
25. Russell, A.; Dennis, R. NARSTO critical review of photo-chemical models and modelling. *Atmos. Environ.* **2000**, *34*, 2283–2324. [CrossRef]
26. Gilliam, R.C.; Hogrefe, C.; Rao, S.T. New methods for evaluating meteorological models used in air quality applications. *Atmos. Environ.* **2006**, *40*, 5073–5086. [CrossRef]
27. Borge, R.; Alexandrov, V.N.; Del Vas, J.J.; Lumbreras, J.; Rodríguez, E. A comprehensive sensitivity analysis of the WRF model for air quality applications over the Iberian Peninsula. *Atmos. Environ.* **2008**, *42*, 8560–8574. [CrossRef]
28. Kumar, A.; Patil, R.S.; Dikshit, A.K.; Kumar, R. Application of WRF Model for Air Quality Modelling and AERMOD—A Survey. *Aerosol Air Qual. Res.* **2017**, *17*, 1925–1937. [CrossRef]

29. Kesarkar, A.P.; Dalvi, M.; Kaginalkar, A.; Ojha, A. Coupling of the Weather Research Forecasting Model with AERMOD for pollutant dispersion modelling. A case study for PM10 dispersion over the Pune City. *Atmos. Environ.* **2006**, *41*, 1976–1988. [CrossRef]

30. Kumar, A.; Patil, R.S.; Dikshit, A.K.; Kumar, R. Application of AERMOD for short-term air quality prediction with forecasted meteorology using WRF model. *Clean Technol. Environ. Policy* **2017**, *19*, 1955–1965. [CrossRef]

31. Myers-Cook, T.; Mallard, J.; Mao, Q. Development of a WRF-AERMOD Tool for Use in Regulatory Application. In Proceedings of the 16th Conference on Air Pollution Meteorology, American Meteorological Society, Atlanta, GA, USA, 16–21 January 2010; pp. 1–10.

32. *Ground Water Information, Greater Mumbai District, Maharashtra*; 1618/DB/20; Ministry of Water Resources, Central Ground Water Board: New Delhi, India, 2009.

33. Gupta, I.; Kumar, R. Trends of particulate matter in four cities in India. *Atmos. Environ.* **2006**, *40*, 2552–2566. [CrossRef]

34. *Comprehensive Environmental Pollution Index (CEPI)*; Central Pollution Control Board, Ministry of Environment and Forests: New Delhi, India, 2010.

35. Skamarock, W.C.; Klemp, J.B.; Dudhia, J.; Gill, D.O.; Barker, D.M.; Duda, M.G.; Huang, X.-Y.; Wang, W.; Powers, J.G. *A Description of the Advanced Research WRF Version 3*; NCAR Technical Note NCAR/TN-45; National Center for Atmospheric Research: Boulder, CO, USA, 2008.

36. Henmi, T.; Flanigan, R.; Padilla, R. *Development and Application of an Evaluation Method for the WRF Mesoscale Model*; ARL-TR3657; Army Research Laboratory: Adelphi, MD, USA, 2005.

37. National Center for Atmospheric Research (NCAR). University Corporation for Atmospheric Research (UCAR), Research Data Achieve (RDA), USA. Available online: https://rda.ucar.edu/datasets/ds083.2/ (accessed on 15 January 2020).

38. Kumar, P.; Kishtawal, C.M.; Pal, P.K. Impact of satellite rainfall assimilation on Weather Research and Forecasting model predictions over the Indian region. *J. Geophys. Res. Atmos.* **2014**, *119*, 2017–2031. [CrossRef]

39. Kumar, P.; Bhattacharya, B.K.; Nigam, R.; Kishtawal, C.M.; Pal, P.K. Impact of Kalpana-1 derived land surface albedo on short-range weather forecasting over the Indian subcontinent. *J. Geophys. Res. Atmos.* **2014**, *119*, 2764–2780. [CrossRef]

40. Kumar, P.; Kishtawal, C.M.; Pal, P.K. Skill of regional and global model forecast over Indian region. *Theor. Appl. Clim.* **2016**, *123*, 629–636. [CrossRef]

41. Sax, T.; Isakov, V. A case study for assessing uncertainty in local-scale regulatory air quality modeling applications. *Atmos. Environ.* **2003**, *37*, 3481–3489. [CrossRef]

42. Singh, N.P.; Gokhale, S. A method to estimate spatiotemporal air quality in an urban traffic corridor. *Sci. Total. Environ.* **2015**, *538*, 458–467. [CrossRef] [PubMed]

43. Thunis, P.; Georgieva, E.; Pederzoli, A. A tool to evaluate air quality model performances in regulatory applications. *Environ. Model. Softw.* **2012**, *38*, 220–230. [CrossRef]

44. *User's Guide for Industrial Source Complex (ISC3) Dispersion Models, Volume I—User Instructions*; EPA-454/B-95-003a; United State Environmental Protection Agency, Office of Air Quality Planning and Standards, Emissions, Monitoring, and Analysis Division, Research Triangle Park: Piedmont, NC, USA, 1999.

45. Bandyopadhyay, A. Prediction of ground level concentration of sulfur dioxide using ISCST3 model in Mangalore industrial region of India. *Clean Technol. Environ. Policy* **2009**, *11*, 173–188. [CrossRef]

46. Cimorelli, A.J.; Perry, S.G.; Venkatram, A.; Weil, J.C.; Paine, R.J.; Wilson Robert, B.; Lee, R.F.; Peters, W.D.; Brode, R.W.; Paumier, J.O. *AERMOD: Description of Model Formulation*; EPA-454/R-03-004; USEPA: Washington, DC, USA, 2004.

47. Bhardwaj, K.S. Examination of the Sensitivity of Land Use Parameters and Population on the Performance of the AERMOD Model for and Urban Area. Civil Engineering. Master's Thesis, University of Toledo, Toledo, OH, USA, 2005.

48. Venkatram, A.; Brode, R.; Cimorelli, A.; Lee, R.; Paine, R.; Perry, S.; Peters, W.; Weil, J.; Wilson, R. A complex terrain dispersion model for regulatory applications. *Atmos. Environ.* **2001**, *35*, 4211–4221. [CrossRef]

49. Venkatram, A.; Isakov, V.; Yuan, J.; Pankratz, D. Modeling dispersion at distances of meters from urban sources. *Atmos. Environ.* **2004**, *38*, 4633–4641. [CrossRef]

50. Zou, B.; Zhan, F.B.; Wilson, J.G.; Zeng, Y. Performance of AERMOD at different time scales. *Simul. Model. Pract. Theory* **2010**, *18*, 612–623. [CrossRef]

51. Zou, B.; Wilson, J.G.; Zhan, F.B.; Zeng, Y. Spatially differentiated and source-specific population exposure to ambient urban air pollution. *Atmos. Environ.* **2009**, *43*, 3981–3988. [CrossRef]

52. *Air Quality Assessment, Emission Inventory and Source Apportionment Studies: Mumbai*; National Environmental Engineering Research Institute, CPCB: New Delhi, India, 2010.

53. Air Quality Monitoring Project-Indian Clean Air Programme (ICAP). *Emission Factor development for Indian Vehicles*; Automotive Research Association of India CPCB-MoEF: New Delhi, India, 2008.

54. Fedoseeva, E.V. An Estimate of the Error of Measurements of Radio Brightness Temperature in Radio-Heat Meteorological Parameters with Background Noise Compensation. *Meas. Tech.* **2015**, *57*, 1463–1468. [CrossRef]

55. Finkelstein, P.L.; Sims, P.F. Sampling error in eddy correlation flux measurements. *J. Geophys. Res.* **2001**, *106*, 3503–3509. [CrossRef]

56. Ulbrich, C.W.; Lee, L.G. Rainfall Measurement Error by WSR-88D Radars due to Variations in *Z–R* Law Parameters and the Radar Constant. *J. Atmos. Ocean. Technol.* **1999**, *16*, 1017–1024. [CrossRef]
57. Sonawane, N.V.; Patil, R.S.; Sethi, V. Health benefit modelling and optimization of vehicular pollution control strategies. *Atmos. Environ.* **2012**, *60*, 193–201. [CrossRef]

 *sustainability*

Article

Can Carbon Trading Policies Promote Regional Green Innovation Efficiency? Empirical Data from Pilot Regions in China

Baoliu Liu, Zhenqing Sun * and Huanhuan Li

School of Economics and Management, Tianjin University of Science & Technology, Tianjin 300457, China; 1748192699@mail.tust.edu.cn (B.L.); lhh9598@mail.tust.edu.cn (H.L.)
* Correspondence: sunzq@tust.edu.cn

Abstract: Although the emission reduction and innovation effects of carbon emissions trading have attracted considerable interest among academics and policy makers, there is a lack of empirical research on how carbon trading pilots in China promote regional green innovation. Therefore, we measured the green innovation efficiency of 30 provinces and cities in mainland China from 2005 to 2018, using selected panel data within a super-efficient SBM model that incorporated undesirable outputs. We used a double differential model to evaluate the impacts of carbon trading policies on the green innovation efficiency of seven carbon trading pilot regions. These impacts were confirmed using the double differential propensity score matching method. Our findings were as follows. (1) The implementation of carbon trading policies has a significant and continuous effect of promoting and improving green innovation efficiency in the pilot areas. (2) Carbon trading induces technological innovation effects, enabling green innovation potential to be realized. Regional green innovation efficiency is further improved through energy substitution and structural upgrading effects and subsequently through all three of the above effects. (3) The synergy between the three major effects of carbon trading policies amplifies regional green innovation efficiency. Therefore, the Chinese government should accelerate its efforts to develop and improve carbon markets, promote carbon trading policies, and actively foster synergy among the three effects to achieve green and sustainable regional development.

Keywords: carbon trading; green innovation efficiency; propensity score matching; difference in differences model

Citation: Liu, B.; Sun, Z.; Li, H. Can Carbon Trading Policies Promote Regional Green Innovation Efficiency? Empirical Data from Pilot Regions in China. *Sustainability* **2021**, *13*, 2891. https://doi.org/10.3390/su13052891

Academic Editor: José Carlos Magalhães Pires

Received: 9 February 2021
Accepted: 3 March 2021
Published: 7 March 2021

Publisher's Note: MDPI stays neutral with regard to jurisdictional claims in published maps and institutional affiliations.

1. Introduction

In recent years, while extensive economic growth has yielded economic dividends, it has also generated various problems, notably, excessive energy consumption and insufficient innovation. The question of how a win-win situation of economic growth and carbon dioxide (CO_2) emission reduction can be achieved, while making full use of essential resources, has emerged as a common societal concern. At the same time, the goal of establishing a global governance system for achieving green, low-carbon, and sustainable development is shared by countries globally. However, the task of achieving the reduction targets stipulated in the Paris Agreement for limiting the rise in the global temperature to no more than 2 °C above preindustrial levels and to reach the global peak as soon as possible is a daunting one [1]. In addition, through the implementation of large-scale carbon dioxide storage projects, the goals and requirements of the Paris Agreement can also be achieved. On the one hand, accelerated adjustment of the industrial structure to redress structural imbalances is required, and on the other hand, emission reduction technologies should be prioritized, and development should be propelled by innovation.

The question of whether a win-win situation for regional economic growth and carbon dioxide (CO_2) emission reduction can be achieved through the adoption of rational economic policies and stringent environmental regulation is therefore a critical one. Efforts to explore and establish different carbon trading systems have been initiated in various regions worldwide. The European Union Emissions Trading Scheme is distinctive among these initiatives, as it is not only politically feasible but also environmentally effective, as well as cost-effective [2]. This system was rapidly extended to cover about 12,000 industrial and power facilities in Europe that were responsible for almost 50% of the EU's greenhouse gas emissions. Between 2005 and 2012, the EU-ETS accounted for 85% of the total global volume of carbon trading [3]. Within the United States, Chicago was the first city to participate in transactions to reduce greenhouse gas emissions, providing a foundation for national efforts to implement activities to reduce greenhouse gas emissions, such as RGGI, WCI, and the California Plan. Australia too launched a domestic carbon trading market after introducing relevant legislation in 2015, gradually developing a regional carbon market with extensive coverage, a carbon price compensation mechanism, and an improved monitoring mechanism over time [4].

In 2011, the Chinese government launched pilot carbon trading projects in seven provinces and cities, namely, Beijing, Tianjin, Shanghai, Chongqing, Guangdong, Hubei, and Shenzhen. The implementation of policies for establishing a carbon trading market in China has occurred relatively late compared with their implementation by other countries or organizations. Consequently, the carbon trading system is still evolving and requires further refinement, and there are information gaps on transactions. The Chinese government has adopted a series of market-oriented measures for achieving emission reduction targets through the development of a carbon trading system for promoting coordinated efforts to stimulate economic growth and environmental improvements. The government is simultaneously actively pursuing a path of green innovation and development framed through a series of concepts, notably, "innovative country", "wild China", and the "five concepts for development". "Green innovation" can also take diverse forms and include the promotion of economic growth [5]. Technical progress and environmental improvement initiatives in China are aimed at achieving the dual goals of "green mountains and clear water" and "mountains of gold and silver". Green innovation simultaneously entails a new process technology, system, and products aimed at reducing environmental pollution and damage and improving energy efficiency [6]. The goal of improving the efficiency of green innovation not only conforms to the concept of green development but it also encourages the implementation of innovation-driven development strategies.

To sum up, carbon trading has gradually gained popularity within most countries worldwide as a mechanism for promoting energy saving, emissions reduction, and low-carbon economic transformation through the use of market-oriented and mandatory measures to reduce energy consumption intensity. This approach also promotes CO_2 emissions reduction, which in turn drives the improvement of regional technological innovations. As the world's largest developing country and a major energy consumer, China's emission base is large, crucially impacting on the global carbon trading volume. Therefore, studies focusing on the operation of China's carbon trading mechanism and status quo would provide valuable inputs that could contribute to the realization of China's low-carbon economic transformation as well as the development of low-carbon technologies in other countries. The achievement of green and sustainable development of regional economies hinges on whether the technical effects of scale, and thus the level of regional technological innovation can be fully realized through the implementation of China's carbon trading policy. Using panel data extracted for 30 provinces and cities in mainland China for the period 2005–2018, we examined the green innovation potential of carbon trading. Specifically we explored the dynamic relationship between regional carbon emissions reduction and green innovation efficiency under the influence of China's carbon trading policy. Moreover, we sought to determine the mechanism by which carbon trading promotes regional green innovation efficiency. Elucidation of this mechanism contributes to advancing research on China's carbon trading regime and yields insights that can be applied to formulate guidelines for promoting green, high-quality regional development. Figure 1 shows the distribution of China's pilot and non-pilot regions involved in the thesis research. Among them, the pilot regions mainly include the seven provinces and cities: Beijing, Tianjin, Shanghai, Hubei, Chongqing, Guangdong, and Shenzhen.

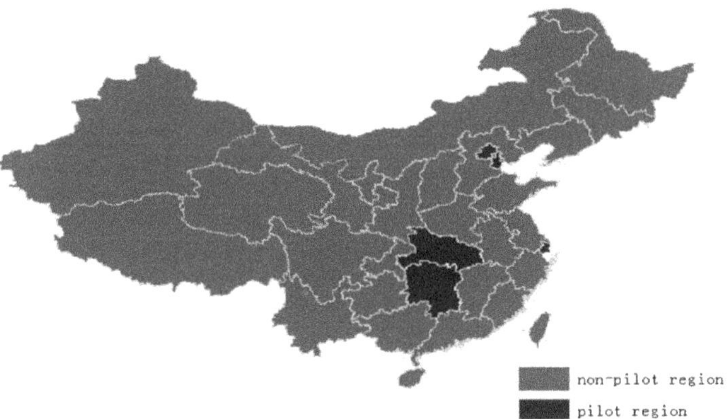

Figure 1. Comparative analysis of pilot and non-pilot regions.

2. Literature Review and Hypothesis

2.1. Literature Review

Scholars in China and abroad have conducted extensive research in the fields of carbon trading and green innovation. The research on carbon trading has mainly focused on impact effects, methods of measurement, carbon quotas, and carbon prices. This paper primarily focuses on studies on the impact effects of carbon trading. Dan et al. applied a differences-in-differences (DID) model to examine the policy effects of carbon trading. Their findings provided support for the "Porter hypothesis", revealing that the carbon trading mechanism encourages technological innovation to a certain extent. However, these authors did not find a strong policy effect on total factor productivity [7]. The trading mechanism of carbon emission rights is required to realize low-carbon transformation of the Chinese economy. Chuanming et al. applied a synthetic control method to investigate the carbon emissions reduction effect of China's carbon trading pilot provinces. They found that the carbon reduction effects of different pilot provinces were heterogeneous because of differences in their levels of economic development and in their industrial structures [8]. Other studies have examined the effects of carbon trading on economic growth. Zhengge et al. investigated whether the SO_2 emission trading mechanism exerts the Porter effect of inducing efficiency and innovation in China. They emphasized that strengthening market development and environmental regulation are necessary conditions for inducing the Porter effect [9]. Chunmei et al., who applied the directional distance function to calculate emissions reduction costs for China's industrial sector, confirmed that the carbon trading market has a significant impact on industries' emissions reduction costs and carbon intensity [10]. Studies that have explored synergistic effects include those of Cheng et al. and Ren Yayun. Their findings indicate that a carbon trading policy that encourages synergistic emissions reduction not only promotes the reduction of CO_2 emissions but it also promotes the reduction of other pollutants, thus playing a role in coordinated emissions reduction [11,12]. Last, a study by Jing and others examined the effect of upgrading the industrial structure, using the synthetic control method to evaluate the impacts of carbon trading on the upgrading of China's industrial structure. Its findings indicated that carbon trading compels the upgrading of the industrial structure [13].

Research on the efficiency of green innovation has primarily entailed the use of two approaches for measuring the values of efficiency and influencing factors. In the first approach, the parameter-based stochastic frontier model and the non-parametric data envelopment analysis methods are used to measure efficiency. In 1997, Chung and others proposed the concept of a directional distance function, which provides methodological support for measuring total factor productivity, including "unexpected outputs" in a region. Watanabe et al. used the directional distance function to assess the impacts of China's inter-provincial initiative to eliminate bad industrial outputs on industrial efficiency during the period 1994–2002. They found that undesirable industrial outputs play an important role in improving industrial efficiency [14]. Neng et al. applied a hybrid DEA model to measure the efficiency of green innovation in China. The findings of their analysis of key factors affecting the efficiency of green innovation revealed that a good industrial structure, a free technology trading market, and basic environmental criteria had a greater impact on green innovation than did other factors [15]. Considering environmental pollution and innovation failure as undesirable outputs, Yanwei and others constructed an SBM–DEA model and alpha and beta convergence models to measure and converge the efficiency of green innovation in China [16]. The second approach entails the study of influencing factors. Scholars adopting this approach have largely focused on the following dimensions: the level of economic development [17], environmental regulation [18], R&D investments [19], and the level of interaction with the outside world [20]. These studies have advanced knowledge of the factors influencing green innovation efficiency within regions, while providing a conceptual foundation for initiatives aimed at improving levels of regional green innovation.

An examination of the recent literature reveals a paucity of studies that have investigated the relationship between carbon trading and green innovation efficiency and a lack of in-depth analysis of the internal mechanism driving this relationship. Moreover, few studies have examined the correlation between carbon emissions reduction and technological innovation despite the effectiveness of carbon trading, as a market-based policy tool, in reducing carbon intensity on the one hand and the importance of regional green innovation efficiency on the other hand. We aimed to address this gap by examining the green innovation potential of carbon trading. Our study makes the following contributions to the literature. First, it shows that the super-efficient SBM model, incorporating the impacts of undesirable outputs (carbon emissions), is an appropriate model for measuring green innovation efficiency in a region. Second, whereas several studies have examined the emissions reduction effects of carbon trading, they have not investigated the potential of carbon trading to promote regional green innovation. Therefore, we examined the diversity of regional green innovation efficiency within a carbon trading framework. Third, we argue that it is necessary to identify the intermediary transmission mechanism whereby carbon trading promotes improvements in a region's green innovation efficiency. At the same time, it is important to analyze differences in the technology innovation effects of carbon trading along with energy substitution and structural upgrading effects on green innovation efficiency. As we show, a study that entails both approaches can contribute useful insights for improving regional energy conservation and emission reduction capabilities and guiding inputs for the exploration of appropriate green innovation paths.

2.2. Research Hypothesis

As an environmental regulation tool, carbon trading influences the costs, benefits, and operating efficiency of the regional economy and promotes green, low-carbon regional development. Therefore, an in-depth study to examine the ways in which carbon trading contributes to the efficiency of green innovation within regions would shed light on their internal linkages.

(1) Technological innovation effect. Academic research has confirmed that technological innovation has a carbon emission reduction effect and that effective technological progress can significantly reduce carbon emissions [21,22]. Technological advances can help companies to control their emissions reduction costs and reduce production, thereby reducing carbon emissions. At the same time, companies located in pilot regions can use technological innovations to reduce emissions, and the technology spillover effect can further promote regional green innovation efficiency [12].

Hypothesis 1 (H1). *Carbon trading contributes to enhancing technological innovation in pilot areas, thereby increasing the level of regional green innovation efficiency.*

Energy substitution effect. The energy substitution effect of carbon trading is evidenced by increased production costs for traditional high-emission enterprises because of the need to expand their production scales. Moreover, CO_2 emission rights relating to ultra-carbon quotas have compelled enterprises to reduce their CO_2 emissions and increase the extent of their clean technology research and development [8]. Thus, the establishment of a carbon market can effectively stimulate the optimization and upgrading of the energy structures of enterprises, thereby increasing the proportion of clean energy that contributes to a reduction of carbon emissions. Clean energy is mainly used in the power sector, which has contributed more than 50% of carbon emissions [11]. Therefore, it is necessary to improve the energy utilization efficiency of this industrial sector through the prioritization of energy-saving and emission-reducing technologies and equipment that promote clean, low-carbon development in the pilot areas.

Hypothesis 2 (H2). *Carbon trading has led to the modification of the energy structure of enterprises in the pilot regions, and the substitution of fossil energy by clean energy has enabled the coordinated development of energy consumption and carbon emission reduction, thereby promoting regional green innovation efficiency.*

(2) Structural upgrading effect. An advanced industrial structure that is optimized and upgraded reflects the changing relationship between different industrial ratios and improved labor productivity within various industries and exerts "structural benefits" [23]. Carbon trading can contribute to the advancement of the industrial structure in two ways. The first entails changing the proportion of different industries and the second entails improving labor productivity within various industries [7]. At the same time, the optimization and upgrading of the industrial structure contributes to the flow of factors between industries, resulting in a gradual reduction in the proportion of the three industries associated with high carbon emissions, which leads to a reduction in industrial carbon emissions while simultaneously promoting green and low-carbon regional development.

Hypothesis 3 (H3). *Carbon trading drives the process of upgrading the industrial structure in the pilot regions, thereby improving the efficiency of regional green innovation.*

A carbon trading policy has three major effects that contribute to promoting green regional development: technological innovation, energy substitution, and structural upgrading effects. Figure 2 shows the specific mechanism whereby these effects are achieved.

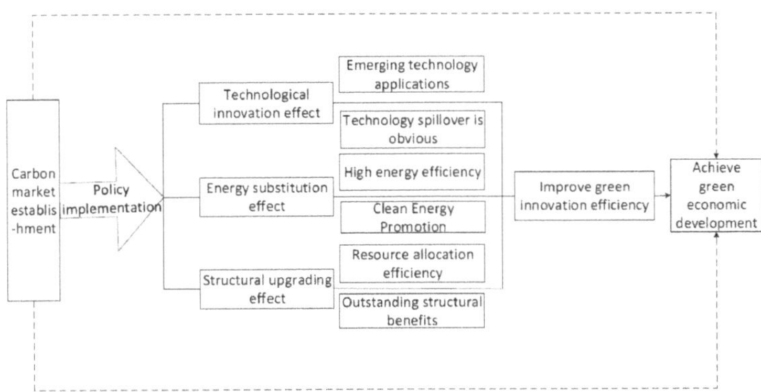

Figure 2. The mechanism whereby carbon trading policies influence the efficiency of green innovation.

3. Methodology and Models Applied

3.1. Super-Efficient SBM Model

Following our review of the relevant literature, we selected the non-parametric DEA method and the parametric SFA method for measuring green technology efficiency [24]. The DEA model can accommodate the relationship between multiple input and output variables. Moreover, it is more aligned with actual situations, as it does not require the setting of a specific function form. However, the traditional DEA model enables a unilateral assessment in which only the expected output is considered, while the effects of the undesirable output are ignored. Relaxation of inputs and outputs leads to a higher value for the measured efficiency. At the same time, the traditional DEA model is radial and angular, which results in some deviation in the calculated results. By contrast, non-radial and non-angle SBM models are better aligned with the study's requirements and address the above-mentioned issues. Consequently, we adopted the approach used in previous studies and incorporated undesirable outputs into the super-efficient SBM model [20,25]. We developed the following model for evaluating green innovation efficiency based on the assumptions that n kinds of decision-making units (DMUs) have m kinds of element inputs, Z_1 kinds of expected outputs, and Z_2 kinds of undesirable outputs:

$$
\min\rho* = \frac{1-\frac{1}{m}\sum\limits_{i=1}^{m}\frac{zi^-}{xik}}{1+\frac{1}{z1+z2}\left(\sum\limits_{r=2}^{z1}\frac{z_r^g}{y_{rk}^g}+\sum\limits_{q=1}^{z2}\frac{z_q^b}{y_{qk}^g}\right)}
$$

$$
\text{s.t.}\begin{cases} xik = \sum\limits_{j=1}^{n} xij\lambda j + zi^-\ i = 1,\cdots,m \\[6pt] y_{rk}^g = \sum\limits_{j=1}^{n} yrj\lambda j - z_r^{g-}\ r = 1,\cdots,z1 \\[6pt] y_{qk}^g = \sum\limits_{j=1}^{n} y_{qj}^b\lambda j + z_q^{b-}\ q = 1,\cdots,z2 \end{cases}
$$

$$
\lambda j > 0\ j = 1,\cdots,n
$$
$$
zi^- \geq 0,\ z_r^g \geq 0,\ z_q^b \geq 0
$$

(1)

Model (1) indicates that when the invalid DMU_k and the effective DMU_k of $\rho*$ are transformed into each other, there is a corresponding correlation between a reduction in input variable z_i^-, an increase in the expected output z_r^{g-}, and a decrease in the undesirable output z_q^{b-}. When the values of z^-, z^g, and z^b are larger, the efficiency value $\rho*$ of DMU_k is correspondingly smaller, and when the values of z^-, z^g, and z^b all have a value of 0, then $\rho* = 1$, indicating that DMU_k is effective, there is no shortage in the expected output, and that the undesirable output is not in excess. DMU_k in the super-efficient SBM model was deemed effective based on a consideration of the factors of the corresponding relaxation variables for the constraints. The specific model used for ranking DMUs was as follows:

$$
\min\varphi* = \frac{1-\frac{1}{m}\sum\limits_{i=1}^{m}\frac{\bar{x}}{xik}}{1+\frac{1}{z1+z2}\left(\sum\limits_{r=2}^{z1}\frac{\bar{y}^g}{y_{rk}^g}+\sum\limits_{q=1}^{z2}\frac{s_q^b}{y_{qk}^g}\right)}
$$

$$
\text{s.t.}\begin{cases} \bar{x} \geq \sum\limits_{j=1,\neq k}^{n} xij\lambda j\ i = 1,\cdots,m \\[6pt] \bar{y}^g = \sum\limits_{j=1,\neq k}^{n} yrj\lambda j\ r = 1,\cdots,z1 \\[6pt] \bar{y}^b = \sum\limits_{j=1,\neq k}^{n} y_{qj}^b\lambda j\ q = 1,\cdots,z2 \end{cases}
$$

$$
\lambda j > 0\ j = 1,\cdots,n\ j \neq 0
$$

$$
\bar{x} \geq xik,\ \bar{y}^g \leq y_{rj}^g,\ \bar{y}^b \leq y_{qj}^b
$$

(2)

We derived and calculated model (2) on the basis of model (1). Using both of these models, we calculated the regional green innovation efficiency values of 30 provinces and cities in China (excluding the Tibet Autonomous Region) for the period 2005–2018 as follows:

$$GIE = \begin{cases} \theta *_{ik} & \theta*_{ik} < 1 \\ \omega *_{ik} & \omega*_{ik} = 1 \end{cases} \quad i = 1, \cdots 30, k = 2005, \cdots 2018 \tag{3}$$

where GIE denotes the green innovation efficiency value for area i during year k.

3.2. DID Model

Currently, many methods exists for evaluating policy effects, such as synthetic control method and DID method [26]. In recent years, this DID model has been used for quantitative evaluations of public policy or for assessing project implementation effects within econometrics. The actual impacts of a policy can be assessed by comparing the amount of change for a specific indicator before and after the policy's implementation using an experimental group and a control group. The DID method has been widely used by scholars because of its ability to reduce endogeneity to some extent [26]. Therefore, we applied this model to investigate the regional emission reduction potential and the green innovation efficiency trend under conditions of the implementation of a carbon trading policy and to examine key factors that influence changes in efficiency.

There are currently seven pilot provinces and cities in China where the carbon trading policy has been implemented: Beijing, Tianjin, Chongqing, Shanghai, Hubei, Guangdong, and Shenzhen. To simplify the analysis, the city of Shenzhen was merged with Guangdong Province, and following Guangming et al., we set 2014 as the year demarcating the separation of the pilot and non-pilot periods, with the pilot period commencing from 2014 [27]. Accordingly, the following basic model was constructed:

$$GIEit = \alpha 0 + \alpha 1 Ci + \alpha 2 Yt + \alpha 3 (Ci \times Yt) + \lambda i + \gamma t + \mu it \tag{4}$$

where GIE denotes green innovation efficiency, i denotes area, t denotes time, and C_i denotes regional dummy variables. If province i is a carbon trading pilot province or city, then the values of C_i are 1 and 0, respectively, for the experimental and control groups. Y_t denotes a time dummy variable. In 2014, which is the year of implementation of the policy, $t \geq 2014$ and $Y_t = 1$; otherwise 0. The estimated coefficients $\alpha 1$, $\alpha 2$, and $\alpha 3$ of the multiplication term $C_i \times Y_t$ are double-difference estimators, indicating the net impact of carbon trading policies. λ_i denotes the individual fixed effects of provinces and cities, γ_t denotes the fixed effect of time, and μ_{it} is the random interference term.

The explanatory use of model (4) on its own could result in the influence of other variables being discounted. Therefore, it is necessary to add control variables to the model to account for the influence of objective factors on the explanatory variables. With reference to previous studies, we transformed the basic model, selecting GDP per capita, R&D investment, carbon intensity, energy structure, population size, and R&D investment as the control variables:

$$GIEit = \alpha 0 + \alpha 1 Ci + \alpha 2 Yt + \alpha 3 (Ci \times Yt) + \sum \alpha j Xj + \lambda i + \gamma t + \mu it \tag{5}$$

where X_j denotes the control variable, and the meaning of other variables is consistent with the above.

3.3. Selection of Variables and Data Sources

3.3.1. Interpreted Variables

Green innovation efficiency differs from other forms of innovation efficiency because it considers the impact of changes in energy consumption and carbon emissions on regional low-carbon development potential. Accordingly, drawing on the findings of previous studies that measured regional green innovation efficiency, we applied the super-efficient SBM model, incorporating undesirable outputs. Input elements, selected with reference to the existing literature, were full-time R&D personnel, R&D funding inputs, and energy resource inputs.

Two categories of output elements of green innovation activities were defined: expected and undesirable outputs. Expected outputs were the value of economic growth, the number of authorized invention patent applications, and revenue from sales of new products. Undesirable outputs were innovation failures and the environmental pollution index. According to Schumpeter's definition of innovation, the success or failure of innovation is reflected in the generation (or not) of economic profits. Failure to innovate affects companies' ability to repay their business loans and their regular cash flows, making it impossible for them to generate profits. Consequently, they are left with non-performing loans. Accordingly, the undesirable output was calculated as the ratio of the amount of non-performing loans of commercial banks to the previous year. The environmental pollution index relates to the discharge of waste water, gas, and solid waste in various regions and was calculated using the entropy weight method to measure the weight of each indicator. The specific index system applied in this study is shown in Table 1.

Table 1. Evaluation index system used to measure green innovation efficiency.

Index	Category	Index Composition	Specific Measurement
Input indicators	Factor input	R&D expenses	R&D expenditure (ten thousand yuan)
		R&D staff	R&D personnel full-time equivalent (person, year)
		Energy resources	Total energy consumption (10,000 tons of standard coal)
Output indicators	Expected output	The level of economic development	GDP per capita (ten thousand yuan, constant price in 2005)
		Knowledge and technology output	Invention patent application authorization volume (pieces)
		Product output	New product sales revenue (ten thousand yuan)
	Unexpected output	Innovation failure	Year-on-year ratio of non-performing loans of commercial banks (%)
		Environmental Pollution Index	The entropy weight method is used to calculate the discharge of waste water, waste gas and solid waste

3.3.2. Core Explanatory Variables

$C_i \times Y_t$ was the core explanatory variable. For a low-carbon city or province, when $Y \geq 2014$, the virtual variable $C_i \times Y_t$ corresponding to the city had a value of 1, and its coefficient indicated the net effect of the carbon trading policy and the strength of its emission reduction effect

3.3.3. Control Variables and Measuring Indicators

Referring to the literature, we selected per capita GDP, R&D investments, carbon intensity, energy structure, and foreign capital dependence as the control variables relating to the level of regional green innovation efficiency.

GDP Per Capita

In general, improvements in levels of regional economic development can drive technological innovation and enhance the level of low-carbon technological innovation. The regional GDP (constant price in 2005) and the proportion of permanent residents at the end of the year were used to express GDP per capita.

R&D Investment Intensity

The level of scientific research has a crucial impact on the extent of regional technological innovation, thereby influencing energy usage. Regional R&D expenditure was used to reflect the amount of regional R&D investment.

Carbon Intensity

Carbon intensity was measured as the ratio of the total CO_2 emissions from fossil combustion in the region to the GDP.

Energy Structure

Coal is the main source of energy consumed. The ratio of total coal to total energy consumption was used to reflect changes in the energy structure, enabling a more intuitive understanding of the trend of energy ratio changes in various regions.

Foreign Capital Dependency

The amount of foreign direct investments significantly influences regional economic growth and the capabilities for controlling environmental pollution. Therefore, in this study, we considered the ratio of foreign direct investment to GDP to reflect the degree of foreign capital dependence.

3.4. Data Sources

Panel data were selected from the available raw data for 30 provinces in China (excluding Tibet, Hong Kong, Macau, and Taiwan) for the period 2005–2018. The data were extracted from the *China Statistical Yearbook*, the *China Energy Statistical Yearbook*, the *China Environmental Statistical Yearbook*, and the *China Science and Technology Statistical Yearbook*.

4. Empirical Results and Analysis

4.1. The DID Method of Regression Analysis

The DID method was applied in a further investigation of the effects of carbon trading policies on regional green innovation efficiency. Accordingly, we performed a regression analysis of the green innovation efficiency of 30 provinces and cities in China (excluding Tibet) for the period 2005–2018. In turn, the analysis was carried out for uncontrolled variables, using a two-way fixed-effect model that incorporates control variables, adds control variables, and controls regional and time effects, and analyzes the differences in impact under different circumstances.

Table 2 presents the results of an analysis to evaluate the impacts of carbon trading policies on green innovation efficiency using the DID method. Model (1) is a benchmark model for analyzing green innovation efficiency without control variables, whereas in model (2), control variables, such as per capita GDP, scientific research inputs, carbon intensity, foreign capital dependence, and energy structure were sequentially added. Model (3) is a model with time effects added to model (2). Overall, with the increase in control variables, the significance of the core explanatory variables and the signs of the coefficients did not change appreciably, and were significantly positive at the 5% level, indicating that the model results were robust. The impact of GDP per capita on green innovation efficiency relating to the control variable changed from negative to positive at a 10% significance level. This result indicates that after the implementation of the carbon trading policy, the pilot regions paid more attention to the coordinated development of economic growth and environmental protection, thereby promoting improvements in regional levels of green innovation. R&D investment was positive at a 1% significance level, indicating that regional R&D investments play a definitive role in promoting technological innovation. The impacts of carbon intensity and the energy structure on the efficiency of green innovation changed from negative to positive, indicating that the reduction effect of carbon trading policies propelled the energy structure's optimization and enhanced green innovation efficiency within a region. Dependence on foreign capital had a positive effect on the efficiency of green innovation, but this effect was not significant. This finding indicates the importance of expanding the proportion of foreign investment and improving the level of technological progress to improve the level of green innovation in a region.

Table 2. Regression results for the impacts of carbon trading policies on green innovation efficiency.

Variable	Green Innovation Efficiency		
	(1)	(2)	(3)
Ci × Yt	0.1332 ** (2.39)	0.0899 ** (2.01)	0.0498 *** (3.45)
GDP per capita		−0.3653 * (−1.73)	0.4929 * (1.86)
Research investment		0.3692 *** (4.62)	0.3678 *** (4.27)
Carbon intensity		−0.4379 * (−1.94)	0.6479 ** (2.17)
Foreign capital dependency		0.0525 (1.29)	0.0635 (1.42)
energy structure		−0.1103 (−0.43)	0.0509 * (1.80)
Control variable	NO	YES	YES
Province fixed	YES	YES	YES
Fixed year	NO	NO	YES
Constant term	0.1756 *** (8.10)	−1.2177 *** (−4.54)	−2.6662 *** (−6.44)
N	420	420	420
R^2	0.0374	0.3850	0.4274

Notes: t-values are shown in brackets; ***, **, and * indicate statistical significance at the 1%, 5%, and 10% levels, respectively.

4.2. Analysis Using the PSM Method

The DID method must meet the requirements of the parallel trend assumption, and whether to implement the carbon trading policy as a virtual variable for overall regression, the parameters may be biased. Therefore, the PSM–DID method was used for the estimation. As required for the PSM method, the Logit model was used to estimate the propensity scores of per capita GDP, scientific research investment, carbon intensity, foreign capital dependence, and the energy structure. These scores are shown in Table 3. Subsequently, provinces within the experimental and control groups were matched using the kernel matching method. At the end of this procedure, 10 samples that did not match the cost were deleted. Regression analysis was performed again using the matching data, and the results are shown in Table 4.

Table 3. Logit regression estimation results using the PSM method.

Variable	Coefficient	Standard Error	T	P
GDP per capita	0.2843 **	0.1435	1.98	0.021
Research investment	0.3737 ***	0.1341	2.79	0.005
Carbon intensity	−0.0215 ***	0.0070	−3.06	0.002
Foreign capital dependency	0.0062 **	0.0002	2.14	0.030
energy structure	−0.7002 ***	0.2574	−2.72	0.003
-cons	1.8016 ***	0.6656	2.71	0.000

Notes: *** and ** indicate statistical significance at the 1%, 5%, and 10% levels, respectively.

Table 3 shows estimation results for the control variables obtained using the Logit regression model. The significance level was high for each variable, which is consistent with the actual situation. Advanced regional economic development along with increased proportions of R&D investments and emissions reduction efforts were indicative of an increased proportion of foreign direct investment and of an optimized and upgraded energy structure. Instead, it is easier to enter the experimental group to ensure the reliability of the regression results. Table 4 depicts the results of continued testing of the carbon trading policy using the DID after deleting unmatched samples. The regression results reveal that overall, the matching carbon trading policy resulted in enhanced green innovation efficiency, which increased by 0.01 units. The increase in GDP per capita was also caused by changes in the matched samples, and the economic dividend phenomenon was clearly apparent, contributing significantly to enhanced green innovation efficiency. At the same time, investments in scientific research, reduced carbon intensity, and green innovation efficiency within a region's energy structure evidently had positive effects. Moreover, with the implementation of the carbon trading policy, this effect was amplified. Thus, the question of whether a mechanism whereby carbon trading policies influence green innovation efficiency in the pilot areas exists, and, if so, how it operates, required an in-depth analysis. This investigation to elucidate and verify such a mechanism is described next.

Table 4. Regression results showing the role of carbon trading policies before and after the propensity score matching process.

	Green Innovation Efficiency					
	Before Matching	After Matching	Before Matching	After Matching	Before Matching	After Matching
$Ci \times Yt$	0.1332 ** (2.39)	0.2063 *** (2.67)	0.0899 ** (2.01)	0.0149 *** (3.11)	0.0498 *** (3.45)	0.0676 ** (2.22)
GDP per capita			−0.3653 * (−1.73)	0.0891* (1.91)	0.4929 * (1.86)	0.2844 ** (2.30)
Research investment			0.3692 *** (4.62)	0.4073 *** (3.44)	0.3678 *** (4.27)	0.3737 ** (3.79)
Carbon intensity			−0.4379 * (−1.94)	−0.2137 (−0.01)	0.6479 ** (2.17)	0.0215 ** (2.06)
Foreign capital dependency			0.0525 (1.29)	0.0006 * (1.81)	0.0635 (1.42)	0.0063 * (1.92)
energy structure			−0.1103 (−0.43)	0.7981 (1.04)	0.0509 * (1.80)	0.7002 ** (2.32)
Control variable	NO	NO	YES	YES	YES	YES
Province fixed	YES	YES	YES	YES	YES	YES
Fixed year	NO	NO	NO	NO	YES	YES
Constant term	0.1756 *** (8.10)	0.2558 *** (6.35)	−1.2177 *** (−4.54)	−1.1822 *** (−2.76)	−2.6662 *** (−6.44)	−1.8016 *** (−2.71)
N	420	260	420	260	420	260
R^2	0.4274	0.8901	0.3850	0.9225	0.4274	0.3532

Notes: *t* values are shown in brackets; ***, **, and * indicate statistical significance at the 1%, 5%, and 10% levels, respectively.

4.3. Testing and Verification of an Intermediary Influence Mechanism

The above empirical results reveal that the pilot provinces and cities where carbon trading policies have been implemented could significantly improve their levels of green innovation efficiency. However, a macroscale analysis of the impact of carbon trading policies on green innovation efficiency lacks sufficient depth for exploring the impact mechanism behind the policy effect. As previous research has shown, carbon trading policies have a significant role to play in promoting low-carbon development of provinces and cities, which, in turn, propels technological innovation, energy substitution, and structural upgrading and improves the efficiency of green innovation. To elucidate the impact mechanism and verify the existence of these three effects, the following steps were implemented, applying the formulas described by Baron and Kenny and Daqian [28,29].

The first step entailed verification of the three major effects of the carbon trading policy in pilot regions using the following formula:

$$TIit(ESit, SUit) = \beta0 + \beta1Ci + \beta2Yt + \beta3(Ci \times Yt) + \sum \beta jXj + \lambda i + \gamma t + \mu it \quad (6)$$

The second step entailed verifying the impacts of carbon trading policies on green innovation efficiency as follows:

$$GIEit = \beta0 + \beta1Ci + \beta2Yt + \beta3(Ci \times Yt) + \sum \beta jXj + \lambda i + \gamma t + \mu it \quad (7)$$

In the third step, the multiplier term and the three major effects were simultaneously inputted into the model and returned to the green innovation efficiency:

$$GIEit = \theta0 + \theta1Ci + \theta2Yt + \theta3(Ci \times Yt) + \theta4TIit(ESit, SUit)\sum \theta jXj + \lambda i + \gamma t + \mu it \quad (8)$$

where TI denotes the effect of technological innovation, expressed by the number of patent applications for energy conservation and emission reduction technologies in various regions. The specific data for the calculation, which were acquired using the method described by Ye Qin et al., indicated that carbon trading enhances the level of regional technological innovation, inducing a technological innovation effect and improving the efficiency of green innovations [30]. ES is the energy substitution effect, entailing the replacement of a proportion of the total amount of regional electricity that is consumed by clean energy. A higher proportion of clean energy used as a substitute within a region corresponds to a stronger effect of enhanced green development. SU is the structural upgrading effect, expressed as the ratio of the tertiary industry to secondary industry within a region. Carbon trading evidently promotes the advancement of the regional industrial structure, and this structural upgrading effect improves the level of green innovation.

Table 5 shows the regression results for the impacts of carbon trading policies relating to the three major effects. The results show that the regression coefficients of the three major effects were all positive at the 1% significance level, indicating that carbon trading policies influence technological innovation, energy substitution, and structural upgrading in the process of applying market-oriented strategies to achieve emission reduction goals. Table 6 shows the regression results for the impacts of carbon trading policies on green innovation efficiency after including the difference factors. The results indicate that the regression coefficients of the effects of technological innovation, energy substitution, and structural upgrading were significantly positive. Specifically, the effect of technological innovation on green innovation efficiency was positive at a 1% significance level, highlighting the important role of the development and use of energy-saving and emission-reducing technologies for advancing green and sustainable regional development. The effects of energy substitution and structural upgrading on green innovation efficiency were positive at the 10% and 5% significance levels, indicating that both of these effects play a role in promoting efficient green innovation, but these effects are not particularly significant. Moreover, the results indicate that China's current energy structure continues to be unviable, as it is still overly dependent on traditional fossil energy sources and requires further optimization and upgrading. These results also confirm the three hypotheses of the study, namely that carbon trading policies improve the efficiency of green innovation within regions through their effects on technological innovation, energy substitution, and structural upgrading.

Table 5. The regression results for carbon trading policies on the three major effects.

Variable	Technological Innovation Effect	Energy Substitution Effect	Structural Upgrading Effect
$C_i \times Y_t$	0.0162 *** (3.13)	0.2025 *** (5.21)	0.0266 *** (4.33)
N	420	420	420
R^2	0.8043	0.9012	0.8624

Notes: *t* values are shown in brackets; *** indicate statistical significance at the 1%, 5%, and 10% levels, respectively.

Table 6. The regression results for carbon trading policies on green innovation efficiency after adding the multiple differences.

Variable	Green Innovation Efficiency		
$Ci \times Yt$	0.2308 *** (3.12)	0.0923 ** (2.10)	0.0831 *** (4.02)
Technological innovation	0.3425 *** (2.90)		
Energy substitution		0.0565 * (1.93)	
Structural upgrade			0.2602 ** (2.08)
N	420	420	420
R^2	0.8913	0.8265	0.9031

Notes: *t* values are shown in brackets; ***, **, and * indicate statistical significance at the 1%, 5%, and 10% levels, respectively.

4.4. Robustness Test for Changing the Sample Interval

The implementation effect of carbon trading policy needs time to test. Considering the timeliness issues before and after the policy, a more balanced data sample for the period

2010–2018 was selected for a further regression conducted to test the robustness of the main regression.

Table 7 shows the regression results of the robustness test during the change sample interval. Columns (1), (2), and (3) show the results of the carbon trading policy on the regressions for the technological innovation, energy substitution, and structural upgrading effects, and column (4) shows the regression results for the three major effects and for green innovation efficiency. The regression results indicate that the coefficients of the interaction terms were all significant at the 1% and 5% levels, and the coefficients of the three major effects were also significant at the 5% and 10% levels. These findings are consistent with those of the main regression, described above, confirming that the regression results presented in this paper are robust. Thus, the findings indicate that the carbon trading policy has produced three major effects and that it has enhanced regional green innovation efficiency through the mechanism of these three effects.

Table 7. Results of the regression to test the robustness of the main regression and to change the window period.

Variable	Technological Innovation Effect	Energy Substitution Effect	Structural Upgrading Effect	Green Innovation Efficiency
	(1)	(2)	(3)	(4)
$C_i \times Yt$	0.5566 *** (5.78)	0.2119 ** (2.56)	0.2572 *** (8.04)	
Technological innovation effect				0.2522 ** (2.27)
Energy substitution effect				0.1051 * (1.90)
Structural upgrading effect				0.1296 * (1.78)
N	270	270	270	270
R^2	0.7890	0.8248	0.8932	0.9023

Notes: *t* values are shown in brackets; ***, **, and * indicate statistical significance at the 1%, 5%, and 10% levels, respectively.

5. Conclusions and Policy Implications

5.1. Conclusions

Panel data obtained for 30 provinces and cities in China (excluding Tibet) were applied within a super-efficient SBM model that included undesirable outputs to measure these regions' green innovation efficiency for the period 2005–2018. A dual difference model and the PSM–DID method were simultaneously used in an empirical examination of the trend in regional green innovation efficiency as it has been impacted by the implementation of carbon trading policies. The conclusions of the study can be summarized as follows.

First, the implementation of carbon trading policies can significantly improve the efficiency of green innovation in pilot areas and promote green, low-carbon regional development. Second, we investigated the influence mechanism of carbon trading policies as a driver of increased efficiency of green innovation. Our findings indicated that the implementation of carbon trading policies improves the efficiency of green innovation through its effects relating to technological innovation, energy substitution, and structural upgrading. The technological innovation effect was positively significant at the 1% level, while the effects of energy substitution and structural upgrading were positively significant, though only at the 10% and 5% levels. Last, our empirical findings revealed the overall synergistic effect of the three individual effects of carbon trading policies in amplifying regional green innovation efficiency.

5.2. Policy Implications

In light of the above conclusions, our findings have the following policy implications for improving the development of China's carbon market. First, the successful experiences in the pilot regions can be replicated by promoting the carbon trading policy on a wider scale and through the advancement of the national carbon market. The implementation of carbon trading policies can play a major role in reducing carbon emissions and driving transformational, low-carbon regional development. On the one hand, further exploration of low-carbon development is needed in the carbon trading pilot areas, focusing on technological innovation and the development of greener production and lifestyles that incorporate, for example, green transportation, green buildings, and green consumption. On the other hand, further development of the pilot regions would enable the expanded scope of market transactions in terms of points and areas and the use of market-oriented strategies to compel companies to upgrade their levels of technological innovation, thereby promoting China's overall green development.

Second, continued efforts are required to increase the proportion of R&D investment in technology, while simultaneously increasing the proportion of clean energy use and adjusting and upgrading the industrial structure. Specifically, the patent incentive system requires improvement, and enterprises or innovation-focused entities should be provided with guidance. Such guidance should focus on the invention and application of energy-saving and emission-reduction patented technologies, reducing the costs for enterprises of investing in new technologies, and providing innovation subsidies for carbon quotas to facilitate green production processes. Further, the process of transforming the energy structure to promote improved green innovation efficiency should be accelerated along with the development of new clean energy and efforts to accelerate reforms of the electricity market system to promote the greening of the power generation process. Finally, efforts to increase the proportion of high-quality service industries, while simultaneously focusing on rationalizing and gradually optimizing the industrial structure, would contribute to its advancement and ensure a gradual rise in levels of regional green innovation.

A final implication of this study relates to the need to develop an institutional system and mechanism for promoting low-carbon development, building on the synergistic effects of technological innovation, energy substitution, and structural upgrades under carbon trading policies. The development of green innovations within regions requires the promotion of advanced green technology and industrial upgrading within various industrial sectors. It simultaneously requires innovations within the macro-control energy system that facilitate the phasing out of energy-intensive and carbon-intensive industries, and the dismantling of the "locking effect" of energy-driven development. By ensuring the coordinated and balanced development of different policy effects, this approach will lead to enhanced efficiency of green innovation within regions and high-quality economic development.

Author Contributions: Conceptualization, B.L. and Z.S.; methodology, H.L.; software, H.L.; validation, B.L., Z.S. and H.L.; formal analysis, B.L.; investigation, H.L.; resources, H.L.; data curation, H.L.; writing—original draft preparation, B.L.; writing—review and editing, H.L.; visualization, Z.S.; supervision, Z.S.; project administration, Z.S.; funding acquisition, Z.S. All authors have read and agreed to the published version of the manuscript.

Funding: This research was funded by the National Social Science Foundation of China, grant number (16AGL002); the Ministry of Education Philosophy and Social Sciences Research Fund, grant number (16JZD014); the Tianjin College Innovation Team Training Program, grant number (TD13-5012/5045).

Institutional Review Board Statement: Ethical review and approval was not required for the study as the research does not involve humans.

Informed Consent Statement: Not applicable.

Data Availability Statement: Data available on request.

Conflicts of Interest: The authors declare no conflict of interest.

References

1. He, J.K.; Lu, L.L.; Wang, H.L. A study on the win-win pathway of economic growth and CO_2 emission reduction. *China Popul. Resour. Environ.* **2018**, *28*, 9–17. [CrossRef]
2. Goulder, L.H.; Parry, I.W. Instrument Choice in Environmental Policy. *Discuss. Pap.* **2008**, *21*, 152–174.
3. Kossoy, A.; Guigon, P. State and trends of the carbon market 2012. *World Bank Other Oper. Stud.* **2013**, *1*, 3–16.
4. Chen, J.; Li, H.; Wang, X. Characteristics Analysis and Enlightenment of Carbon Emission Trading System in Australia. *Ecol. Econ.* **2013**, *12*, 15–22.

5. Schiederig, T.; Tietze, F.; Herstatt, C. Green innovation in technology and innovation management–an exploratory literature review. *R&D Manag.* **2012**, *42*, 180–192.
6. Kemp, R.; Arundel, A. Survey Indicators for Environmental Innovation. *Indic. Data Eur. Anal.* **1998**, *18*, 22–28.
7. Fan, D.; Wang, W.; Liang, P. Analysis of the performance of carbon emissions trading right in China—The evaluation based on the difference-in-difference model. *China Environ. Sci.* **2017**, *37*, 2383–2392.
8. Liu, C.; Sun, Z.; Zhang, J. Research on the effect of carbon emission reduction policy in China's carbon emissions trading pilot. *China Popul. Resour. Environ.* **2019**, *29*, 49–58.
9. Tu, Z.; Shen, R. Can Emissions Trading Scheme Achieve the Porter Effect in China? *Econ. Res. J.* **2015**, *50*, 160–173.
10. Liu, C.M.; Gao, Y. Research on the Cost of Carbon Emissions among Chinese Industrial Sectors Based on Carbon Trading. *Soft Sci.* **2016**, *30*, 85–88.
11. Cheng, B.; Dai, H.; Wang, P.; Zhao, D.; Masui, T. Impacts of carbon trading scheme on air pollutant emissions in Guangdong Province of China. *Energy Sustain. Dev.* **2015**, *27*, 174–185. [CrossRef]
12. Ren, Y.; Fu, J. Research on the effect of carbon emissions trading on emission reduction and green development. *China Popul. Resour. Environ.* **2019**, *29*, 11–20.
13. Tan, J.; Zhang, J. Does China's ETS Force the Upgrade of Industrial Structure—Evidence from Synthetic Control Method. *Res. Econ. Manag.* **2018**, *39*, 104–119.
14. Watanabe, M.; Tanaka, K. Efficiency analysis of Chinese industry: A directional distance function approach. *Energy Policy* **2007**, *35*, 6323–6331. [CrossRef]
15. Shen, N.; Zhou, J. A study on china's green innovation efficiency evaluation and functional mechanism based on hybrid DEA and SEM model. *J. Ind. Eng. Eng. Manag.* **2018**, *32*, 46–53.
16. Lv, Y.; Xie, Y.; Lou, X. Study on the Convergence of China's Regional Green Innovation Efficiency. *Sci. Technol. Prog. Policy* **2019**, *36*, 37–42.
17. Peng, W. Green Creation and Chinese Economic High-Quality Development. *Jianghan Tribune* **2019**, *9*, 36–43.
18. Du, L.; Zhao, Y.; Tao, K.; Lin, W. Compound Effects of Environmental Regulation and Governance Transformation in Enhancing Green Competitiveness. *Econ. Res. J.* **2019**, *54*, 106–120.
19. Wang, D.; Li, S.; Sueyoshi, T. DEA environmental assessment on U.S. Industrial sectors: Investment for improvement in operational and environmental performance to attain corporate sustainability. *Energy Econ.* **2014**, *45*, 254–267. [CrossRef]
20. Gong, X.; Li, M.; Zhang, H. Has OFDI Promoted the Industrial Enterprises' Green Innovation Efficiency in China—Evidence based on Agglomeration Economic Effect. *J. Int. Trade* **2017**, *11*, 127–137.
21. Goodchild, A.; Toy, J. Delivery by drone: An evaluation of unmanned aerial vehicle technology in reducing CO_2 emissions in the delivery service industry. *Transp. Res. Part D Transp. Environ.* **2018**, *61*, 58–67. [CrossRef]
22. Park, C.; Xing, R.; Hanaoka, T.; Kanamori, Y.; Masui, T. Impact of Energy Efficient Technologies on Residential CO_2 Emissions: A Comparison of Korea and China. *Energy Procedia* **2017**, *111*, 689–698. [CrossRef]
23. Liu, W.; Zhang, H.; Huang, Z. Investigation on the Height of China's Industrial Structure and the Process of Industrialization and Regional Differences. *Econ. Perspect.* **2008**, *11*, 4–8.
24. Xiao, R.; Wang, Z.; Qian, L. Research on the Industrial Enterprise's Innovation Efficiency in China Considering Environmental Factor. *Manag. Rev.* **2014**, *26*, 56–66.
25. Tone, K. *Dealing with Undesirable Outputs in DEA: A Slacks-Based Measure (SBM) Approach*; GRIPS Research Report Series; Tokyo, National Graduate Institute for Policy Studies: Tokyo, Janpan, 2003.
26. Song, H.; Sun, Y.; Chen, D. Assessment for the Effect of Government Air Pollution Control Policy: Empirical Evidence from "Low-carbon City" Construction in China. *Manag. World* **2015**, *35*, 95–108+195.
27. Li, G.M.; Zhang, W.J. Research on industrial carbon emissions and emissions reduction mechanism in China's ETS. *China Popul. Resour. Environ.* **2017**, *27*, 141–148.
28. Baron, R.M.; Kenny, D.A. The moderator–mediator variable distinction in social psychological research: Conceptual, strategic, and statistical considerations. *J. Pers. Soc. Psychol.* **1986**, *51*, 1173–1182. [CrossRef] [PubMed]
29. Shi, D.; Ding, H.; Wei, P.; Liu, J. Can Smart City Construction Reduce Environmental Pollution. *China Ind. Econ.* **1987**, *6*, 117–135.
30. Ye, Q.; Zeng, G.; Dai, S.Q. Research on the effects of different policy tools on China's emissions reduction innovation: Based on the panel data of 285 prefectural-level municipalities. *China Popul. Resour. Environ.* **2018**, *28*, 115–122.

 *sustainability*

Article

Investigating the Emission of Hazardous Chemical Substances from Mashrabiya Used for Indoor Air Quality in Hot Desert Climate

Chuloh Jung * and Nahla Al Qassimi

Department of Architecture, College of Architecture, Art and Design, Healthy and Sustainable Buildings Research Center, Ajman University, Ajman P.O. Box 346, United Arab Emirates; n.alqassimi@ajman.ac.ae
* Correspondence: c.jung@ajman.ac.ae

Abstract: Dubai has the reputation of a continuously growing city, with skyscrapers and mega residential projects. Many new residential projects with poor choices of material and ventilation have led to a faster rise in sick building syndrome (SBS) in Dubai than in any other country, and the IAQ (indoor air quality) has become more critical. Volatile organic compounds (VOCs) and formaldehyde (HCHO) affect the health of residents, producing the phenomenon known as SBS (sick building syndrome). It has been reported that wood materials used for furniture and wooden windows and doors are a significant source of indoor air pollution in new houses. This paper aims to identify the factor elements emitting harmful chemical substances, such as VOCs and HCHO, from wooden mashrabiya (traditional Arabic window) by examining the characteristics of the raw and surface materials through test pieces. As a methodology, a small chamber system was used to test the amount of hazardous chemicals generated for each test piece. For Total volatile organic compounds (TVOC) and HCHO, the blank concentration before the injection and the generation after seven days were measured. The results showed that to reduce TVOC, it is necessary to secure six months or more as a retention period for raw materials and surface materials. The longer the retention period, the smaller the TVOC emission amount. In the case of mashrabiya, an HCHO low-emitting adhesive and maintenance for one month or more are essential influencing factors. It was proven that using raw materials with a three-month or more retention period and surface materials with a one-month or more retention period is safe for indoor mashrabiya. This study is the first study in the Middle East to identify factors and characteristics that affect the emission of hazardous chemicals from wood composite materials, such as wood mashrabiya, that affect indoor air quality in residential projects in Dubai. It analyzes the correlation between emission levels and the retention period of raw and surface materials, in order to provide a new standard for indoor air pollutants.

Keywords: mashrabiya; total volatile organic compounds (TVOC); formaldehyde (HCHO); retention period; hot desert climate

Citation: Jung, C.; Al Qassimi, N. Investigating the Emission of Hazardous Chemical Substances from Mashrabiya Used for Indoor Air Quality in Hot Desert Climate. *Sustainability* **2022**, *14*, 2842. https://doi.org/10.3390/su14052842

Academic Editor: Diego Pablo Ruiz Padillo

Received: 3 January 2022
Accepted: 14 February 2022
Published: 28 February 2022

Publisher's Note: MDPI stays neutral with regard to jurisdictional claims in published maps and institutional affiliations.

1. Introduction

From the point of view of selecting building materials in the design stage of a house, it is helpful to use data on the amount of hazardous chemical substances generated from each building material [1,2]. In the case of using construction materials with a large amount of hazardous chemical substances, more ventilation is required to meet a certain concentration standard than when using low-generation building materials. In a hot desert climate, this leads to an increased energy consumption [3]. Therefore, it is desirable not only in terms of comfort and health, but also in an energy-saving aspect to use materials that generate low-toxicity substances [4].

In the United Arab Emirates (UAE), the Emirates Green Building Council launched the Energy Efficiency Program (EEP) in 2013 to reduce the carbon footprint for existing energy-inefficient buildings [5]. The EEP database was launched in 2014 to facilitate building retrofit projects, and the Technical Guidelines for Retrofitting Existing Buildings was published in 2015 [6]. After the guidelines, Building Retrofit Training (BRT) was launched in 2017 by the Dubai Supreme Council of Energy and Masdar to increase the fundamental knowledge of retrofits specific to the MENA (Middle East/North Africa) region [6].

During this process, the recent construction of energy-efficient houses in Dubai has become more air-tight [7]. Accordingly, the importance of air quality in the indoor environment of a house is increasing [8]. In particular, it is known that indoor air pollution caused by chemicals such as volatile organic compounds (VOCs) and HCHO affects the health of residents [9].

According to the Dubai Healthcare City report, an estimated 15% of Dubai residents have suffered sick building syndrome (SBS) symptoms, such as fatigue, headache, red eyes, eye/nose/throat irritation, dry cough, dry or itchy skin, dizziness, and difficulty in focusing on work [10,11]. Due to this SBS phenomenon, the Dubai Municipality initiated the indoor air quality (IAQ) concentration standards, requiring less than 0.08 ppm (parts per million) of HCHO, less than 300 micrograms/m^3 of total volatile organic compounds (TVOCs), and less than 150 micrograms/m^3 of particulate matter (PM$_{10}$) (less than 10 microns) in 8 h of continuous monitoring prior to occupancy for new houses [12].

In previous studies, it has been reported that wood materials used for furniture and wooden windows and doors are a significant source of indoor air pollution in new houses [13]. It should be noted that wood products emit hazardous chemical substances continuously for an extended time, since the emission characteristics of hazardous chemicals from wood over time are different from those of hazardous chemical substances generated from paints and adhesives [14,15].

In particular, in the case of new houses, the amount of built-in furniture is increasing [16]. As the risk of indoor contamination by hazardous chemical substances generated in wood materials during construction increases, the need for the control of this has also increased [17]. These wood products are applied indoors as a composite material, rather than a single wood product, formed by applying different surface materials, such as a particle board (PB) or medium-density fiberboard (MDF), and going through various processing methods [18]. In addition, the installation of these wood products follows a process in which partially finished products are brought into the room, with a time gap according to the process, and are installed and finished onsite [19]. Therefore, it is possible to reduce, to some extent, the natural emission of toxic chemical substances from wood materials by making adjustments to such factors as the retention period on site during the installation process [20].

Indoor wood products are molded from composite materials rather than single materials, and identifying the emission characteristics of hazardous chemical substances could be helpful in improving indoor air quality [21,22].

Currently, wood materials mainly used for new houses in Dubai can be classified into mashrabiya (Figure 1) and built-in furniture materials.

The influencing factors can be derived by reviewing the previous literature on the emission characteristics of hazardous chemical substances from single materials or composite materials from furniture and wooden windows/doors.

Several researchers have conducted research on the emission characteristics of chemical substances emitted from wood materials, such as furniture and wooden windows/doors [23]. Adamová et al. (2020) [24] and Ulker et al. (2021) [25] report that, unlike the emission process of paints or adhesives, chemicals in wood materials are continuously emitted into the air in trace amounts for a very long time. Their research focuses on single materials such as plywood, PB, and MDF.

Richter et al. (2021) [26] reports the effect of a single material combination on the emission of pollutants from composite finishing materials, since the building finishing materials mainly used in houses are composite finishing materials. For the composition of the test piece, a composite material composed of wallpaper and adhesive or floorboard and adhesive was used in the experiment. The release mechanism was estimated based on the results showing the difference in the emission amount of the composite material according to the difference in the contaminant content of the single material constituting the composite material. The amount of composite materials emitted differs depending on the composition of the single material and the degree of contamination.

Figure 1. Wooden mashrabiya window at Dubai Hill Estate.

Cao et al. (2019) [27] reports that MDF is treated with low-pressure melamine-impregnated paper (LPM), which is a surface method for furniture materials, on both sides, and those that were not treated were targeted. The experiment was carried out with differences in the exposure of the cut surface in the case of the MDF-treated material on both sides. As a result, the emission intensity tended to increase as the ratio of the cut surface increased (the surface finish area decreased). It was reported that most of the wood materials showed a difference in the amount of emission depending on the surface finishing method.

Qi et al. (2019) [28] produced six types of built-in furniture by selecting furniture composed of raw materials and surface finishing materials commonly used in apartments and identified the chemical pollutant emission characteristics by raw material grade and surface finish. This study reported that the concentration of HCHO was different depending on the core material in the same surface finishing material.

Xiong et al. (2019) [29] reports that the emission of TVOC is not affected by the grade of the core material, but is affected by the finish of the surface material. In the TVOC case, the raw material grade influences and the difference in the quantity of emission by the surface material were established.

Through the above literature review, in the case of a wood composite finishing material, very diverse factors affect the emission of hazardous chemical substances. The degree of contamination of single raw materials or surface materials, differences in surface finishing methods, and the retention period of raw materials constituting composite finishing materials were identified as factors affecting the release of hazardous chemical substances [30].

The purpose of this study is to identify the factors emitting harmful chemical substances from wooden mashrabiya by examining the characteristics of the raw materials and surface finishing materials of the test pieces.

2. Materials and Methods

2.1. Types and Characteristics of Raw Materials

2.1.1. Fiberboard

Fiberboard is a generic term for plate products formed by fiberizing wood or other plants and is classified according to density [31]. Since fiberboard has a dense structure and excellent machinability, it can be used instead of general wood for construction sites that require precise dimensions or for frames and moldings that require accurate angles [32]. The Environmental Protection Agency (EPA) in the United States has classified the above materials according to the amount of HCHO emission since 2018 [33] (Table 1).

Table 1. Classification of fiberboards and EPA HCHO emission standards.

Types	Density	EPA HCHO Emission Standards
Hard Fiberboard (HB) - Veneer Core		
HB - Composite Core	More than 0.85 g/cm^3	0.05 ppm
Medium-Density Fiberboard (MDF)	0.35 g/cm^3–0.85 g/cm^3	0.11 ppm

It is classified as MDF with a density of less than 0.35 g/cm^3–0.85 g/cm^3, and HB with a density of 0.85 g/cm^3 or more. MDF is the most used and can be produced from 3.0 mm to 30 mm thick [34] (Figure 2).

Figure 2. HB (**left**), MDF (**middle**), and glued laminated timber (**right**).

2.1.2. Glued Laminated Timber

Glued laminated timber refers to a material that is laminated and adhered in the length, width, and thickness directions by paralleling each other in the fiber direction using sawmills, boards, or small square materials [35] (Figure 2). Additionally, it refers to a material that is decorated on the surface of the aggregated material for aesthetics.

Dissimilar to plywood by orthogonal and odd-sheet lamination methods, glued laminated timber is not manufactured in a plate shape, and it is manufactured in various shapes and sizes depending on the purpose of use. It is mainly used as a wooden frame material for windows and doors and as a core material for doors, and one of the most popular glued laminated timbers is Lauan laminated timber (LLT).

2.2. Types and Characteristics of Surface Finishes
2.2.1. Veneer

Veneer refers to a product manufactured by processing raw wood as thin as paper [36]. It is used as a decorative material to express the naturalness and splendor of wood. It is a material that can express the texture of wood by finishing the primary structural frame with gypsum board or plywood, applying adhesive to the surface, and attaching it by heating and pressing methods [37]. In most cases, it is usually composed of a thin plate of 0.2 mm, and it is rarely used as an exterior material.

For the sample used in this experiment, samples prepared by wrapping veneer around raw materials of LLT, MDF, and HB, used for mashrabiya, were used with hot-melt-type adhesive [38].

2.2.2. Polyvinylchloride (PVC) Sheet

PVC sheet is manufactured by mixing polyvinyl chloride as a raw material with stabilizer, plasticizer, and pigment [39]. Since it is easy to process and mass production is possible, it is currently widely used as a surface material for overlay or wrapping of interior materials, furniture, and wooden windows and doors.

For the sample used in this experiment, a sample piece wrapped around PB was used with a water-based liquid adhesive.

2.2.3. Finishing Foil (F/F)

It is a patterned paper in which paper is impregnated with thermosetting resin (melamine, urea, and acrylic), and then the surface is painted [40]. This material was developed to reduce the cost of LPM. Dissimilar to LPM, an additional adhesive must be used to bond the finishing foil. In the Middle East, it is commonly used as an interior building material and furniture surface finishing material [41].

For the sample used in this experiment, the sample was wrapped in raw materials of LLT, MDF, and HB, and the finishing foil with a hot-melt adhesive precoated on the backside was used.

2.3. Hazardous Chemical Emission Test Method for Mashrabiya

2.3.1. Test Piece Overview

The test piece used in this experiment was from the same composite material in the interior of the Dubai Hill house [42]. The raw materials, adhesives, and surface materials of actual mashrabiya in Dubai Hill were used. Figure 3 shows the configuration of the test piece. It was composed of a surface finishing method of wrapping using an adhesive between the raw and surface materials.

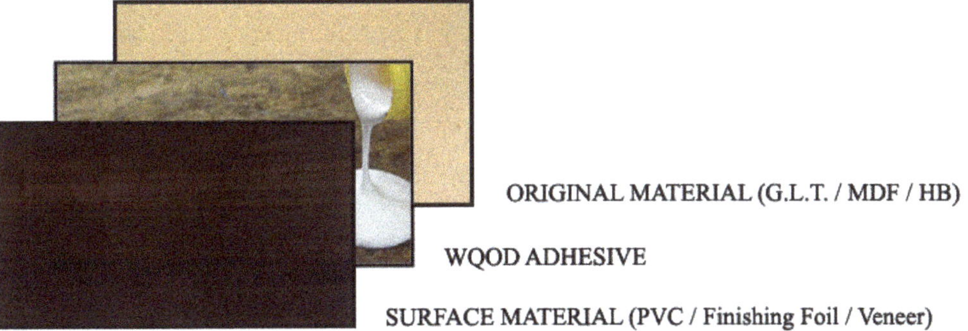

Figure 3. Material composition of the test piece.

Table 2 shows the classification of test pieces according to the composition of raw materials, adhesives, surface materials of the test pieces, the production date of the test pieces, and the test start date after standing for a certain time.

Among the test specimens, in the case of test pieces TP-01, TP-02, and TP-03, PVC, F/F, and veneer were wrapped in LLT, using materials of traditional style mashrabiya, and PVC, F/F, and veneer were wrapped in MDF and HB, respectively. The test pieces of TP-04, TP-05, TP-06, TP-07, TP-08, and TP-09 were manufactured assuming the material of laser-cut mashrabiya [43].

Table 2. Test piece composition and production/test start date.

Name	Material Composition for Test Piece	Production Date	Test Start Date
TP-01	LLT + Wood Adhesive + PVC	04/09/2021	28/09/2021
TP-02	LLT + Wood Adhesive + F/F	04/09/2021	28/09/2021
TP-03	LLT + Wood Adhesive + Veneer	04/09/2021	28/09/2021
TP-04	MDF + Wood Adhesive + PVC	04/10/2021	12/10/2021
TP-05	MDF + Wood Adhesive + F/F	04/10/2021	12/10/2021
TP-06	MDF + Wood Adhesive + Veneer	04/10/2021	12/10/2021
TP-07	HB + Wood Adhesive + PVC	02/11/2021	05/11/2021
TP-08	HB + Wood Adhesive + F/F	02/11/2021	05/11/2021
TP-09	HB + Wood Adhesive + Veneer	02/11/2021	05/11/2021

Table 3 shows the composition of each test piece manufactured with the difference between the different retention periods of raw materials and surface materials [44]. For the composition of the test piece, PVC sheet, F/F, and veneer were processed for LLT, MDF, and HB, each having a different retention period, respectively. In the case of raw materials, three-month-old LLT, six-month-old MDF, and HB were used, and in the case of surface materials, 30–75-day-old F/F, veneer, and PVC were used. In addition, all 9 test pieces had different retention periods. The retention period after production refers to the time elapsed before being used for testing after completing the production of materials with different retention periods of raw materials and surface materials by different processing methods. Specific test samples were required because, in general, when a factory product is brought into the construction field, its usual retention period is 10 to 30 days. In particular, in the case of 500 households or more, after installation, the retention time of one month is required. Therefore, to secure the emission data of furniture material considering the effect of the actual indoor air environment, the factor of the retention period after production should be considered.

Table 3. Test piece composition according to the retention period of raw materials.

Surface Material	Raw Material	LLT	MDF	HB	Test Piece Retention Period after Production (Days)
	Retention Period (Days)	90	180	180	
	30	TP-01			14
PVC	45		TP-04		8
	70			TP-07	3
	30	TP-02			14
F/F	45		TP-05		8
	70			TP-08	3
	30	TP-03			14
Veneer	45		TP-06		8
	70			TP-09	3

In this experiment, the retention period after production for 3, 6, and 14 days was used proportionally to create the same environment as in the construction field.

The manufacturing process of the test piece used in this experiment was the same as the manufacturing process for the villa construction [45].

It was manufactured with raw materials that passed the same process as the production period of mashrabiya, which was installed in the villa, and kept at room temperature [46].

2.3.2. Experiment Conditions and Calculation of Emission Intensity

A small chamber system according to the International Organization for Standardization (ISO) 16,000 was used to test the amount of hazardous chemicals generated by each test piece [47]. The blank concentration of seven days before injecting the sample into the chamber and the quantity of generation after injecting the sample were assessed for TVOC and HCHO [48]. The chamber's indoor air conditions were maintained at 25 °C of the temperature, 50% of humidity, and 0.5 times/hour the number of ventilation. The area of the test piece was 0.044 m^{2}, and the sample load rate was 2.2 m^2/m^3. Emission intensity was calculated according to ISO 10580.

2.3.3. Sample Collection and Analysis Methods

For the collection of VOCs, an adsorption tube filled with Tenax TA 200 mg was connected to a flow sampling pump to collect a total of 3.5 L of air in the chamber at 167 mL/min, since Tenax TA is a widespread polymer sorbent recommended for retaining VOCs according to ISO 16000-6. The absorption procedure was as follows: VOCs were collected on the Tenax TA sorbent tube by pulling an air sample across the sorbent bed with the aid of a portable mechanical pump. VOCs were thermally desorbed from the sorbent tube within the split/splitless injection port of a GC and focused onto the head of a capillary column. GC analysis with negative ion chemical ionization MS allowed the quantification of desorbed VOCs.

To analyze the VOCs and target substances contained in standard samples and test pieces, ATD-400 (PerkinElmer, Waltham, MA, USA) directly connected to the GC column by GC/MS (gas chromatography/mass spectrometry, Agilent 6890/5973N, Santa Clara, CA, USA) was used. The analysis conditions of GC/MS were an HP-1 capillary column. The column flow was 1 mL/min, and the column temperature rate reached 60 °C within 5 min, and, thereafter, it was allowed to rise to 260 °C by 5 °C every 5min. The ion source temperature of MS was set to 260 °C and used for analysis (Table 4).

Table 4. Overview of measuring instruments.

Measuring Items	Measuring Instruments	Measurement Range/Accuracy
VOCs	• Gas Chromatograph (Agilent 6890 GC, Santa Clara, CA, USA) • Mass Spectrometer (Agilent 5973N MSD, Santa Clara, CA, USA)	• Pressure sensors accuracy: ±2% full scale • Repeatability: ±0.05 psi • Temperature coefficient: ±0.01 psi/°C • Drift: ±0.1 psi/6 months • Flow sensors accuracy: <±5% • Repeatability: ±0.35% of setpoint • Temperature coefficient: ±0.20 mL/min normalized temperature
HCHO	• High-Performance Liquid Chromatography (Shimadzu 10AVP Series HPLC System, Kyoto, Japan)	• 0.01–51,200 µS cm−1 FS • 0.01, 0.1, 1, 10, 100 $\mu S\ cm^{-1}$/mV
Room Temperature Relative Humidity	• Digital Thermo-Hygrometer (TR-72U, Taipei, Taiwan)	• −20 to 80 °C, 10 to 95% RH • Average +/− 0.3 °C (−20 to 80 °C) • Average +/− 5% RH (At 25 °C 50 % RH)

HCHO was collected using an LpDNPH S10L cartridge (Supelco Inc., Bellefonte, PA, USA) to collect carbonyl compounds. The sample was collected by connecting it to a flow sampling pump to remove the interference caused by ozone (O_3), and an O_3 scrubber was connected in front of the LpDNPH S10L cartridge.

At this time, a total of 7.0 L of air was collected in the chamber at 167 mL/min, and the collected samples were stored in a cool and dark place until extraction. For the extraction of the analytical sample, the DNPH-carbonyl derivative formed by reacting with DNPH was extracted with 5 mL of HPLC-grade acetonitrile, and analysis was performed immediately. The analysis for HCHO was performed using high-performance liquid chromatography (HPLC, Shimadzu, Kyoto, Japan) and fixed at a maximum wavelength of 360 nm.

3. Results

In this experiment, the difference in the emission amount was analyzed for composite material test pieces. The characteristics of the emission quantity due to changes in the raw material and surface material properties, and the retention duration after molding, were investigated. In addition, to understand the emission level of hazardous chemicals emitted from these test pieces, they was compared and evaluated using the Environmental Product Declaration (EPD) standard in Dubai [49].

A total of nine test pieces, prepared by the finishing method of wrapping PVC sheets, F/F, and veneer surface materials around raw materials of HB, MDF, and LLT, were used. Figures 4 and 7 show the results of the emission for TVOC and HCHO from three different test pieces, such as TP-01 using a PVC sheet with a retention period of 30 days in LLT with a retention period of 3 months, TP-04 using a PVC sheet on 45-day-old MDF with a 6-month retention period, and TP-07 using a PVC sheet with a retention period of 70 days in HB with a retention period of 6 months.

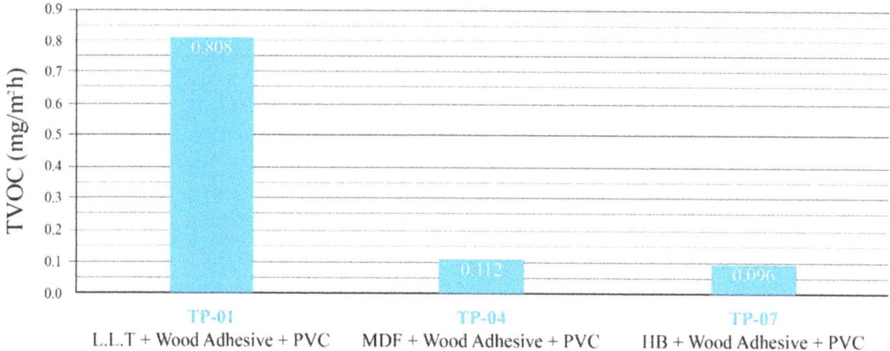

Figure 4. TVOC emission of PVC-wrapped test pieces.

Figures 5 and 8 show the results of the emission for TVOC and HCHO from three different test pieces, such as TP-02 using F/F sheets with a retention period of 30 days in LLT with a retention period of 3 months, TP-05 using F/F sheets on 45-day-old MDF with a retention period of 6 months, and TP-08 using F/F sheets with a retention period of 70 days in HB with a retention period of 6 months.

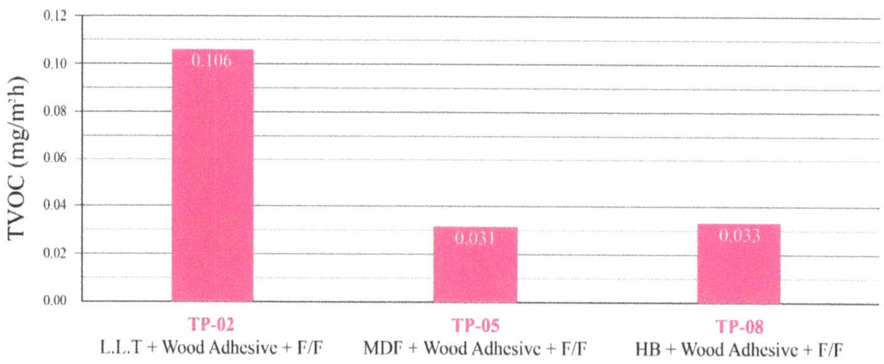

Figure 5. TVOC emission of F/F-wrapped test pieces.

Additionally, Figures 6 and 9 show the results of the emission for TVOC and HCHO from three different test pieces, such as TP-03 using veneer with a retention period of 30 days on an LLT with a retention period of 3 months, TP-06 using veneer on MDF aged for 45 days with a retention period of 6 months, and TP-09 using veneer with a retention period of 70 days in HB with a retention period of 6 months.

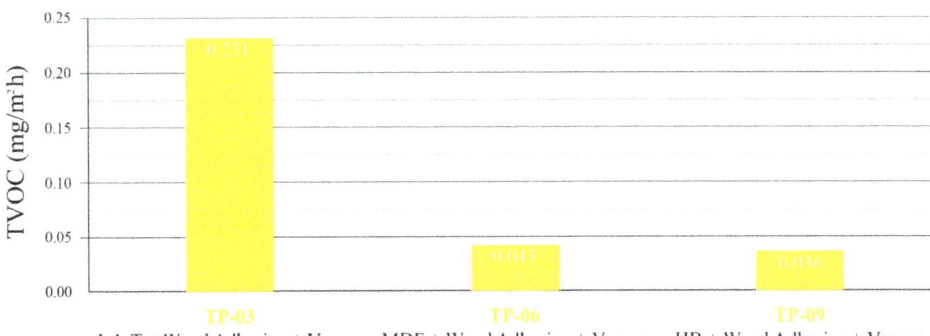

Figure 6. TVOC emission of veneer-wrapped test pieces.

First of all, it was a characteristic of the degree of contamination of raw materials and surface materials. As for raw materials, MDF and HB used the same grade of formaldehyde-dissipating raw materials. However, as LLT is currently excluded from classification in the EPD standards, it had the prerequisite that the degree of emission was uncertain.

3.1. The Characteristics of TVOC Emission

3.1.1. Emission Characteristics Due to Differences in Raw Materials and Surface Materials

Figure 4 shows the emission of TVOC for the specimens composed of different types of raw materials when the PVC sheet had the same finish. It was found that TP-01 using LLT raw materials showed a significantly higher emission than TP-04, which was MDF, and TP-07, which was HB. This trend was also shown in Figures 6 and 7. Therefore, to identify the cause of the difference in the emission, the effect of the characteristics of raw materials and surface materials was analyzed.

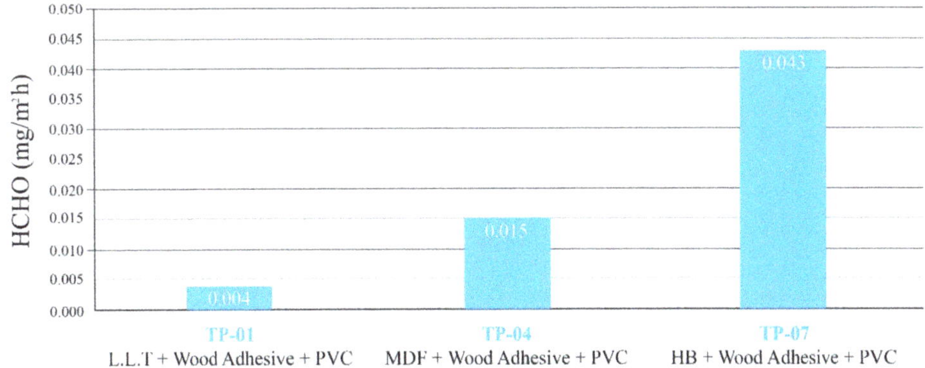

Figure 7. HCHO emission of PVC-wrapped test pieces.

In the case of the surface material, as described above in Section 2.1, the PVC sheet emitted more TVOC than the veneer and F/F. In the case of adhesives, most of the adhesives were environmentally friendly hot-melt-based or water-based.

In order to understand the emission characteristics due to differences in surface materials, different surface materials with the same retention period of 30 days in the same LLT, PVC sheet (Figure 4, TP-01), and F/F (Figure 5, TP-02) were compared with the case finished with veneer (Figure 6, TP-03). PVC showed the highest emission at 0.808 mg/m^{2}h. In the case of veneer and F/F, the emission amounts were 0.231 mg/m^{2}h and 0.106 mg/m^{2}h, respectively, showing a slight difference. Additionally, to show the same trend, the raw materials of MDF were dried with different surface materials with the same retention period of 45 days, such as the PVC sheet (Figure 4, TP-04), F/F (Figure 5, TP-05), and veneer (Figure 6, TP-06). The same trend was also observed in the case of using different surface materials that had passed the same 70 days for the raw materials of HB, such as the PVC sheet (Figure 4, TP-07), F/F (Figure 5, TP-08), and veneer (Figure 6, TP-09)

As a result of comparing and analyzing the TVOC emission in the case of finishing with different finishing materials that passed the same time on the same raw material, the difference due to the type of surface material was shown. The case of finishing with a PVC sheet showed a relatively high level, and F/F and veneer showed almost a slight difference and were detected to be low. It was judged that these results had a relatively significant effect on the type of surface material.

In what follows, we compared and evaluated the difference in the amount of emission for the retention period among the characteristics of raw materials and surface materials. In the case of raw materials in Figure 4, LLT (TP-01), which had a retention period of 3 months after production, showed a higher emission than MDF (TP-04) and HB (TP-07), which had a retention period of 6 months. In particular, PVC, which was the same surface material, was used, but by using sheets that passed 30 days for LLT, 45 days for MDF, and 70 days for HB, the PVC sheets applied to MDF and HB had a more extended retention period than LLT.

Therefore, the TVOC emission was affected by the difference between the retention periods of raw materials as the MDF+PVC and HB+PVC test pieces with relatively long retention periods of raw materials and surface materials were significantly lower than those of LLT+PVC with short retention periods. In addition, in the case of HB+PVC and MDF+PVC, HB and MDF raw materials had the same retention period, but the retention period of the surface material PVC had a difference of more than one month. When raw materials with a relatively long retention period of 6 months or more were used, the effect on the retention period of the surface material was judged to be weak. The same trend was observed in the test pieces finished with F/F and veneer (Figures 5 and 6).

3.1.2. Emission Characteristics Due to Differences in Retention Period

As shown in Table 3, the retention period of the test pieces was 14 days for the LLT+PVC (TP-01), LLT+F/F (TP-02), and LLT+veneer (TP-03) composites, eight days for the MDF+PVC (TP-04), MDF+F/F (TP-05), MDF+veneer (TP-06) composites, and three days for the HB+PVC (TP-07), HB+F/F (TP-08), HB+veneer (TP-09) composites. The characteristics of TVOC emissions were identified by classifying them into composites with different elapsed days.

Concerning Figure 4, the TVOC release was observed for the test piece of LLT+PVC (Figure 4, TP-01), which had an retention period after molding (14 days), compared to MDF+PVC (Figure 4, TP-04), with a relatively short retention period after molding, and HB+PVC (Figure 4, TP-07) was found to be higher than that of the specimen. The results shown in Figures 5 and 6 were also the same, and it was confirmed that the samples of LLT+ F/F (Figure 5, TP-02) and LLT and veneer (Figure 6, TP-03) showed high levels. In addition, it can be seen that the case where the retention period was short for eight days and three days did not show a clear difference, and showed a relatively low release amount compared to the case of fourteen days.

It was analyzed that the effect of TVOC emission from the composite material on the retention period after molding was small. TVOC emission had little effect on the retention period due to the component characteristics of the molding material, the adhesive with HCHO.

3.2. The Characteristics of HCHO Emission

3.2.1. Emission Characteristics Due to Differences in Raw Materials and Surface Materials

As shown in Figure 7, the difference in the amount of HCHO emitted between the test pieces was visible, and Figures 8 and 9 also showed the same trend. In order to find out the factors that indicated the difference in the amount of HCHO emitted by each test piece, the effects of the characteristics of raw materials and surface materials were analyzed.

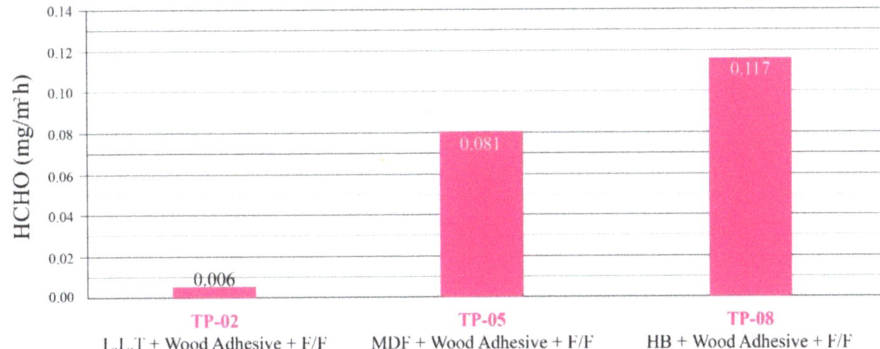

Figure 8. HCHO emission of F/F-wrapped test pieces.

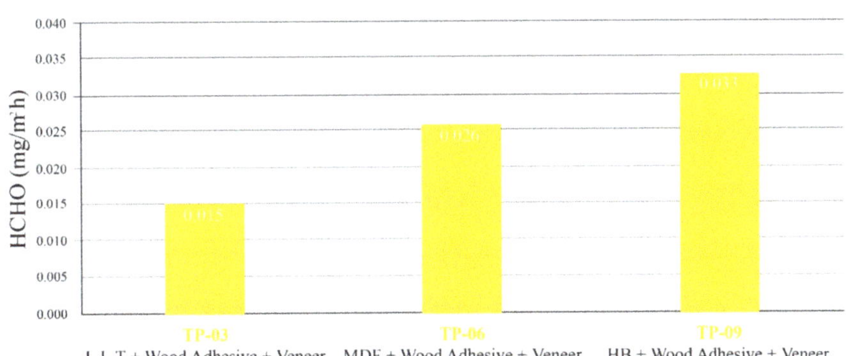

Figure 9. HCHO emission of veneer-wrapped test pieces.

The degree of contamination by raw materials, surface materials, and adhesives constituting the composite material test piece was the same as described above. On the premise of this, as shown in Figure 7, when the same surface material was applied to different raw materials, it was confirmed that the emission amount differed depending on the type of raw material. Figures 8 and 9 also showed the same trend, confirming a difference in the amount of emission depending on the type of raw material. However, it cannot be concluded that this result was a difference in the type of raw material, because it showed a significant difference in the emission amount even in the case of MDF and HB with the same HCHO emission grade. As can be seen from Figures 8 and 9, this trend was the same in all cases.

Meanwhile, the PVC sheets with different surface materials (Figure 7, TP-01), F/F (Figure 8, TP-02), and veneer (Figure 9, TP-03) had the same retention period of 30 days in the same LLT. In comparison, the PVC sheet showed 0.004 mg/m²h, F/F 0.006 mg/m²h, and veneer 0.015 mg/m²h, indicating that veneer showed a relatively high emission.

However, when applying different surface materials with 45 days elapsed to the same MDF and applying the same 70 days elapsed surface material to HB, the previous case was veneer, whereas F/F had a relatively high emission. Although it was possible to confirm the difference in the amount of emission depending on the surface material, it was judged that there was no general trend limited to a specific surface material.

The following compared and analyzed the characteristics of HCHO emission for the retention period of raw materials and surface materials. As shown in Figure 7, in the case of raw materials, LLT with a retention period of 3 months after production was 0.004 mg/m^{2}h, MDF with a retention period of 6 months was 0.015 mg/m^{2}h, and HB was 0.043 mg/m^{2}h, confirming that HB with a more extended retention period showed a higher emission than LLT with a shorter retention period.

Additionally, in the PVC sheet, which was a surface material, 30 days for LLT, 45 days for MDF, and 70 days for HB were used, and HB had the most prolonged retention period. F/F in Figures 8 and 9 and the surface material of veneer showed the same trend. The emission of HCHO from the composite material was high in the case where the retention period of raw materials and surface materials was long. It was confirmed that the retention period of raw materials and surface materials was not a factor significantly affecting the amount of formaldehyde emitted from the composite material.

Therefore, the amount of HCHO emitted from the test piece of the composite material did not clearly show a general tendency and consistent difference depending on the characteristics, such as the degree of contamination, type, and retention period of raw materials and surface materials.

3.2.2. Emission Characteristics Due to Differences in Retention Period

HCHO emission characteristics for specimens with different retention periods after molding of composite material specimens were investigated. Figure 7 shows the amount of HCHO emitted from the composite material after 14 days of retention after molding for LLT+PVC specimens, eight days for MDF+PVC, and three days for HB+PVC.

It was confirmed that the release of HCHO was the highest in HB+PVC, which had the shortest retention period after molding. However, this was the test piece with the most prolonged retention period of raw materials and surface materials, and it was judged that the release of HCHO from the composite materials was strongly affected by the retention period after molding by applying them rather than the retention period of the raw materials and surface materials. It was confirmed that Figures 8 and 9 showed the same trend.

It can be seen that the composite material test piece applied in this study was a sample of a method using an adhesive during molding, so it can be seen that the adhesive component was related to the HCHO emission factor. It was analyzed that the retention period after molding dramatically affected the amount of HCHO emitted from the composite material.

4. Discussion

Dubai has the reputation for being a continuously growing city with skyscrapers and mega residential projects. Many new residential projects with poor material, and ventilation choices led Dubai to SBS faster than any other country.

This study was the first study in the Middle East to identify factors and characteristics that affect the emission of hazardous chemicals from wood composite materials such as wood mashrabiya that affect the IAQ at residential projects in Dubai.

Many previous studies on indoor air quality in Dubai set up apartments or villas as units to study on-site measurement and reduction methods for IAQ. However, in this study, the TVOC and HCHO emission factors and characteristics of wood mashrabiya, which have been traditionally used a lot in the Middle East but have not been scientifically measured, were analyzed for the correlation with the retention period of raw materials and surface materials to provide the new standard for indoor air pollutants.

The EPD standard in Dubai was applied to evaluate the emission level of hazardous chemicals for the test pieces used in this study, and it was compared and evaluated with the EPD standard in Dubai [50].

EPD does not explicitly stipulate wooden windows [51]. However, since standards for wooden office furniture were established, this chapter evaluated them according to this standard. According to the EPD Small Chamber Act, if HCHO emission after seven days was 0.125 mg/m^{2}h or less, it was considered to meet the standard.

In addition, in regulating the amount of VOC emissions from wood materials, all surfaces of wood materials must be paved to suppress the emission of VOCs [52]. In particular, it was stipulated that the rest, except for the edge, of the wood material should be packed with a sheet composed of a thermosetting resin material such as melamine resin, and the emission amount of VOCs after seven days according to the small chamber method was stipulated to be less than 0.4 mg/m^{2}h.

HCHO was evaluated for 0.125 mg/m^{2}h or less, in this study, and the TVOC emission level was 0.4 mg/m^{2}h or less for nine test pieces. As shown in Table 5, it was found that the evaluation results exceeded the TVOC emission standard only when the PVC sheet was wrapped in LLT, and the remaining eight test pieces were all suitable.

Table 5. Emissions of hazardous substances from long-term storage test pieces of raw materials.

Name	Material Composition for Test Piece	Emission Intensity (mg/m^{2}h)		Evaluation
		TVOC	HCHO	
TP-01	LLT + Wood Adhesive + PVC	0.808	0.004	Unsuitable
TP-02	LLT + Wood Adhesive + F/F	0.112	0.015	Suitable
TP-03	LLT + Wood Adhesive + Veneer	0.096	0.043	Suitable
TP-04	MDF + Wood Adhesive + PVC	0.106	0.006	Suitable
TP-05	MDF + Wood Adhesive + F/F	0.031	0.081	Suitable
TP-06	MDF + Wood Adhesive + Veneer	0.033	0.117	Suitable
TP-07	HB + Wood Adhesive + PVC	0.231	0.015	Suitable
TP-08	HB + Wood Adhesive + F/F	0.043	0.026	Suitable
TP-09	HB + Wood Adhesive + Veneer	0.036	0.033	Suitable

Therefore, it was confirmed that wood materials mainly used for mashrabiya showed a stable emission of harmful substances in the case of raw materials with a shelf life of three months or more, and surface finishing materials of one month or more, i.e., with a long storage period.

Results show that the retention period after importing or processing raw materials was the most critical factor. It showed the importance of the grade of raw materials and a certain period of action and sufficient curing in product manufacturing and management.

5. Conclusions

In this study, the TVOC and HCHO emission factors and characteristics of wood mashrabiya were analyzed to correlate with the retention period of raw materials and surface materials to provide the new IAQ standard for the Dubai municipality.

In the case of raw and surface materials of mashrabiya with a retention period of 1 month or more, the retention period was the most influential factor for the emission of TVOC. It was confirmed that the longer the retention period of raw materials and surface materials, the smaller the TVOC emission amount. It was required to ensure that raw materials and surface materials had a more extended retention period than a particular amount of time to induce a TVOC decrease.

It was confirmed that the emission of HCHO from the raw and surface materials of mashrabiya with a retention period of one month or more was significantly affected by the retention period after production and was considered to be influenced by the processing method using the adhesive for molding.

As a result of evaluating the indoor air emission of the test piece, raw materials with three months or more and surface materials with one month or more should be used. In the case of indoor mashrabiya, it is essential to secure the retention period of raw and surface materials and enough of a retention period after production.

This study targeted mashrabiya with a relatively long retention period. It was necessary to compare and evaluate the short-term case with a retention period of less than one month to secure the reliability of the above conclusion.

Author Contributions: Conceptualization, C.J. and N.A.Q.; methodology, C.J.; software, C.J.; validation, C.J. and N.A.Q.; formal analysis, N.A.Q.; investigation, N.A.Q.; resources, C.J. and N.A.Q.; data curation, C.J.; writing—original draft preparation, C.J.; writing—review and editing, N.A.Q.; visualization, C.J.; supervision, N.A.Q.; project administration, N.A.Q.; All authors have read and agreed to the published version of the manuscript.

Funding: This research received no external funding.

Institutional Review Board Statement: Not applicable.

Informed Consent Statement: Not applicable.

Data Availability Statement: New data were created or analyzed in this study. Data will be shared upon request and consideration of the authors.

Acknowledgments: The authors would like to express their gratitude to Ajman University for APC support and Healthy and Sustainable Buildings Research Center at Ajman University for providing great research environment.

Conflicts of Interest: The authors declare no conflict of interest.

References

1. Almusaed, A.; Almssad, A.; Homod, R.Z.; Yitmen, I. Environmental profile on building material passports for hot climates. *Sustainability* **2020**, *12*, 3720. [CrossRef]
2. Awad, J.; Jung, C. Evaluating the Indoor Air Quality after Renovation at the Greens in Dubai, United Arab Emirates. *Buildings* **2021**, *11*, 353. [CrossRef]
3. Jung, C.; Awad, J. The Improvement of Indoor Air Quality in Residential Buildings in Dubai, UAE. *Buildings* **2021**, *11*, 250. [CrossRef]
4. Tsai, W.T. Overview of green building material (GBM) policies and guidelines with relevance to indoor air quality management in Taiwan. *Environments* **2018**, *5*, 4. [CrossRef]
5. Emirates GBC. Energy Efficiency Program. 2020. Available online: https://emiratesgbc.org/technical-programs/energy-efficiency-program/ (accessed on 12 August 2021).
6. Emirates GBC. Emirates Green Building Council Launches First 'Technical Guidelines for Retrofitting Existing Buildings' in the UAE. 2015. Available online: https://emiratesgbc.org/press_releases/emirates-green-building-council-launches-first-technical-guidelines-for-retrofitting-existing-buildings-in-the-uae/ (accessed on 14 August 2021).
7. Krarti, M.; Dubey, K. Review analysis of economic and environmental benefits of improving energy efficiency for UAE building stock. *Renew. Sustain. Energy Rev.* **2018**, *82*, 14–24. [CrossRef]
8. Ghauri, A.; Hameed, M.; Hughes, A.J.; Nazarinia, M. Numerical Analysis of a Zero Energy Villa in the UAE. In *International Sustainable Buildings Symposium*; Springer: Berlin/Heidelberg, Germany, 2017; pp. 183–197.
9. Jung, C.; Awad, J. The Analysis of Indoor Air Pollutants Emission from New Apartments at Business Bay in Dubai, UAE. *Front. Built Environ.* **2021**, *7*, 765689. [CrossRef]
10. The Nationals Sick Buildings Are Leading to Sick UAE Office Workers, Doctors Say. Available online: https://www.thenationalnews.com/uae/health/sick-buildings-are-leading-to-sick-uae-office-workers-doctors-say-1.175866 (accessed on 16 August 2021).
11. Gulf News. Let's Not Forget Indoor Air Quality as Well. 2020. Available online: https://gulfnews.com/business/analysis/lets-not-forget-indoor-air-quality-as-well-1.1589873286956 (accessed on 3 September 2021).
12. DEWA. Green Building Regulations & Specifications. 2021. Available online: https://www.dewa.gov.ae/~{}/media/Files/Consultants%20and%20Contractors/Green%20Building/Greenbuilding_Eng.ashx (accessed on 4 September 2021).
13. Mannan, M.; Al-Ghamdi, S.G. Indoor Air Quality in Buildings: A Comprehensive Review on the Factors Influencing Air Pollution in Residential and Commercial Structure. *Int. J. Environ. Res. Public Health* **2021**, *18*, 3276. [CrossRef]
14. Joseph, P.; Tretsiakova-McNally, S. Sustainable non-metallic building materials. *Sustainability* **2010**, *2*, 400–427. [CrossRef]
15. Karunarathna, M.S.; Smith, R.C. Valorization of lignin as a sustainable component of structural materials and composites. *Sustainability* **2011**, *12*, 734. [CrossRef]

16. Kuys, J.; Al Mahmud, A.; Kuys, B. A Case Study of University–Industry Collaboration for Sustainable Furniture Design. *Sustainability* **2021**, *13*, 10915. [CrossRef]
17. Chiesa, G.; Cesari, S.; Garcia, M.; Issa, M.; Li, S. Multisensor IoT platform for optimising IAQ levels in buildings through a smart ventilation system. *Sustainability* **2019**, *11*, 5777. [CrossRef]
18. May, N.; Guenther, E.; Haller, P. Environmental indicators for the evaluation of wood products in consideration of site-dependent aspects: A review and integrated approach. *Sustainability* **2017**, *9*, 1897. [CrossRef]
19. Piasecki, M.; Kozicki, M.; Firląg, S.; Goljan, A.; Kostyrko, K. The approach of including TVOCs concentration in the indoor environmental quality model (IEQ)—Case studies of BREEAM certified office buildings. *Sustainability* **2018**, *10*, 3902. [CrossRef]
20. Son, Y.S.; Lim, B.A.; Park, H.J.; Kim, J.C. Characteristics of volatile organic compounds (VOCs) emitted from building materials to improve indoor air quality: Focused on natural VOCs. *Air Qual. Atmos. Health* **2013**, *6*, 737–746. [CrossRef]
21. Salthammer, T.; Schripp, T.; Wientzek, S.; Wensing, M. Impact of operating wood-burning fireplace ovens on indoor air quality. *Chemosphere* **2014**, *103*, 205–211. [CrossRef]
22. Yu, C.W.; Kim, J.T. Long-term impact of formaldehyde and VOC emissions from wood-based products on indoor environments; and issues with recycled products. *Indoor Built Environ.* **2012**, *21*, 137–149. [CrossRef]
23. Bourmaud, A.; Beaugrand, J.; Shah, D.U.; Placet, V.; Baley, C. Towards the design of high-performance plant fibre composites. *Prog. Mater. Sci.* **2018**, *97*, 347–408. [CrossRef]
24. Adamová, T.; Hradecký, J.; Pánek, M. Volatile organic compounds (VOCs) from wood and wood-based panels: Methods for Evaluation, Potential Health Risks, and Mitigation. *Polymers* **2020**, *12*, 2289. [CrossRef]
25. Ulker, O.C.; Ulker, O.; Hiziroglu, S. Volatile organic compounds (VOCs) emitted from coated furniture units. *Coatings* **2021**, *11*, 806. [CrossRef]
26. Richter, M.; Horn, W.; Juritsch, E.; Klinge, A.; Radeljic, L.; Jann, O. Natural Building Materials for Interior Fitting and Refurbishment—What about Indoor Emissions? *Materials* **2021**, *14*, 234. [CrossRef]
27. Cao, T.; Shen, J.; Wang, Q.; Li, H.; Xu, C.; Dong, H. Characteristics of VOCs released from plywood in airtight environments. *Forests* **2019**, *10*, 709. [CrossRef]
28. Qi, Y.; Shen, L.; Zhang, J.; Yao, J.; Lu, R.; Miyakoshi, T. Species and release characteristics of VOCs in furniture coating process. *Environ. Pollut.* **2019**, *245*, 810–819. [CrossRef]
29. Xiong, J.; Chen, F.; Sun, L.; Yu, X.; Zhao, J.; Hu, Y.; Wang, Y. Characterization of VOC emissions from composite wood furniture: Parameter determination and simplified model. *Build. Environ.* **2019**, *161*, 106237. [CrossRef]
30. Liu, X.; Mason, M.A.; Guo, Z.; Krebs, K.A.; Roache, N.F. Source emission and model evaluation of formaldehyde from composite and solid wood furniture in a full-scale chamber. *Atmos. Environ.* **2015**, *122*, 561–568. [CrossRef]
31. El-Kassas, A.M.; Mourad, A.I. Novel fibers preparation technique for manufacturing of rice straw based fiberboards and their characterization. *Mater. Des.* **2013**, *50*, 757–765. [CrossRef]
32. EPA Sources of Greenhouse Gas Emissions. Available online: https://www.epa.gov/ghgemissions/sources-greenhouse-gas-emissions (accessed on 18 October 2021).
33. Mantanis, G.I.; Athanassiadou, E.T.; Barbu, M.C.; Wijnendaele, K. Adhesive systems used in the European particleboard, MDF and OSB industries. *Wood Mater. Sci. Eng.* **2018**, *13*, 104–116. [CrossRef]
34. Dietsch, P.; Tannert, T. Assessing the integrity of glued-laminated timber elements. *Constr. Build. Mater.* **2015**, *101*, 1259–1270. [CrossRef]
35. Bekhta, P.; Salca, E.A. Influence of veneer densification on the shear strength and temperature behavior inside the plywood during hot press. *Constr. Build. Mater.* **2018**, *162*, 20–26. [CrossRef]
36. Zerbst, D.; Affronti, E.; Gereke, T.; Buchelt, B.; Clauß, S.; Merklein, M.; Cherif, C. Experimental analysis of the forming behavior of ash wood veneer with nonwoven backings. *Eur. J. Wood Prod.* **2020**, *78*, 321–331. [CrossRef]
37. Wang, W.; Zammarano, M.; Shields, J.R.; Knowlton, E.D.; Kim, I.; Gales, J.A.; Li, J. A novel application of silicone-based flame-retardant adhesive in plywood. *Constr. Build. Mater.* **2018**, *189*, 448–459. [CrossRef]
38. Petrović, E.K.; Hamer, L.K. Improving the healthiness of sustainable construction: Example of polyvinyl chloride (PVC). *Buildings* **2018**, *8*, 28. [CrossRef]
39. Yan, X.; Han, Y.; Yin, T. Synthesis of urea-formaldehyde microcapsule containing fluororesin and its effect on performances of waterborne coatings on wood surface. *Polymers* **2021**, *13*, 1674. [CrossRef]
40. Salca, E.A.; Hiziroglu, S. Hardness and Roughness of Overlaid Wood Composites Exposed to a High-Humidity Environment. *Coatings* **2019**, *9*, 711. [CrossRef]
41. Suethao, S.; Shah, D.U.; Smitthipong, W. Recent progress in processing functionally graded polymer foams. *Materials* **2020**, *13*, 4060. [CrossRef]
42. Construction Business News Dubai Hills Vista Project by Emaar and Inspired by Automobili Lamborghini Is Sold Out. Available online: https://www.cbnme.com/news/dubai-hills-vista-project-by-emaar-and-inspired-by-automobili-lamborghini-is-sold-out/ (accessed on 20 October 2021).
43. Eltawahni, H.A.; Olabi, A.G.; Benyounis, K.Y. Investigating the CO₂ laser cutting parameters of MDF wood composite material. *Opt. Laser Technol.* **2011**, *43*, 648–659. [CrossRef]
44. Santaniello, F.; Djupström, L.B.; Ranius, T.; Weslien, J.; Rudolphi, J.; Sonesson, J. Simulated long-term effects of varying tree retention on wood production, dead wood and carbon stock changes. *J. Environ. Manag.* **2017**, *201*, 37–44. [CrossRef]

45. Ramage, M.H.; Burridge, H.; Busse-Wicher, M.; Fereday, G.; Reynolds, T.; Shah, D.U.; Scherman, O. The wood from the trees: The use of timber in construction. *Renew. Sustain. Energy Rev.* **2017**, *68*, 333–359. [CrossRef]
46. Bribián, I.Z.; Capilla, A.V.; Usón, A.A. Life cycle assessment of building materials: Comparative analysis of energy and environmental impacts and evaluation of the eco-efficiency improvement potential. *Build. Environ.* **2011**, *46*, 1133–1140. [CrossRef]
47. Wi, S.; Park, J.H.; Kim, Y.U.; Kim, S. Evaluation of environmental impact on the formaldehyde emission and flame-retardant performance of thermal insulation materials. *J. Hazard. Mater.* **2021**, *402*, 123463. [CrossRef]
48. Liu, Z.; Little, J.C. Materials responsible for formaldehyde and volatile organic compound (VOC) emissions. *Toxic. Build. Mater.* **2012**, *17*, 76–121.
49. EnviroLink Environmental Product Declaration (EPD Certification). Available online: https://www.envirolink.me/https-www-envirolink-me-environmental-product-declarations-epd/ (accessed on 24 October 2021).
50. EnviroLink. Environmental Product Declarations EPD LEED Product Certification. 2021. Available online: https://www.envirolink.me/environmental-product-declarations-epd/ (accessed on 2 November 2021).
51. Takano, A.; Hafner, A.; Linkosalmi, L.; Ott, S.; Hughes, M.; Winter, S. Life cycle assessment of wood construction according to the normative standards. *Eur. J. Wood Prod.* **2015**, *73*, 299–312. [CrossRef]
52. Kim, S. Control of formaldehyde and TVOC emission from wood-based flooring composites at various manufacturing processes by surface finishing. *J. Hazard. Mater.* **2010**, *176*, 14–19. [CrossRef]

Article

Air Quality Modeling of Cooking Stove Emissions and Exposure Assessment in Rural Areas

Yucheng He [1], Sanika Ravindra Nishandar [1], Rufus David Edwards [2,*] and Marko Princevac [1]

[1] Department of Mechanical Engineering, Marlan and Rosemary Bourns College of Engineering, University of California Riverside, Riverside, CA 92521, USA; yucheng.he@email.ucr.edu (Y.H.); sanika.nishandar@email.ucr.edu (S.R.N.)

[2] Department of Epidemiology and Biostatistics, Program in Public Health, University of California Irvine, Irvine, CA 92697, USA

* Correspondence: edwardsr@hs.uci.edu

Abstract: Cooking stoves produce significant emissions of $PM_{2.5}$ in homes, causing major health impacts in rural communities. The installation of chimneys in cooking stoves has been documented to substantially reduce indoor emissions compared to those of traditional open fires. Majority of the emissions pass through chimneys to the outdoors, while some fraction of the emissions leak directly into the indoor air, which is defined as fugitive emission. Indoor $PM_{2.5}$ concentrations are then the result of such fugitive emissions and the infiltration of outdoor neighborhood pollutants. This study uses a combination of the one-contaminant box model and dispersion models to estimate the indoor $PM_{2.5}$ household concentration. The results show that the contributions of outdoor infiltration to indoor $PM_{2.5}$ concentrations increase with higher packing densities and ventilation rates. For a case study, under WHO recommended ventilation conditions, the 24 h average mass concentration is ~21 µg/m^3, with fugitive concentration accounting for ~90% of the total exposure for highly packed communities. These results help to identify the potential benefits of intervention strategies in regions that use chimney stoves.

Keywords: dispersion model; health risk assessment; particulate matter; indoor air quality; cook stove; biomass burning

Citation: He, Y.; Nishandar, S.R.; Edwards, R.D.; Princevac, M. Air Quality Modeling of Cooking Stove Emissions and Exposure Assessment in Rural Areas. *Sustainability* **2023**, *15*, 5676. https://doi.org/10.3390/su15075676

Academic Editors: José Carlos Magalhães Pires, Álvaro Gómez-Losada and Vincenzo Torretta

Received: 13 January 2023
Revised: 11 March 2023
Accepted: 15 March 2023
Published: 24 March 2023

1. Introduction

Globally, many rural communities rely on traditional biomass burning stoves to meet household energy demands, both indoors [1] and outdoors [2], which result in a significant burden of disease [2–4]. WHO announced that an estimated 3.2 million deaths per year were attributed to household air pollution in 2020 [5]. In addition, many studies have reported high health risk associated with traditional open-fire cooking [6,7]. A large portion of the rural population utilizes traditional stoves for household needs because of socioeconomic factors including availability of fuel, the cost of the stove, and a lack of alternative energy sources such as LPG [3]. The necessity of reducing pollutants associated with stove burning has led to the development of technologies to improve combustion efficiency. The change of behavior in the kitchen can also reduce $PM_{2.5}$ exposure. Although there are many different stove types, the combination of a burning chamber and a flue/chimney are commonly used in many areas of the world, in which majority of the emissions are exhausted from the chimney to the outdoor environment. Indoor $PM_{2.5}$ concentrations are the result of the fraction of the emissions that leak directly into the indoor air [8], combined with the outdoor infiltrated pollutants. The outdoor neighborhood pollutants quantified in this study are from chimney emissions of an individual home and upstream neighborhood homes [9]. Often, the stacks for household chimneys installed in these regions are short, resulting in neighborhood pollution, and substantial emissions accumulate in the vicinity of homes, which enables infiltration back indoors [10,11]. The current study focuses on the contributions of neighborhood pollution to indoor air pollution associated with different housing packing densities, which have not previously been well characterized.

Epidemiological studies have indicated that $PM_{2.5}$ exposure can cause adverse health impacts through heart, respiratory, and other chronic diseases [12]. Biomass fuel burning is recognized as a major cause of chronic obstructive pulmonary disease, especially for individuals in developing countries [13]. Estimating the risk of exposure that can lead to health problems is vital to inform risk abatement strategies. The current analysis evaluates the outdoor neighborhood pollution distributions, outdoor to indoor infiltration, indoor stove fugitive contamination, and the associated $PM_{2.5}$ exposure risk [14]. Subsequently, EPA health risk assessment [15] was applied to quantify the inhalation risk of $PM_{2.5}$ to the female population of different age groups to contribute to the development of control strategies for air quality management in and around rural communities where solid fuels are used for cooking.

2. Materials and Methods

2.1. Background

The disability-adjusted life years (DALYs) represent the sum of years of populations living in a status of less than good health resulting from specific causes. Figure 1 presents the household attributed DALYs per 100,000 people according to WHO released data for the year 2019 [16]. The DALYs are substantially higher in developing countries, where biomass combustion supplies the majority of household primary energy [17].

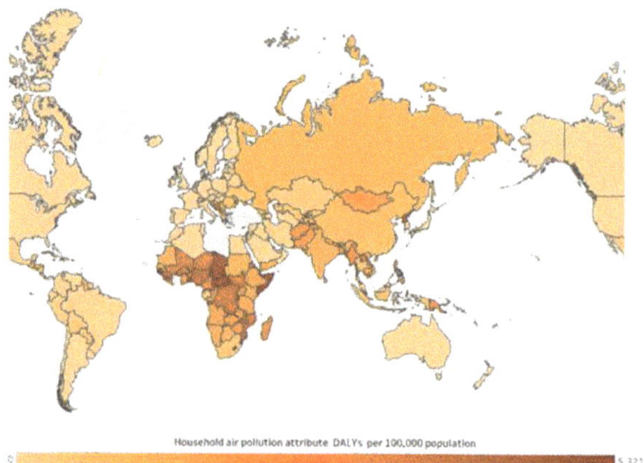

Figure 1. Household air pollution attributed DALYs per 100,000 people [16].

2.2. Evaluation Framework

The schematic of the study process is shown in Figure 2. The study is divided into the estimation of outdoor and indoor air quality. For the outdoor air quality study, meteorological parameters such as temperature, wind speed, direction, and cloud cover are required. These required micrometeorological parameters for the dispersion model inputs are derived from routine weather data [18,19]. Integrating the dispersion model with a meteorological preprocessing approximation is a good alternative when the field measurements are not available. The approximation details, together with field validation, are described in the Supplementary Material. Figure S1 shows a good agreement of the measured friction velocity with the approximation model output. A dispersion model is then deployed to quantify outdoor pollutant distribution. Such outdoor pollutants near the household are the infiltration source to the indoor environment. The indoor generated $PM_{2.5}$ is from indoor fugitive emissions. The infiltrated and the indoor-generated $PM_{2.5}$, combine to form indoor pollution. Finally, a US EPA health risk assessment methodology [20] is utilized to quantify the potential dose and risk quotient for long-term exposure to such pollutants.

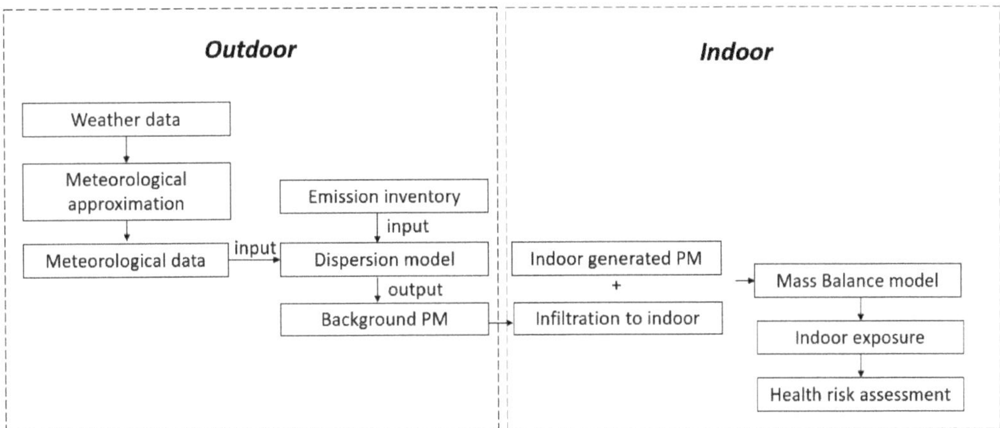

Figure 2. Analysis flow.

2.3. Dispersion Model

Dispersion models are effective, widely utilized tools to evaluate atmospheric pollution level when field measurements are not available. Such models have been utilized to quantify pollution emitted from different sources, such as cooking, traffic, and industry [17,21,22]. The ability to isolate emission sources, thus targeting the sole effect of one possible source, can inform source management and relevant policymaking. Most of the plume models do not consider complex effects of obstacles such as buildings on the dispersion of pollutants in urban or suburban areas [23]. The current study compares AERMOD with the Quick Urban and Industrial Complex (QUIC) results to examine the influence of building morphology. QUIC accommodates building influences, rapidly enables detailed modeling of the flow field around buildings, and applies this generated wind field in a particle dispersion model. The simulation results are utilized to quantify the neighborhood infiltration because it estimates pollution dispersion in the vicinity of the buildings. AERMOD is extensively used for regulatory purposes and plays a substantial role in decision making [24]. AERMOD does not explicitly solve the flow features in the vicinity of obstacles, but accounts for obstacles through a Plume Rise Model Enhancements (PRIME) model. The current analysis compares neighborhood pollution results using both approaches. The comparison of point source dispersion among Gaussian, QUIC, and water channel evaluation is presented in Figure S2. And the comparison of contours for the outdoor emission estimation of AERMOD output with QUIC output is given in Figure S3.

QUIC has broad applications, primarily in modeling wind flow and dispersion patterns in urban or suburban areas, providing building-scale results that can show pollutant concentrations and their interaction with eddies in the built environment [23,25]. QUIC is a fast-response dispersion model that is comprised of a wind field model QUIC-URB and a dispersion model QUIC-PLUME. The flow patterns modeled by QUIC-URB have been validated with USEPA wind tunnel measurements [26]. QUIC-PLUME has also been validated through many experiments and modeling comparisons. For example, Zajic et al. compared QUIC-PLUME output to a Gaussian plume model output at different atmospheric stability, with the results being in good agreement for unstable and neutral atmospheric stabilities [27].

AERMOD was developed by the U.S. Environmental Protection Agency (EPA) in conjunction with the American Meteorological Society (AMS) to incorporate scientific advances into a dispersion model for regulatory applications [28]. AERMOD is a regulatory model with superior performance to other models in a 17 field study databases [29]. Many studies have integrated meteorological preprocessing to estimate AERMOD required inputs. Kumar et al. integrated a weather forecast model, using data-driven predicted meteorological data as inputs [30]. The current meteorological approximation has been validated for different built environments including relatively uniform terrain, as well as urban canopies [18,31].

2.4. Indoor PM$_{2.5}$

The indoor fugitive emission and infiltration of outdoor pollution are the two major sources of household pollutants (please see the schematic in Figure 3). Such external infiltration from the upstream community is determined by ventilation type, room volume, flow direction, and nature and size of openings in walls, windows, and doors. The ventilation in rural households is usually natural, wherein the leakage of airflow is through the openings in the building walls, windows, and doors. The dimensionless infiltration factor of outdoor pollution to the indoors is described by [14,32]:

$$F_{inf} = \frac{P * a}{a + k} \tag{1}$$

where a is the air changes per hour (ACH), P is the penetration coefficient that indicates the fraction of outdoor pollutant passing indoors [33], and k is the pollutant deposition rate per hour. ACH is the rate of indoor air replacement by outdoor air. It is an important parameter that determines air ventilation in microenvironments, thus affecting indoor air exposure [34].When particles penetrate through the building envelope, gravity, diffusion, and inertial interaction are the major determinants of P. P shows a hill-shape distribution with respect to particle size, and it is assumed to be 0.8 in the study case [32]. K describes the flux of particulate matter deposited on the frames of windows, doors, walls, and other surfaces when traveling indoors and it is adapted to be 0.53 h^{-1} for particle size at 2.5 μm, based on experiments in six naturally ventilated houses [32].

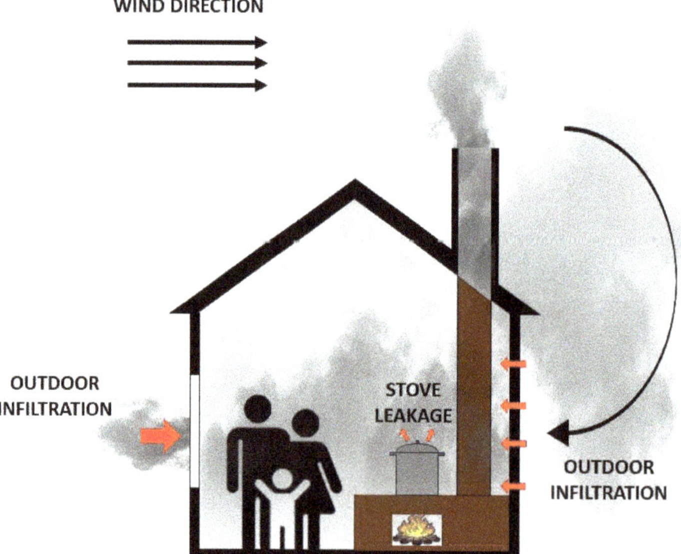

Figure 3. Household pollution sources from indoor fugitive emission and outdoor infiltration.

The ACH in a region can vary substantially based on the local wind speed, location, and the closing/opening of the door and windows. Figure 4 shows different ACH measured in rural areas in India [7], Bangladesh [34], China [35], and Nepal [36]. When the windows are closed, the ventilation is solely dependent on the leakage through gaps. The Literature states that the ACH in such conditions is around 1–2 for houses with solid walls. WHO suggested a default ACH of 21, for kitchens using biomass for cooking, to achieve the indoor air quality standard [37]. For current analysis, indoor pollution concentration and residents' potential dose were estimated for ACH range of 1–25.

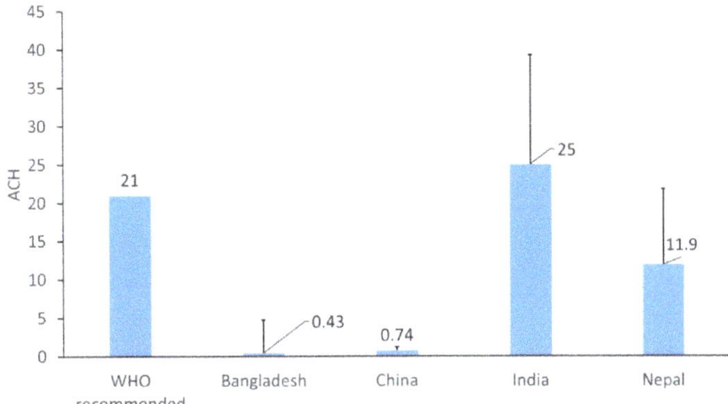

Figure 4. Reference rural air changes per hour in different regions.

Using the calculated dimensionless infiltration factor, the outdoor infiltration is then described by:

$$C_b = F_{inf} \times C_{out} \tag{2}$$

where C_{out} is the outdoor $PM_{2.5}$ concentration in $\mu g/m^3$, which is obtained from the dispersion model output.

Computational Fluid Dynamic (CFD) models are at times utilized for indoor air quality modeling [38,39]. In contrast, the current study deployed a computationally inexpensive single zone mass balance model [40] to estimate the indoor generated $PM_{2.5}$ concentration, which is described by:

$$C(t) = C_b + \frac{S}{V(a+k)} + (C_{ini} + C_b + \frac{S}{V(a+k)})e^{-(a+k)t} \tag{3}$$

where V is the kitchen volume, which is ~40 m^3 for multiple regions [41]. S is the indoor source emission rate in $\mu g/h$. In reality, the emission rate measurements vary widely, based on factors such as experimental methodology, combustion facilities, and fuel properties [8]. The indoor fugitive emission is around 2–5% of the total emission, based on the laboratory experiments, as well as field measurements [8,10,41]. Household energy needs are met mainly by biomass fuels, including crop residues, wood, coal, etc. Usage of different types of fuels impact the overall emission levels. Figure 5 presents the $PM_{2.5}$ emission factor (EF) of different fuel types, in the laboratory [42] and in field, in different regions including China [8], Nepal [42], Mexico [43], and India [42]. For the same type of fuel, the EF differences may be due to factors such as fuel shape, moisture content, and the stove differences. With the different emission factors, one can estimate the emission inventory based on the fuel consumption rate from the various cooking events [44].

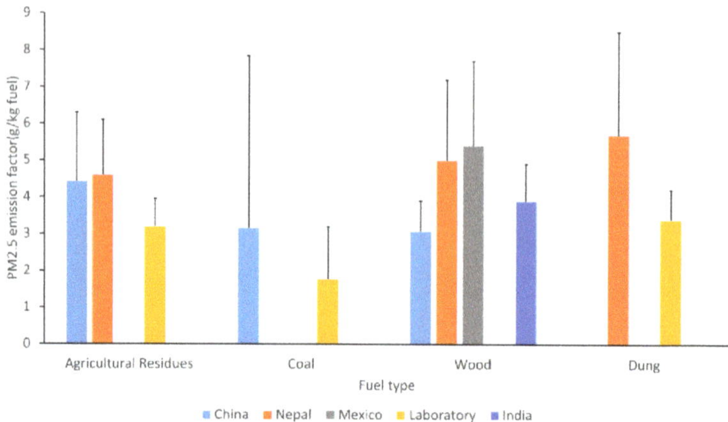

Figure 5. Reference emission factor for different fuel types.

2.5. Health Risk Assessment

The potential dose of stove-induced $PM_{2.5}$ for individuals under long-term exposure to cooking emissions is assessed using US EPA risk assessment [20]. The potential dose (I) in µg/kg·day of $PM_{2.5}$ can be quantified as [15,45,46]:

$$I = C_A \frac{IR \times ET \times EF \times ED}{BW} \times \frac{1}{AT} \quad (4)$$

where C_A is the concentration of $PM_{2.5}$ in µg/m^3. ET is the exposure time in h/day, which in this study, is assumed to be 3 h exposure per day [47]. EF is the exposure frequency (days/year). ED is the exposure duration in the study period. AT is the average time of exposure in a day, which is $ED \times 365$. BW is the body weight (kg). IR is the inhalation rate (m^3/h), which represents the volume of air inhaled over a specified timeframe. The inhalation rates are typically indexed to activity levels. The inhalation rate for different age groups, segregated by gender and the average body weight, is referred from the EPA Exposure Factors Handbook [48]. Note that for the male gender, the potential dose results are very similar, within 2–7%. For brevity, here we present only results calculated using available female parameters, as in many rural locations, the primary coking activities are carried out by women. Under moderate activity levels, for age groups from 0.5–3, 3–10, and 10–18 years old, the average body weight is 11 kg, 23 kg, and 50 kg, and the average inhalation rate is 0.6 m^3/h, 0.9 m^3/h, and 1.26 m^3/h, respectively, while for the adult age groups from 18–30, 30–60, and above 60 years old, the average body weight is 62 kg, 68 kg, and 67 kg, and the average inhalation rate is 1.32 m^3/h, 1.32 m^3/h, and 1.2 m^3/h, respectively.

The risk quotient is often used to inform the health implications due to pollutant exposure. The risk quotient is described by:

$$RQ = \frac{I}{RfD} \quad (5)$$

where RfD is the reference dose of $PM_{2.5}$ (µg/kg·day) and represents the safe average daily dose. RfD is calculated from Equation (4), with a reference concentration of 5 µg/m^3. If RQ < 1, the exposure is not considered adverse to public health; if RQ > 1, the exposure is considered detrimental to public health.

3. Results

3.1. Outdoor Pollution Level

A case study of neighborhood pollution attributed to chimney vented $PM_{2.5}$ emissions is conducted using QUIC model, with a wind speed of 2 m/s (at 10 m height) and a south-southwest wind direction. The QUIC modeling results of ambient $PM_{2.5}$ pollution during the steady cooking state is shown in Figure 6. The channeling effect caused by wind flow encountering the building obstacles will lead to accumulation of $PM_{2.5}$ in the building wakes [23]. The QUIC model enables the detection of severely impacted regions, considering the building morphology. A typical rural village consists of regions with different building densities. In each of these building densities, the neighborhood pollution level varies because the effects of trapping $PM_{2.5}$ near buildings differ. As indicated in Figure 6, the maximum ground level concentration occurred in the high building density region. Although the prevailing wind is from the south-southwest, the buildings in the high-density region substantially disturb the flow and consequently, distribute the emissions in different directions, while in the low-density region, the rarely disturbed flow quickly dilutes $PM_{2.5}$.

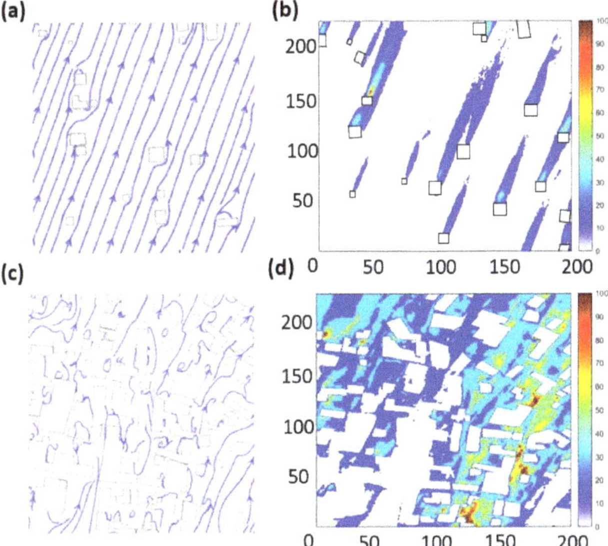

Figure 6. Outdoor dispersion modeling results: (**a,c**) flow trajectory for a low- and high-density region, respectively; (**b,d**). $PM_{2.5}$ mass concentration distribution in a low- and high-density region, respectively.

The current analysis used an emission rate of 52 mg/min for the outdoor pollution modeling based on field measurements, with 96% of total emissions from chimneys [41]. The mean $PM_{2.5}$ in the high building density region is 21.2 ± 4.26 µg/m^3. The mean $PM_{2.5}$ in the low building density region is 4.57 ± 2.8 µg/m^3. The high-density area is more impacted, regardless of the wind direction, due to the building density. Hence, this region has the highest level of $PM_{2.5}$ in the communities.

Other factors, such as seasonal relative humidity, temperature, and precipitation, also play a significant role in outdoor pollution dispersion across seasons [49]. In the Brazilian rainforest, during the dry season, exposures to $PM_{2.5}$ can be 6 times higher than during the rainy season [15]. This lower exposure can be attributed to the leaching of air pollution to the ground as a result of higher precipitation in the rainy season [50].

3.2. Indoor Pollution Level

Instead of completely switching to clean fuel, which may be impractical, many studies have also recommended alternative actions to reduce household air pollution and exposure. Other studies have also recommended alternative actions to reduce household air pollution and exposure, such as increasing the natural ventilation rates [51]. Combining chimneys with improved combustion chambers in stoves can also result in substantially reduced overall emissions, although chimney maintenance is necessary to maintain these reductions [41]. Leakage emissions from well-maintained stoves were shown to be substantially lower than from those in need of repair, such as the filling of cracks and the cleaning of chimneys [52].

Despite the up to 90% reductions in indoor air concentrations of $PM_{2.5}$ associated with the installation of chimneys [53], fugitive concentration, combined with outdoor infiltrations, still contributes to poor indoor air quality. The indoor pollutant concentration is determined by the fugitive emission rates, room volume, particulate matter decay rate, and the infiltration of pollution from outdoors due to ventilation. Figure 7 shows the modeled indoor generated $PM_{2.5}$ mass concentrations during 1 h of cooking under different air exchange rates, incorporating contributions of both fugitive emissions and neighborhood infiltration for high-packing density. The fugitive emission rate ranges from 0.26–2.6 mg/min, based on direct field measurements in different regions [8,10,41,54]. The ACH ranges from 1–25 h^{-1}, from poor ventilated cases, to WHO default ventilation rates for ISO tiers.

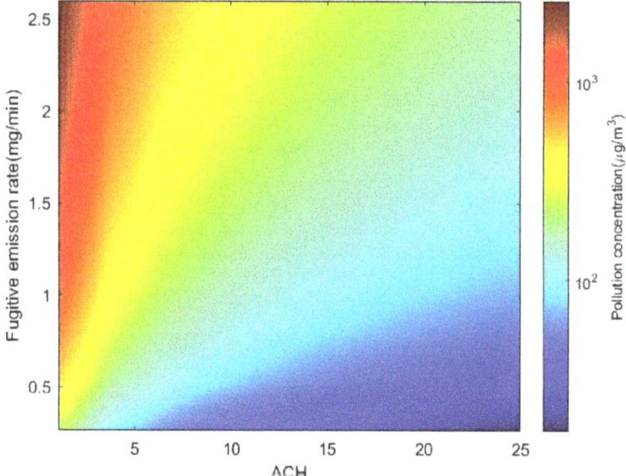

Figure 7. Indoor $PM_{2.5}$ concentrations under different ACH incorporating fugitive emission and neighborhood infiltration for a high-packing density neighborhood.

Infiltration of pollution from outdoors consists of both neighborhood pollution and regional background $PM_{2.5}$. The neighborhood pollution contribution to indoor concentrations depends on the packing density of upstream homes. Figure 8 shows the percentage of fugitive contribution to the total indoor air $PM_{2.5}$ concentrations. At 25 ACH, for homes in the high packing density area, the indoor generated $PM_{2.5}$ accounts for 90% of the total concentration, while for a low packing density region, the number is 98%.

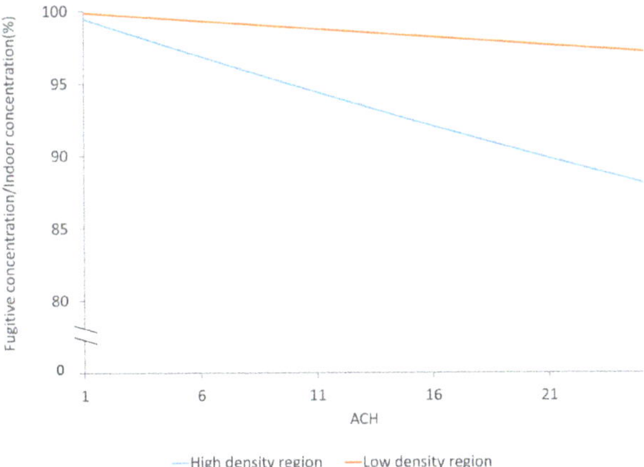

Figure 8. Contribution of fugitive emission to indoor air $PM_{2.5}$ concentrations.

After the cooking concludes, the decay trend of the relative mass concentration $(C(t)/C_{max})$ under ACH = 1, 5, 10, 20, and 25 are shown in Figure 9. Studies indicated that the building characteristics, including ventilation, orientation, the morphology of the streets, wall construction, eave spaces, open–closed windows, etc., dominate the air exchange rate [50,55].

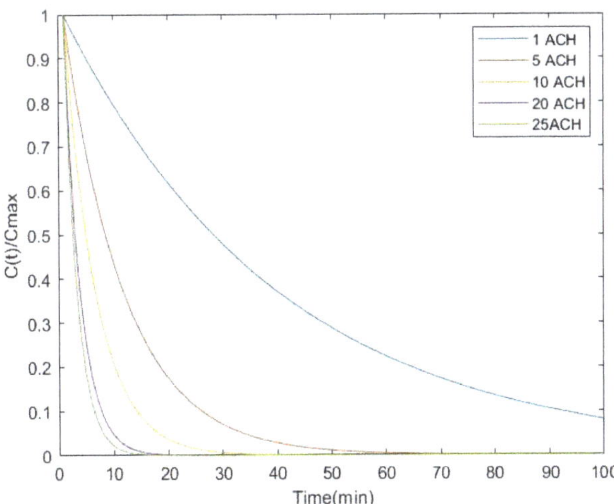

Figure 9. Influence of different ACHs on relative indoor concentration decay trend after steady cooking events.

This case study uses a fugitive emission rate of 2.1 mg/min, which is directly measured in rural Mexico using nested hoods to capture all emissions [41]. Even at the recommended 21 ACH, the indoor pollution concentration during the 1 h steady cooking event is 174 µg/m^3, with fugitive emissions contributing 90% to indoor concentrations. The corresponding 24 h average $PM_{2.5}$ level is ~21 µg/m^3 under the assumption that each household conducts 3 h cooking each day [47]. This exceeds the 2021 WHO air quality guideline of 15 µg/m^3 for 24 h.

$PM_{2.5}$ concentration levels in different rooms can be significantly lower than in other rooms [56] and can vary by season [49]. Zuk et al. found that the kitchen concentrations were two times that of other rooms [56]. In addition, in many homes cooking-generated $PM_{2.5}$ may readily spread to the adjacent rooms in the house [49,57]. Since people spend the most time in bedrooms and living rooms, having a separate kitchen can help reduce exposures, although, room ventilation and location relative to the kitchen have been reported to impact the $PM_{2.5}$ level in the room [58]. Behavior changes, such as opening doors and windows [59] and the use of extraction fans, may also reduce indoor concentrations.

3.3. Potential Dose Assessment

The inhaled dose is determined by individual behaviors and the distance from the emission source. A number of studies have shown that personal exposures are ~50% of indoor kitchen concentrations [52,60], as personal exposures include time spent away from the kitchen in other environments. Figure 10 shows the potential dose estimated under personal exposure. The risk quotient (RQ) for exposed residents is 26, 7.93, 3.96, and 2.19 under 1, 5, 10, and 20 ACH, respectively. This shows that even under high ventilation rates, household emissions moderately contribute to total chronic exposure and may induce respiratory health problems. Besides, the potential intake of pollutants from cooking activity is high, and thus, the long-term exposure can significantly impact individuals who perform the cooking.

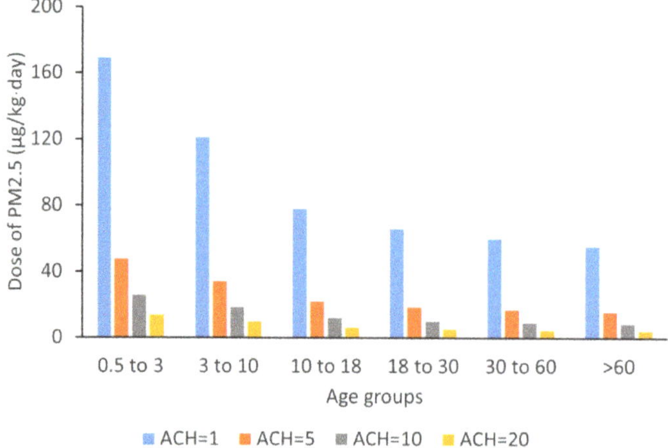

Figure 10. Case study of potential dose of $PM_{2.5}$ for different age groups under different ventilation parameters.

In general, potential $PM_{2.5}$ doses decreased with age groups, and children under 3 years had the highest potential dose, in agreement with other studies [15,50], because younger children are more active and breathe more per unit of body weight.

A study of chimney stove impact conducted over a period of 12 months by Chakraborty et al. found that the median value of $PM_{2.5}$ RQ was 1.63 [61]. Although the utilization of a chimney stove has adverse health effects, the results show a lower $PM_{2.5}$ potential dose compared to that from a traditional open fire, for which the observed RQ can be as high as 5.57 [61].

Single-zone models tend to overestimate concentrations, as the model assumes a well-mixed environment, which may attribute to the discrepancies in the calculated and measured concentration [62]. The overestimation of potential dose can thus be a limitation of this approach.

4. Conclusions

The aim of this study is to provide a full-cycle analysis integrating air quality models, infiltration models, and risk assessment models to better understand the impacts of neighborhood pollution and stove fugitive emissions on the potential dose. The main conclusions of this study are:

1. $PM_{2.5}$ pollutants tend to accumulate in the wake of buildings, and the pollutant infiltration can contribute significantly to poor indoor air quality. The major contributor to indoor pollution, however, is fugitive emissions from cooking stoves.
2. The contribution of chimney stoves to infiltration increases with higher packing densities and may contribute to indoor pollutant concentrations.
3. Indoor concentrations from chimney stoves exceed WHO air quality guideline values for $PM_{2.5}$. The associated health risk assessment shows that the risk quotient (RQ) is 2.19, despite good ventilation conditions.

Supplementary Materials: The following supporting information can be downloaded at: https://www.mdpi.com/article/10.3390/su15075676/s1, Figure S1: Comparison of meteorological approximation model output and field measurement; Figure S2: Comparison of point source dispersion among Gaussian, QUIC, and water channel evaluation; Figure S3: Comparison of (left) AERMOD output with (right) QUIC output for one-hour outdoor emission estimation. References [18,19,63–65] are cited in the supplementary materials.

Author Contributions: Conceptualization, Y.H., S.R.N., R.D.E. and M.P.; data curation, Y.H. and S.R.N.; formal analysis, Y.H. and S.R.N.; investigation, Y.H. and S.R.N.; software, Y.H. and S.R.N.; validation, Y.H. and S.R.N.; visualization, Y.H. and S.R.N.; methodology, Y.H. and S.R.N.; funding acquisition, R.D.E.; project administration, R.D.E. and M.P.; supervision, M.P. and R.D.E.; writing—original draft, Y.H. and S.R.N.; writing—review and editing, Y.H., S.R.N., R.D.E. and M.P. All authors have read and agreed to the published version of the manuscript.

Funding: This work was funded by Clean Stacking Options and Regional IAP Scenarios for Rural Mexico, NIH-5585744.

Institutional Review Board Statement: Not applicable.

Informed Consent Statement: Not applicable.

Data Availability Statement: Not applicable.

Acknowledgments: We are grateful to the Los Alamos National Laboratory for providing the QUIC model under license number: LIC-20-04147. Special thanks to Hannah Lee, Rebecca Albano, and Arthor Bernal for running the QUIC simulation.

Conflicts of Interest: The authors declare no conflict of interest.

References

1. Chafe, Z.A.; Brauer, M.; Klimont, Z.; Van Dingenen, R.; Mehta, S.; Rao, S.; Riahi, K.; Dentener, F.; Smith, K.R. Household cooking with solid fuels contributes to ambient PM2.5 air pollution and the burden of disease. *Environ. Health Perspect.* **2015**, *122*, 1314–1320. [CrossRef] [PubMed]
2. Pilishvili, T.; Loo, J.D.; Schrag, S.; Stanistreet, D.; Christensen, B.; Yip, F.; Nyagol, R.; Quick, R.; Sage, M.; Bruce, N. Effectiveness of six improved cookstoves in reducing household air pollution and their acceptability in rural western Kenya. *PLoS ONE* **2016**, *11*, e0165529. [CrossRef] [PubMed]
3. García-Frapolli, E.; Schilmann, A.; Berrueta, V.M.; Riojas-Rodríguez, H.; Edwards, R.D.; Johnson, M.; Guevara-Sanginés, A.; Armendariz, C.; Masera, O. Beyond fuelwood savings: Valuing the economic benefits of introducing improved biomass cookstoves in the Purépecha region of Mexico. *Ecol. Econ.* **2010**, *69*, 2598–2605. [CrossRef]
4. McCreanor, J.; Cullinan, P.; Nieuwenhuijsen, M.J.; Stewart-Evans, J.; Malliarou, E.; Jarup, L.; Harrington, R.; Svartengren, M.; Han, I.-K.; Ohman-Strickland, P.; et al. Respiratory Effects of Exposure to Diesel Traffic in Persons with Asthma. *N. Engl. J. Med.* **2007**, *357*, 2348–2358. [CrossRef]
5. World Health Organization. Household Air Pollution. Available online: https://www.who.int/news-room/fact-sheets/detail/household-air-pollution-and-health (accessed on 17 December 2022).
6. Parajuli, I.; Lee, H.; Shrestha, K.R. Indoor Air Quality and ventilation assessment of rural mountainous households of Nepal. *Int. J. Sustain. Built Environ.* **2016**, *5*, 301–311. [CrossRef]
7. Johnson, M.; Lam, N.; Brant, S.; Gray, C.; Pennise, D. Modeling indoor air pollution from cookstove emissions in developing countries using a Monte Carlo single-box model. *Atmos. Environ.* **2011**, *45*, 3237–3243. [CrossRef]

8. Shen, H.; Luo, Z.; Xiong, R.; Liu, X.; Zhang, L.; Li, Y.; Du, W.; Chen, Y.; Cheng, H.; Shen, G.; et al. A critical review of pollutant emission factors from fuel combustion in home stoves. *Environ. Int.* **2021**, *157*, 106841. [CrossRef]
9. Ruiz, V.; Masera, O. *Estimating Kitchen PM 2.5 and CO Concentrations out of Stove Emissions: The Case of Mexican Plancha-Type Stoves*; Universidad Nacional Autónoma de México: Morelia, Mexico, 2018.
10. Shen, G.; Du, W.; Luo, Z.; Li, Y.; Cai, G.; Lu, C.; Qiu, Y.; Chen, Y.; Cheng, H.; Tao, S. Fugitive Emissions of CO and PM2.5 from Indoor Biomass Burning in Chimney Stoves Based on a Newly Developed Carbon Balance Approach. *Environ. Sci. Technol. Lett.* **2020**, *7*, 128–134. [CrossRef]
11. Lim, M.; Myagmarchuluun, S.; Ban, H.; Hwang, Y.; Ochir, C.; Lodoisamba, D.; Lee, K. Characteristics of indoor pm2.5 concentration in gers using coal stoves in ulaanbaatar, mongolia. *Int. J. Environ. Res. Public Health* **2018**, *15*, 2524. [CrossRef]
12. Pope, C.A.; Dockery, D.W. Health effects of fine particulate air pollution: Lines that connect. *J. Air Waste Manag. Assoc.* **2006**, *56*, 709–742. [CrossRef]
13. Laumbach, R.J.; Kipen, H.M. Respiratory health effects of air pollution: Update on biomass smoke and traffic pollution. *J. Allergy Clin. Immunol.* **2012**, *129*, 3–11. [CrossRef] [PubMed]
14. Breen, M.S.; Schultz, B.D.; Sohn, M.D.; Long, T.; Langstaff, J.; Williams, R.; Isaacs, K.; Meng, Q.Y.; Stallings, C.; Smith, L. A review of air exchange rate models for air pollution exposure assessments. *J. Expo. Sci. Environ. Epidemiol.* **2014**, *24*, 555–563. [CrossRef] [PubMed]
15. De Oliveira, B.F.A.; Ignotti, E.; Artaxo, P.; Do Nascimento Saldiva, P.H.; Junger, W.L.; Hacon, S. Risk assessment of PM2.5 to child residents in Brazilian Amazon region with biofuel production. *Environ. Health A Glob. Access Sci. Source* **2012**, *11*, 64. [CrossRef] [PubMed]
16. World Health Organization. Household Air Pollution Attributable Dalys. Available online: https://www.who.int/data/gho/data/indicators/indicator-details/GHO/household-air-pollution-attributable-dalys (accessed on 16 December 2022).
17. Edwards, R.; Princevac, M.; Weltman, R.; Ghasemian, M.; Arora, N.K.; Bond, T. Modeling emission rates and exposures from outdoor cooking. *Atmos. Environ.* **2017**, *164*, 50–60. [CrossRef]
18. Qian, W.; Princevac, M.; Venkatram, A. Using temperature fluctuation measurements to estimate meteorological inputs for modelling dispersion during convective conditions in urban areas. *Bound.-Layer Meteorol.* **2010**, *135*, 269–289. [CrossRef]
19. Holtslag, A.A.M.; Van Ulden, A.P. A simple scheme for daytime estimates of the surface fluxes from routine weather data. *J. Appl. Meteorol. Climatol.* **1983**, *22*, 517–529. [CrossRef]
20. Means, B. *Risk-Assessment Guidance for Superfund. Volume 1. Human Health Evaluation Manual. Part A. Interim Report (Final)*; Environmental Protection Agency: Washington, DC, USA, 1989.
21. Venkatram, A.; Isakov, V.; Thoma, E.; Baldauf, R. Analysis of air quality data near roadways using a dispersion model. *Atmos. Environ.* **2007**, *41*, 9481–9497. [CrossRef]
22. López, M.T.; Zuk, M.; Garibay, V.; Tzintzun, G.; Iniestra, R.; Fernández, A. Health impacts from power plant emissions in Mexico. *Atmos. Environ.* **2005**, *39*, 1199–1209. [CrossRef]
23. Brown, M.J.; Williams, M.D.; Nelson, M.A.; Werley, K.A. QUIC Transport and Dispersion Modeling of Vehicle Emissions in Cities for Better Public Health Assessments. *Environ. Health Insights* **2015**, *9s1*, EHI-S15662. [CrossRef]
24. Patiño, W.R.; Duong, V.M. Intercomparison of Gaussian Plume Dispersion Models Applied to Sulfur Dioxide Emissions from a Stationary Source in the Suburban Area of Prague, Czech Republic. *Environ. Model. Assess.* **2021**, *27*, 119–137. [CrossRef]
25. Bowker, G.E.; Gillette, D.A.; Bergametti, G.; Marticorena, B. Modeling flow patterns in a small vegetated area in the northern Chihuahuan Desert using QUIC (Quick Urban & Industrial Complex). *Environ. Fluid Mech.* **2006**, *6*, 359–384. [CrossRef]
26. Bowker, G.E.; Perry, S.G.; Heist, D.K. A comparison of airflow patterns from the QUIC model and an atmospheric wind tunnel for a two-dimensional building array and a multi-city block region near the World Trade Center site. Presented at the 5th Symposium on the Urban Environment, Vancouver, BC, Canada, 23–27 August 2004.
27. Brown, M.J. *Quick Urban and Industrial Complex (QUIC) CBR Plume Modeling System: Validation-Study Document*; Los Alamos National Lab. (LANL): Los Alamos, NM, USA, 2018.
28. Cimorelli, A.J.; Perry, S.G.; Venkatram, A.; Weil, J.C.; Paine, R.J.; Wilson, R.B.; Lee, R.F.; Peters, W.D.; Brode, R.W. AERMOD: A dispersion model for industrial source applications. Part I: General model formulation and boundary layer characterization. *J. Appl. Meteorol.* **2005**, *44*, 682–693. [CrossRef]
29. Perry, S.G.; Cimorelli, A.J.; Paine, R.J.; Brode, R.W.; Weil, J.C.; Venkatram, A.; Wilson, R.B.; Lee, R.F.; Peters, W.D. AERMOD: A Dispersion model for industrial source applications. Part II: Model performance against 17 field study databases. *J. Appl. Meteorol.* **2005**, *44*, 694–708. [CrossRef]
30. Kumar, A.; Dikshit, A.K.; Patil, R.S. Use of simulated and observed meteorology for air quality modeling and source ranking for an industrial region. *Sustainability* **2021**, *13*, 4276. [CrossRef]
31. Wiernga, J. Representative roughness parameters for homogeneous terrain. *Bound.-Layer Meteorol.* **1993**, *63*, 323–363. [CrossRef]
32. Chao, C.Y.H.; Wan, M.P.; Cheng, E.C.K. Penetration coefficient and deposition rate as a function of particle size in non-smoking naturally ventilated residences. *Atmos. Environ.* **2003**, *37*, 4233–4241. [CrossRef]
33. Liu, D.L.; Nazaroff, W.W. Modeling pollutant penetration across building envelopes. *Atmos. Environ.* **2001**, *35*, 4451–4462. [CrossRef]

34. Das, D.; Moynihan, E.; Nicas, M.; McCollum, E.D.; Ahmed, S.; Roy, A.D.; Chowdhury, N.; Hanif, A.A.M.; Babik, K.R.; Baqui, A.H.; et al. Estimating residential air exchange rates in rural Bangladesh using a near field-far field model. *Build. Environ.* **2021**, *206*, 108325. [CrossRef]
35. Zhou, B.; Zhao, B.; Zhou, W. Characterizing PM2.5 concentration and air exchange rates in Chinese rural kitchens: A field study. In Proceedings of the 10th International Healthy Buildings Conference, Brisbane, Australia, 8–12 July 2012; Volume 1, pp. 260–261.
36. Soneja, S.I.; Tielsch, J.M.; Curriero, F.C.; Zaitchik, B.; Khatry, S.K.; Yan, B.; Chillrud, S.N.; Breysse, P.N. Determining particulate matter and black carbon exfiltration estimates for traditional cookstove use in rural nepalese village households. *Environ. Sci. Technol.* **2015**, *49*, 5555–5562. [CrossRef]
37. ISO. *Technical Report ISO/TR Solutions—Harmonized Laboratory Cookstoves Based on Laboratory Testing*; ISO: Geneva, Switzerland, 2018; Volume 2018.
38. Nishandar, S.R.; He, Y.; Princevac, M.; Edwards, R.D. Fate of Exhaled Droplets From Breathing and Coughing in Supermarket Checkouts and Passenger Cars. *Environ. Health Insights* **2023**, *17*. [CrossRef]
39. Mohammadi, M.; Calautit, J. Impact of Ventilation Strategy on the Transmission of Outdoor Pollutants into Indoor Environment Using CFD. *Sustainability* **2021**, *13*, 10343. [CrossRef]
40. Leary, C.O.; Jones, B.; Leary, C.O.; Jones, B. A Method to Measure Emission Rates of PM2.5s from Cooking A Method to Measure Emission Rates of PM 2.5 s from Cooking. In Proceedings of the 38th Air Infiltration and Ventilation Centre Conference, Nottingham, UK, 13–14 September 2017.
41. Ruiz-García, V.M.; Edwards, R.D.; Ghasemian, M.; Berrueta, V.M.; Princevac, M.; Vázquez, J.C.; Johnson, M.; Masera, O.R. Fugitive Emissions and Health Implications of Plancha-Type Stoves. *Environ. Sci. Technol.* **2018**, *52*, 10848–10855. [CrossRef] [PubMed]
42. Edwards, R.; Bond, T.; KR, S. Characterization of emissions from small, variable solid fuel combustion sources for determining global emissions and climate impact. *Final Proj.* **2017**, *83503601*, 1–69.
43. Johnson, M.; Edwards, R.; Alatorre Frenk, C.; Masera, O. In-field greenhouse gas emissions from cookstoves in rural Mexican households. *Atmos. Environ.* **2008**, *42*, 1206–1222. [CrossRef]
44. Du, W.; Zhuo, S.; Wang, J.; Luo, Z.; Chen, Y.; Wang, Z.; Lin, N.; Cheng, H.; Shen, G.; Tao, S. Substantial leakage into indoor air from on-site solid fuel combustion in chimney stoves. *Environ. Pollut.* **2021**, *291*, 118138. [CrossRef]
45. Bixapathi, B.; Reddy, P.V.V.; Rao, M.S.; Raghavendra, T. Health risk assessment of four important ambient air pollutants in Hyderabad. *NVEO-Nat. Volatiles Essent. Oils J.* **2021**, *8*, 9925–9935.
46. de Souza Silva, P.R.; Ignotti, E.; de Oliveira, B.F.A.; Junger, W.L.; Morais, F.; Artaxo, P.; Hacon, S. High risk of respiratory diseases in children in the fire period in Western Amazon. *Rev. Saude Publica* **2016**, *50*, 29. [CrossRef]
47. World Health Organization. Input Data to Run Household Multiple Emission Sources and Performance Target Models. Available online: https://www.who.int/tools/input-data-to-run-household-multiple-emission-sources-and-performance-target-models (accessed on 16 December 2022).
48. U.S. Environmental Protection Agency. *U.S. EPA. Exposure Factors Handbook*; U.S. Environmental Protection Agency: Washington, DC, USA, 2011.
49. Edwards, R.D.; Liu, Y.; He, G.; Yin, Z.; Sinton, J.; Peabody, J.; Smith, K.R. Household CO and PM measured as part of a review of China's National Improved Stove Program. *Indoor Air* **2007**, *17*, 189–203. [CrossRef]
50. Kaewrat, J.; Janta, R.; Sichum, S.; Kanabkaew, T. Indoor air quality and human health risk assessment in the open-air classroom. *Sustainability* **2021**, *13*, 8302. [CrossRef]
51. Pokhrel, A.K.; Bates, M.N.; Acharya, J.; Valentiner-Branth, P.; Chandyo, R.K.; Shrestha, P.S.; Raut, A.K.; Smith, K.R. PM2.5 in household kitchens of Bhaktapur, Nepal, using four different cooking fuels. *Atmos. Environ.* **2015**, *113*, 159–168. [CrossRef]
52. Hartinger, S.M.; Commodore, A.A.; Hattendorf, J.; Lanata, C.F.; Gil, A.I.; Verastegui, H.; Aguilar-Villalobos, M.; Mäusezahl, D.; Naeher, L.P. Chimney stoves modestly improved Indoor Air Quality measurements compared with traditional open fire stoves: Results from a small-scale intervention study in rural Peru. *Indoor Air* **2013**, *23*, 342–352. [CrossRef] [PubMed]
53. Masera, O.; Edwards, R.; Arnez, C.A.; Berrueta, V.; Johnson, M.; Bracho, L.R.; Riojas-Rodríguez, H.; Smith, K.R. Impact of Patsari improved cookstoves on indoor air quality in Michoacán, Mexico. *Energy Sustain. Dev.* **2007**, *11*, 45–56. [CrossRef]
54. Jetter, J. *In Stove 60-Liter Institutional Stove with Wood Fuel—Air Pollutant Emissions and Fuel Efficiency*; US Environmental Protection Agency: Washington, DC, USA, 2016.
55. Howard-Reed, C.; Wallace, L.A.; Ott, W.R. The effect of opening windows on air change rates in two homes. *J. Air Waste Manag. Assoc.* **2002**, *52*, 147–159. [CrossRef] [PubMed]
56. Zuk, M.; Rojas, L.; Blanco, S.; Serrano, P.; Cruz, J.; Angeles, F.; Tzintzun, G.; Armendariz, C.; Edwards, R.D.; Johnson, M.; et al. The impact of improved wood-burning stoves on fine particulate matter concentrations in rural Mexican homes. *J. Expo. Sci. Environ. Epidemiol.* **2007**, *17*, 224–232. [CrossRef]
57. Kim, H.; Kang, K.; Kim, T. Measurement of particulate matter (PM2.5) and health risk assessment of cooking-generated particles in the kitchen and living rooms of apartment houses. *Sustainability* **2018**, *10*, 843. [CrossRef]
58. Estévez-García, J.A.; Schilmann, A.; Riojas-Rodríguez, H.; Berrueta, V.; Blanco, S.; Villaseñor-Lozano, C.G.; Flores-Ramírez, R.; Cortez-Lugo, M.; Pérez-Padilla, R. Women exposure to household air pollution after an improved cookstove program in rural San Luis Potosi, Mexico. *Sci. Total Environ.* **2020**, *702*, 134456. [CrossRef]

59. Iribagiza, C.; Sharpe, T.; Coyle, J.; Nkubito, P.; Piedrahita, R.; Johnson, M.; Thomas, E.A. Evaluating the effects of access to air quality data on household air pollution and exposure—An interrupted time series experimental study in rwanda. *Sustainability* **2021**, *13*, 11523. [CrossRef]
60. Pillarisetti, A.; Alnes, L.W.H.; Ye, W.; McCracken, J.P.; Canuz, E.; Smith, K.R. Repeated assessment of PM2.5 in Guatemalan kitchens cooking with wood: Implications for measurement strategies. *Atmos. Environ.* **2023**, *295*, 119533. [CrossRef]
61. Chakraborty, D.; Kumar, N. Reduction in household air pollution and associated health risk: A pilot study with an improved cookstove in rural households. *Clean Technol. Environ. Policy* **2021**, *23*, 1993–2009. [CrossRef]
62. Mutahi, A.W.; Borgese, L.; Marchesi, C.; Gatari, M.J.; Depero, L.E. Indoor and outdoor air quality for sustainable life: A case study of rural and urban settlements in poor neighbourhoods in kenya. *Sustainability* **2021**, *13*, 2417. [CrossRef]
63. Schlichting, H.; Gersten, K. Boundary Layer Theory, 8th English edn. *J. Fluid Mech.* **2020**, *415*, 346–347.
64. Wang, I.T.; Chen, P.C. Estimations of heat and momentum fluxes near the ground. *Bull. Am. Meteorol. Soc.* **1980**, *61*, 97. [CrossRef]
65. Boarnet, M.G.; Edwards, R.; Princevac, M.; Wu, J.; Pan, H.; Bartolome, C.J.; Ferguson, G.; Fazl, A.; Lejano, R. Near-Source Modeling of Transportation Emissions in Built Environments Surrounding Major Arterials. *Univ. Calif. Transp. Cent. Univ. Calif. Transp. Cent. Work. Papers* **2009**.

 *sustainability*

Article

Impact of Ventilation Strategy on the Transmission of Outdoor Pollutants into Indoor Environment Using CFD

Murtaza Mohammadi and John Calautit *

Department of Architecture and Built Environment, University of Nottingham, Nottingham NG7 2RD, UK; murtaza.mohammadi1@nottingham.ac.uk
* Correspondence: john.calautit1@nottingham.ac.uk; Tel.: +44-7780-3531-63

Abstract: The transition to remote working due to the pandemic has accentuated the importance of clean indoor air, as people spend a significant portion of their time indoors. Amongst the various determinants of indoor air quality, outdoor pollution is a significant source. While conventional studies have certainly helped to quantify the long-term personal exposure to pollutants and assess their health impact, they have not paid special attention to the mechanism of transmission of pollutants between the two environments. Nevertheless, the quantification of infiltration is essential to determine the contribution of ambient pollutants in indoor air quality and its determinants. This study evaluates the transmission of outdoor pollutants into the indoor environment using 3D computational fluid dynamics modelling with a pollution dispersion model. Naturally ventilated buildings next to an urban canyon were modelled and simulated using Ansys Fluent and validated against wind tunnel results from the Concentration Data of Street Canyons database. The model consisted of two buildings of three storeys each, located on either side of a road. Two line-source pollutants were placed in the street, representing traffic emissions. Three internal rooms were selected and modelled on each floor and implemented with various ventilation strategies. Results indicate that for a canyon with an aspect ratio of 1, indoor spaces in upstream buildings are usually less polluted than downstream ones. Although within the canyon, pollution is 2–3 times higher near the upstream building. Cross ventilation can minimise or prevent infiltration of road-side pollutants into indoor spaces, while also assisting in the dispersion of ambient pollutants. The critical configuration, in terms of air quality, is single-sided ventilation from the canyon. This significantly increases indoor pollutant concentration regardless of the building location. The study reveals that multiple factors determine the indoor–outdoor links, and thorough indexing and understanding of the processes can help designers and urban planners in regulating urban configuration and geometries for improved indoor air quality. Future works should look at investigating the influence of indoor emissions and the effects of different seasons.

Keywords: air pollution transmission; indoor–outdoor relation; urban canyon; computational fluid dynamics; factors of transmission

Citation: Mohammadi, M.; Calautit, J. Impact of Ventilation Strategy on the Transmission of Outdoor Pollutants into Indoor Environment Using CFD. *Sustainability* **2021**, *13*, 10343. https://doi.org/10.3390/su131810343

Academic Editors: José Carlos Magalhães Pires and Álvaro Gómez-Losada

Received: 17 June 2021
Accepted: 9 September 2021
Published: 16 September 2021

Publisher's Note: MDPI stays neutral with regard to jurisdictional claims in published maps and institutional affiliations.

1. Introduction and Literature Review

Good air quality is essential for living a healthy life. Approximately 4.9 million deaths connected to indoor and outdoor air pollution occurred in 2017 [1]. As reported by the Global Burden of Disease [2], poor indoor air quality led to about 1.6 million deaths in the same year. The health impacts of polluted air have been an ongoing urban management topic for discussions, which have also been emphasised in several epidemiological studies. Accordingly, pollutants such as CO, NO_2, O_3, etc., can have a short- and long-term impact on both humans and animals' primary and secondary health [3,4]. Past research recognises that the reduction of air pollution will increase life expectancy considerably [5]; additionally, according to [6–8], the exposure to particulate matter (PM) increased the mortality rate of COVID-19 patients during the pandemic. Consequently, governments and institutions are advised to develop guidelines and control pollution levels.

The mode of commuting and work patterns have been greatly altered by the pandemic. Generally, people spent a significant amount of time indoors [9,10] and migration of the work environment to homes due to the lockdown has further accentuated the trend. The shift to working remotely emphasises the necessity of more studies on indoor air quality [11–15]. Therefore, the mechanism of indoor pollution and its effect on building occupants' health and well-being should be well understood and adequately assessed when designing new buildings and spaces. The activities that cause air pollution in a building include cooking, indoor heating, use of mechanical equipment, shedding from skin and/or clothes, smoking, dusting, cleaning products, etc. Outdoor–indoor air interaction also influences the transmission of pollutants between the two spaces [16]. This transmission occurs by infiltration, natural ventilation, and mechanical ventilation [17]. For indoor spaces that are naturally ventilated, one of the crucial IAQ determinants is outdoor pollution [18]. The relationship between the indoor and outdoor environment is important for visual connection, thermal regulation, and ventilation.

Environmental parameters such as meteorology, temperature difference across the building, urban landscape, spatial configuration, indoor activity, furnishings, etc., alter the rate of air exchange and consequent pollution transmission [19]. Natural ventilation relies on thermal and pressure gradients across the building façade to induce air exchange between the two domains, and attempts have been made to correlate the relationship. Multiple studies have observed that indoor PM concentration closely follows the outdoor level when the space is naturally ventilated, especially in the finer particle size range [20–22]. For instance, Zhao et al. [23] showed that average particle number concentration (PNC), in naturally ventilated German houses, followed outdoor variation during the warm season when windows were left open for a longer duration. Similarly, studies [24–26] demonstrated that indoor PM concentration in homes increased when windows were opened, and the PNC distribution curve closely followed the outdoor trend.

Air exchange due to the thermal stack effect is well established, and the buoyancy-driven airflow occurs especially in the case of single-sided ventilation [27]. This can lead to pollutants exhausted from lower levels of a building to re-enter at higher levels, and the phenomenon is amplified in low wind speed conditions [28]. However, there are multiple factors at play, and the relative contribution may significantly vary. Tippayawong [29] established a significant negative correlation between indoor $PM_{2.5}$ concentrations and outside temperature during the day, while, on the contrary, Lv et al. [30] found a positive relation. Similarly, there are contradictory observations related to the association between relative humidity and indoor $PM_{2.5}$ levels [31,32]. This is also the case for wind speed, which may be desirable in some situations and unfavourable in others. Higher wind speeds may resuspend particulate matter, increasing the aerial concentration, while it may also improve ventilation and remove particulate matter from an indoor space. It may also be size-dependent, as was indicated by Orza et al. [33]. They observed that higher wind speeds increased the concentration of particles larger than 7.5 μm and decreased the concentration for sizes smaller than 0.8 μm. Thus, there are various determinants modifying the relationship and attention must be paid to local and regional context, which may determine the dominant factor.

Another important parameter to consider is the urban context and spatial morphology of the surrounding environment. Street canyon design, building footprint, open spaces, etc., affect the dispersion of pollutants in the outdoor environment and consequently impact the indoor environment [34,35]. Ai and Mak [36] reviewed over 150 studies and indicated that deeper street canyons suffered from decoupled air flow at lower wind speeds. This leads to local recirculation in the canyon, preventing the dispersion of pollutants. This was corroborated by [37], who showed a decrease in pollution levels by increasing the permeability of the urban forms, thereby assisting the dispersion of pollutants. Urban elements such as trees, aqueducts, lamp posts, etc., can also modify air flow and alter the pollution levels [38–40]. However, the major focus of urban pollution studies has been on assessing ambient levels and inhalation exposure.

Novelty of the Present Work

While over 450 documents were returned when searched with the following keywords, 'pollution' AND 'modelling' AND 'urban' AND 'canyon' (inclusive) on Scopus, only a few studies were investigating the transmission of outdoor pollution into indoor environments. Knowing the ambient pollution levels and the relationship between indoor and outdoor air conditions, one can estimate the IAQ and take appropriate precautionary measures. Very few studies have modelled both the indoor and outdoor environment as a combined system due to the inherent complexity and difficulty in validating the models. Tong et al. [41] investigated the pollution level inside an office space as a function of its distance from a pollutant source. The study focused only on a single isolated building next to a pollution source, and hence the impact of an upstream building on the airflow distribution and pollution concentration within the urban canyon and indoor spaces should be explored. The geometry under consideration was not an urban canyon design, which disregarded local amplification and interactions. Moreover, the pollutant was modelled as a constant flux from the inlet. Yang et al. [42] examined the impact of window opening percentage along a façade on the indoor quality of a downstream building. The work did not measure the indoor pollutant levels directly but instead chose to calculate the ventilation flux to account for IAQ. Similarly, Peng et al. [43] introduced openings in the façade of the typical canyon model to understand the dispersion characteristics of the pollution. Peng et al. [43] modelled a 2D canyon and indoor environment to assess indoor conditions. However, turbulence is essentially 3D in nature, and the finer nuances of the wind flow are not generated in a 2D flow. Some other studies which have similar objectives are listed in Table 1.

Table 1. Similar CFD studies investigating dispersion of urban air pollutants, and their differences w.r.t the present investigation.

Ref.	Domain	Turbulence Model	Parameter Studied	Differences w.r.t Present Investigation
[44]	An isolated building (auditorium)	Standard k-e	Indoor PM_1, $PM_{2.5}$, and PM_{10} concentrations	A constant ambient concentration was specified at the boundary and the effect of trees was studied in modifying the indoor pollution levels.
[45]	2D canyon with a viaduct	RNG k-e	Indoor normalised tracer gas concentrations	A 2D canyon model including indoor domain was created and transmission of traffic pollutants was studied. Effect of temperature difference on transmission was investigated.
[46]	Several building clusters	RNG k-e	Normalised pollutant concentration	PM10 particles were injected into the domain, and outdoor pollutant concentration was analysed at pedestrians' level. Effect of urban configuration on dispersion of outdoor pollutants was studied.
[47]	Building cluster	LES and Standard k-e	Normalised tracer gas concentration	Outdoor dispersion of pollutant was studied for a group of buildings. Experimental and CFD simulations were carried out to assess the transmission in outdoor environment.
[48]	Building cluster		Tracer gas concentration	Viral load calculations were performed for apartment buildings and to estimate inhalation exposure of coronavirus. The study used smoke dispersion models (ATOR and CFD) to predict outdoor concentrations followed by estimating the indoor exposure by referring to established I/O ratio.

Table 1. *Cont.*

Ref.	Domain	Turbulence Model	Parameter Studied	Differences w.r.t Present Investigation
[49]	Isolated multiroom	SST k-w	H_2S concentration	Indoor pollution transmission was investigated for a multiroom block. Several natural ventilation strategies were studied, and the average concentration was presented for the building.
[50]	Urban canyon	Standard k-e	Normalised tracer gas concentration	Transmission of outdoor pollutants into indoor domain of only the downstream building was studied. Effect of building height and arrangement was tested.
[51]	Isolated building	Baseline k-w	Indoor tracer gas concentration	Few window types were simulated to identify the transmission characteristics of pollutant when released from an adjacent room.
[52]	Consecutive street canyon	RNG k-e	Tracer gas retention time	Ventilation pattern and pollutant retention time were analysed inside a target canyon placed in the middle of a series of urban canyons. 2D and 3D models were analysed for varying street lengths.
[53]	Urban canyon	RNG k-e LES	Normalised tracer gas concentration	Outdoor dispersion characteristics of traffic pollutants were studied for an isolated urban canyon.
[54]	Two industrial workshops	RNG k-e	Normalised tracer gas concentration	Cross-transmission of pollutants from one industrial building to another were studied under various thermal conditions.

To account for the shortcomings in previous studies, the current study developed a CFD model which adopted a modified version of the urban canyon pollution dispersion model. The present work builds upon earlier studies on pollution dispersion in street canyons [55], by integrating an indoor domain in the model. The model was validated against earlier experimental studies. The impact of the window opening pattern on the transmission of outdoor vehicular pollution into the indoor environment was studied. CFD was used to perform the simulations, as it offers a finer spatio-temporal resolution of the fluid interactions, which is often desirable for episodic contaminations, such as during a pandemic [41]. The tool is versatile and can be replicated to test various alternatives. Analytical assessment and measurement studies are unable to account for the diverse physics governing pollution dispersion and the fine resolution of measurement [56]. Unlike the previous works, the present study will be modelling a 3D urban canyon which is necessary to fully understand how the urban geometry, position of the indoor space in the building, and type of natural ventilation impacts the indoor–outdoor mean flow and pollutant transmission. The proposed model can help identify the sources of pollution issues in an urban area and provide useful guidance for future urban planning.

2. Materials and Methods

2.1. CFD Modelling and Choice of Turbulence

The choice of turbulence is often determined by the problem at hand, the focus of the investigation, and the available computational resources. While the RANS model is more often used in building studies because of its low computational cost, it cannot accurately predict the fluctuating flow field in the wake of the structure, such as separation and recirculation. LES (large eddy simulation) is a superior model, which can give better results and the transient nature of the flow field. It, however, requires higher computational resources and modelling of appropriate boundary conditions [57]. For the current study, an average concentration of pollutants within the indoor environment is needed to quantify the transmission in relation to outdoor levels. Therefore, the RANS (Reynolds-averaged Navier–Stokes) model was adopted as the predictions are within acceptable limits while also saving computation cost, and is based on statistical averaging leading to averaged equations of mass and momentum [58]. Three different turbulence models, the k-ω BSL, the RNG k-ϵ, and the RSM, were used in the validation. Further analysis was carried out using RSM, as it provided the most reliable results when compared with wind tunnel results, detailed in Section 2.4.

2.2. Computational Domain, Mesh, and Boundary Conditions

The geometrical configuration was based on the experimental investigation by Gromke et al. [59]. The model is fairly popular and has been applied in multiple studies of pollution dispersion in urban canyons, including [44,60–62]. Detailed information pertaining to model limits and wind tunnel results are also available online [63]. The canyon consists of two cuboidal building blocks of dimensions 10, 10, and 100 m in width, height, and length, respectively. The distance between the two buildings was 10 m, representing an aspect ratio of 1. Two line-sources of pollutant were placed in the canyon. The domain consists of an outer zone and an inner sub-zone near the vicinity of the canyon with a finer mesh.

The outer domain measures 30, 24, and 8 H length, width, and height, respectively, where H is the height of the building. The building was placed 8 H away from the front boundary and 7 H from the side boundaries, leading to a blockage ratio of 5.2%. The inner domain extends 18, 3, and 5 H in width, height, and depth, respectively. The domain was discretised using hexahedral elements with several mesh sizes tested for grid independence. Mesh size was refined until no significant improvement in the results was observed. The smallest dimension of the elements in the vicinity of the canyon was 0.05 H, with at least 15 cells across the length of the smallest edge. The validation model consists of 5.2 million cells, with 1.2 million cells in the canyon. Figure 1 shows the domain description and mesh characteristics.

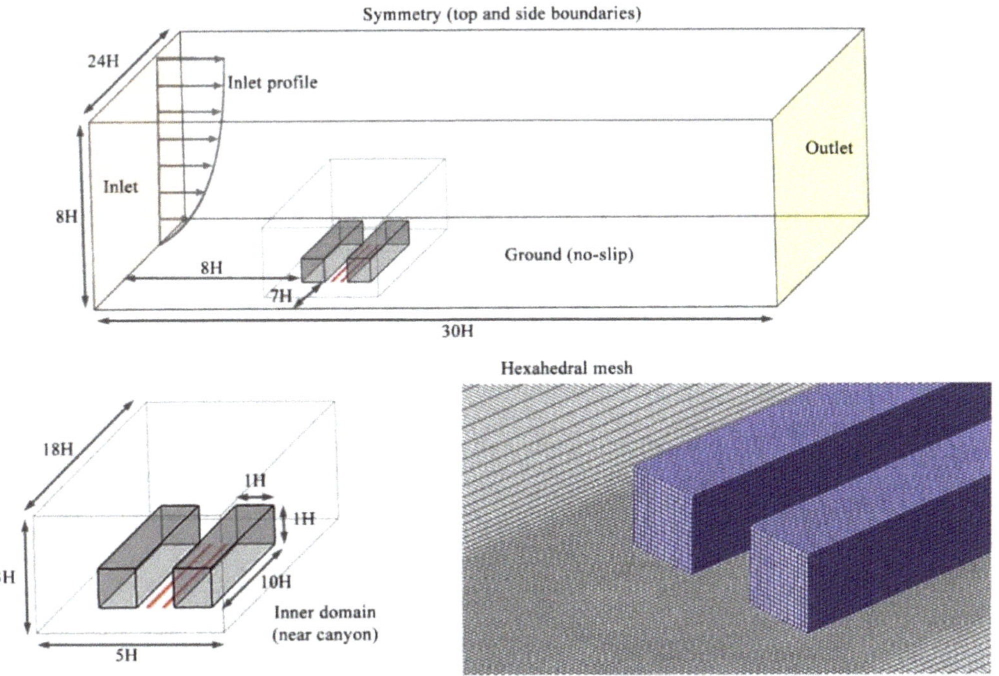

Figure 1. CFD domain, boundary condition, and surface mesh.

Inlet wind profile (speed, turbulent kinetic energy, dissipation rate profiles) was set according to the study described by [64], shown below:

$$u(z) = u_{ref} \left(\frac{z}{z_{ref}} \right)^{\alpha}$$

(1)

$$k = \frac{u_*^2}{\sqrt{C_\mu}} \left(1 - \frac{z}{\delta} \right)$$

(2)

$$\varepsilon = \frac{u_*^3}{kz} \left(1 - \frac{z}{\delta} \right)$$

(3)

where u_{ref} is the reference velocity equal to 4.7 m/s at a reference height z_{ref} = H. δ is the boundary layer depth (0.5 m), u^* is the friction velocity set to 0.54 m/s. The von Karman constant (k), and the velocity field and turbulence function (C_μ) were set to 0.4 and 0.09, respectively. Isothermal conditions are assumed, i.e., only wind-driven ventilation and pollution dispersion were considered. SIMPLE scheme was adopted for the pressure-velocity coupling, and second-order spatial discretisation was selected.

The tracer gas, SF_6, acted as a proxy for particulate matter and was released at the rate of Q = 10 mg/s from the two-line sources (red lines in Figure 1). The measured concentration was normalised according to the formula:

$$c^+ = \frac{c * z_{ref} * u_{ref}}{Q}$$

(4)

2.3. Canyon Model with Various Configurations

The modification to the original canyon model was in the form of an indoor environment at the centre and the ends of the building, representing the extreme cases. The height of each room was H/3, and the room depth was H. The building was naturally ventilated through the windows located on either side of the room. Window dimensions were H/10 × H/10, with sill level at 1 m above the floor. For the present case, H was set to 10 m. Walls and floors were given a thickness of 0.3 m and were accounted for in the geometry. Figure 2 shows the setup of the internal environment and its relation to the external canyon. Although it is possible to simulate all rooms in the building, the present work focused only on 18 key indoor spaces to reduce the computational time. However, additional rooms can be modelled in future works for a more detailed analysis.

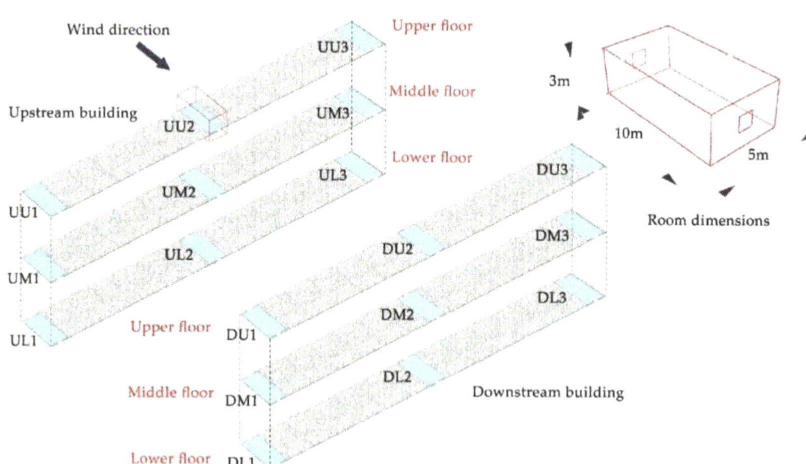

Figure 2. Location of internal spaces in the upstream and downstream buildings.

2.4. Method Verification and Validation

For verification purposes, three different turbulence models, the k-ω BSL, the RNG k-ϵ, and the RSM were tested and compared with experimental results available on the Concentration Data of Street Canyons (CODASC) website [63] and the studies [44,64,65]. Normalised pollutant concentration values were extracted from the canyon facing walls, A and B. While Wall A is located on the upstream building, Wall B is located on the downstream building, as shown in Figure 3. The case with all windows closed was used for verification, as it represents the decoupled indoor and outdoor domain.

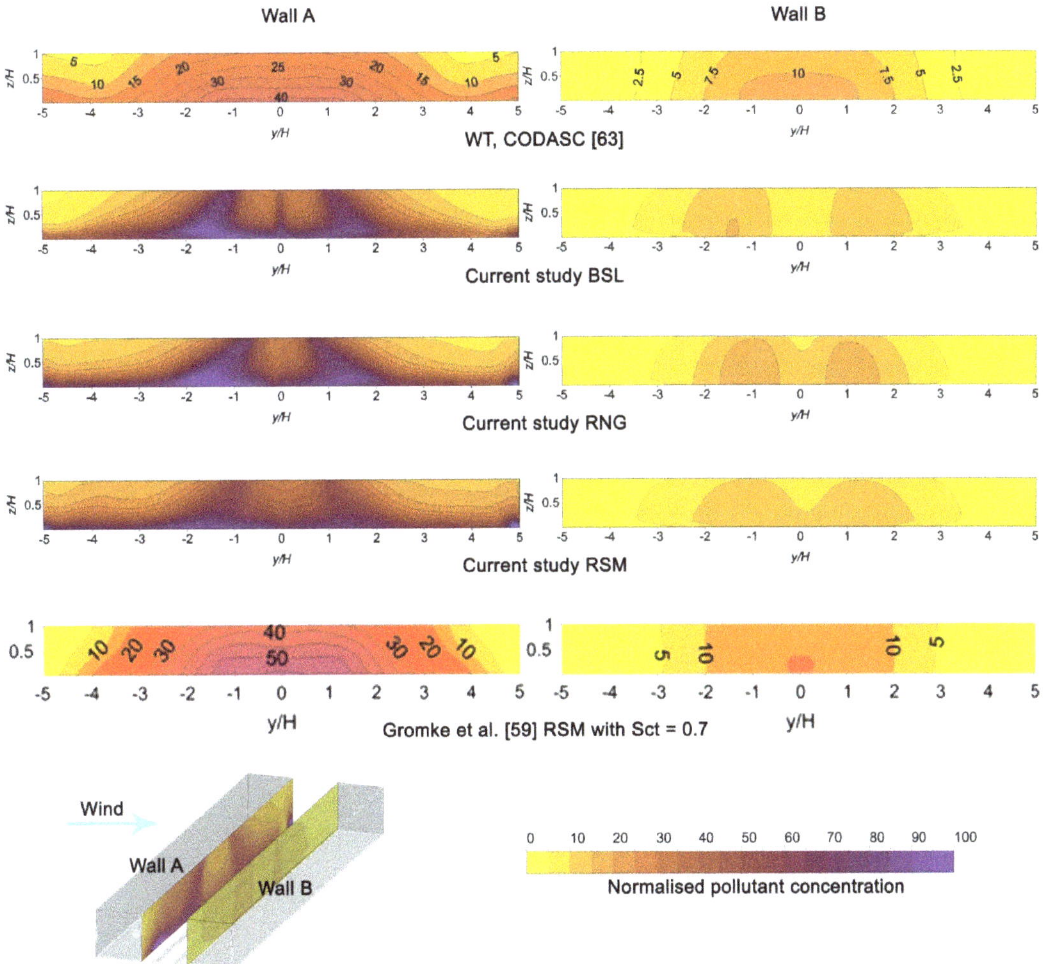

Figure 3. Validation of the pollutant concentration results against previous works [59,63].

The wind tunnel data predict the highest concentration near the centre of Wall A, especially near the bottom. The concentration gradually decreases while moving towards the corner; however, a slight increase was observed near the far end. On Wall B, the concentration decreases towards the sides; however, the magnitude of the pollution is 2–4 times lower as compared to the other wall. All three turbulence models predict similar concentration patterns, albeit with some variation. The BSL and the RNG model overpredict

SF_6 by 150% near the centre of Wall A, and slight underprediction (20%) occurs on the centre of Wall B. RSM also records a high pollution level near the centre and bottom of the walls, with values in the range 70–80 on Wall A. Compared with the numerical results by [55], the RSM model behaves most closely, with slight under prediction of about 33% near the centre.

Additionally, the pollutant concentration along the mid-height of the wall was also extracted, as shown in Figure 4. It was observed that while all three turbulence models overpredict pollutant concentration with respect to the wind tunnel data, the values are comparable to the numerical results by [55]. The RSM model fairly replicates the trend by [55], although studies by [65,66] are closer to the wind tunnel results. The difference in the domain settings could explain the apparent deviation. The pollutant was released at discrete points from pressure taps placed on the floor in the wind tunnel test, which were aligned along 4 parallel lines in the canyon, extending about 0.92 H beyond the building length. In the current numerical simulation, two continuous line-sources were modelled, which extend up to the canyon length.

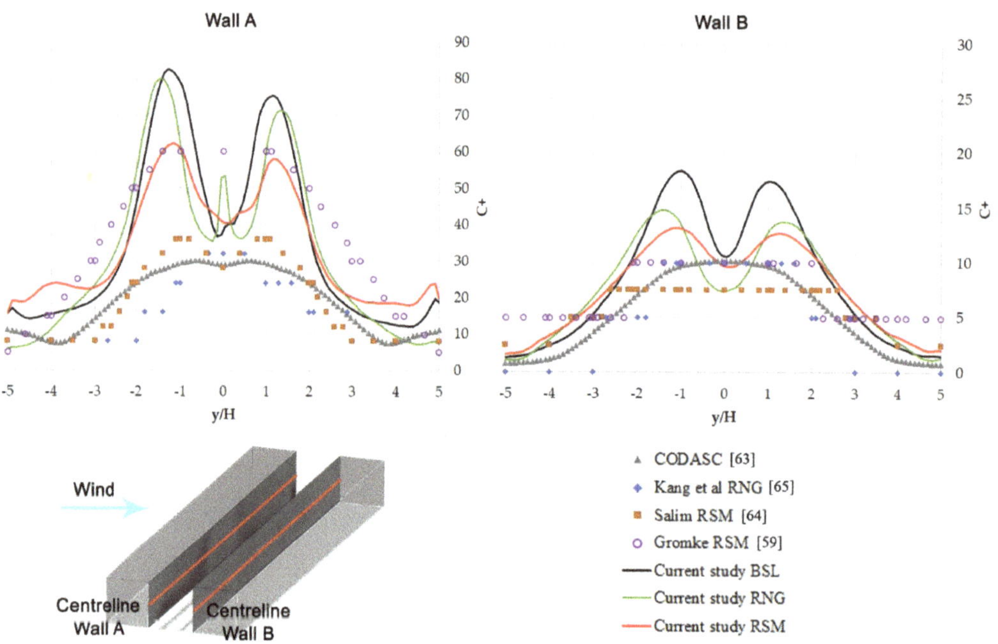

Figure 4. Validation of the pollutant concentration results against previous works [59,63–65].

As the research focused on indoor air pollution and the effect of urban configuration, the deviation was considered within the permissible range. The choice of RSM as the turbulence model was justified as the predictions within the region of interest were most close to the experimental data.

3. Results and Discussion

Pollution concentration data were analysed at a height of 1.5 m above the floors, which represented the average breathing height of indoor occupants. Similarly, cross-sectional planes through the canyon, at $X = 0$ m and $X = 37.5$ m, were also analysed. Table 2 shows the c^+ data along the central ($X = 0$) and side plane ($X = 37.5$). The results were symmetric along the centre of the canyon, and therefore, only one side plane is shown below. It is observed that the indoor c^+ levels are low to almost 0 for all cases, except for Case 4. Higher pollutant concentration occurs along Wall A regardless of the ventilation or position of the room. When cross ventilation was enabled, the upstream building was ventilated from the windward side, ensuring that pollutants from the canyon were not drawn indoors. Likewise in the downstream building, cross ventilation in Cases 1 and 5 ensures that the air is ventilated from the wind shadow side of the building, i.e., away from the polluted canyon. However, there is a slight increase in c^+ values in rooms UU1 and UU3. This suggests that indoor spaces located on the upper corners of the downstream building were generally ventilated from the canyon side, thereby some amount of infiltration is likely.

Table 2. Normalised pollutant concentration along the cross-sectional plane through the canyon; (**A**) central plane, (**B**) side plane.

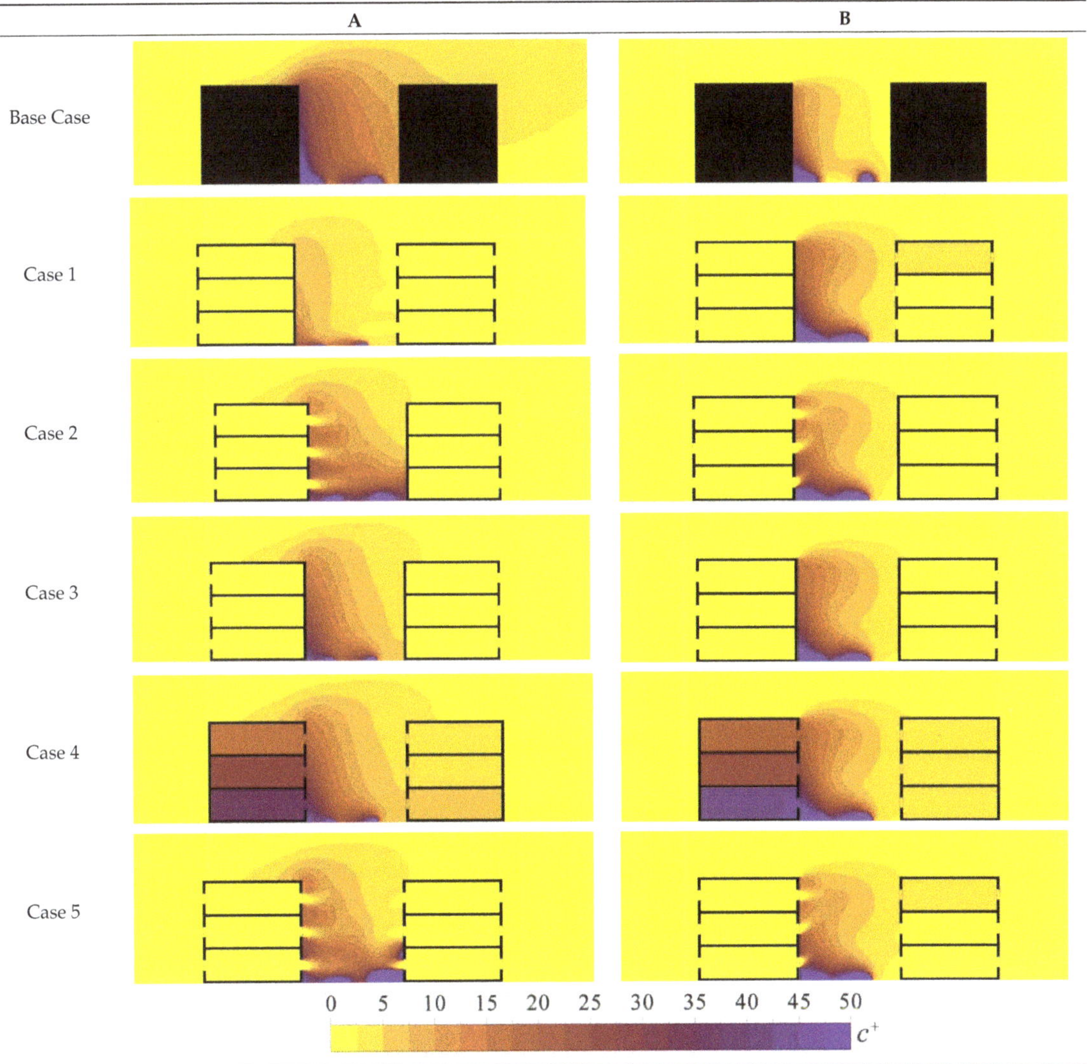

In the case of single-sided ventilation, the quality of indoor air is significantly impacted when air flow occurs from the canyon side. This is evident in Case 4, where high concentration of pollution is observed inside both the buildings. Between the two built masses, the downstream building shows lower concentration of indoor pollutants as compared to the upstream building, contrary to general assumption. This can be explained by the clockwise recirculation generated in the canyon, which forces pollutants to accumulate along Wall A. Internal rooms UL2, UM2, and UU2 show higher c^+ values when compared to DL2, DM2, and DU2. Simultaneously, pollution concentration also varies with the floor height. In the upstream building, the indoor concentration level decreases with floor height, while in the downstream building, this variation is only observed for the centrally located spaces. The rooms located towards the ends of the canyon, however, show a reversal in concentration levels. The lower floors record lower c^+ as compared to upper floors (Table 2)

It is interesting to observe that the downstream building is ventilated from the leeward side when cross ventilation was enabled (Cases 1, 4, and 5). This ensures that no pollutants migrate from the canyon side. Recirculation zones generated in the wake of the buildings force the air to rise along the building walls. The clockwise wind flow in the street forces the pollutants to rise along Wall A and flow above the canyon, some of which is recirculated down Wall B.

Table 3 shows the normalised concentration of pollutants extracted at the height of 1.5 m above the ground floor. In the reference case, the swirls generated in the canyon cause concentration to increase along the leeward and side walls of the upstream building. Higher pollutants are observed near the middle of the canyon. Cross flow of air through the internal rooms, such as in Case 1, causes dilution of pollutants in the canyon, although increase in c^+ is observed for some cases near Wall B, such as in Cases 2 and 5. In general, infiltration of pollutants into indoor spaces remains negligible, as air flow inside the room is not from the canyon side. For Case 4, however, the air is drawn into the indoor space from the canyon side, leading to a high concentration of pollutants inside. c^+ values as high as 45 are observed in the upstream rooms (UL1, UL2, and UL3), while values in the range of 3–5 are observed in the downstream rooms (DL1, DL2 and DL3).

Concentration for other cases remain comparatively low. Although the results showed that the cross ventilation minimised or eliminated the built up of pollutants inside the spaces, in practice incorporating such a strategy might not be practical for all types of rooms and buildings. In most cases, there will be several apartments, rooms, or partitions which will dampen the air flow. Hence, further studies should consider a more realistic indoor space instead of an idealised cross flow ventilated space. The results show that incorporating a natural ventilation strategy may not always lead to healthy indoor spaces, in particular, if the ventilated room or space is facing the urban canyon side and employs single-sided ventilation. Future works should consider evaluating the trade-off between natural ventilation and pollutant transmission for these types of spaces.

Pollutant concentration is comparatively lower on the first floor, as shown in Table 4. Recirculating eddies lead to a higher concentration of pollutants near Wall A and around the corners of the upstream building. c^+ contours are symmetric along the central axis, and the highest values are observed at about a distance of 2.5 H from either edge of the building. In Case 1, ventilation of room DM2 from the leeward side forces air into the canyon, diluting the pollutants near Wall B. Similar observations are also made for Cases 4 and 5. Whereas, no such observation is made for cross flow of air through UM2, in the upstream building. On the contrary, a slight increase is observed near Wall B. Case 4 is the most critical configuration, wherein the indoor space is ventilated from the canyon side. Air is drawn from the canyon into the adjoining rooms and a high level of pollution is observed in the upstream building, with an average c^+ value of about 25. Rooms in the downstream building perform relatively better, with c^+ values less than 5. In general, pollution levels on this floor are the same as the ones measured on the ground floor, except for Case 4. While air flows into the internal spaces from the non-canyon side for most configurations, Case 4 shows an exception with air being drawn from the canyon side. This also leads to the intake of pollutants from the canyon, and significantly impacts the performance of the spaces.

Table 3. Normalised pollutant concentration at occupants' height on ground floor.

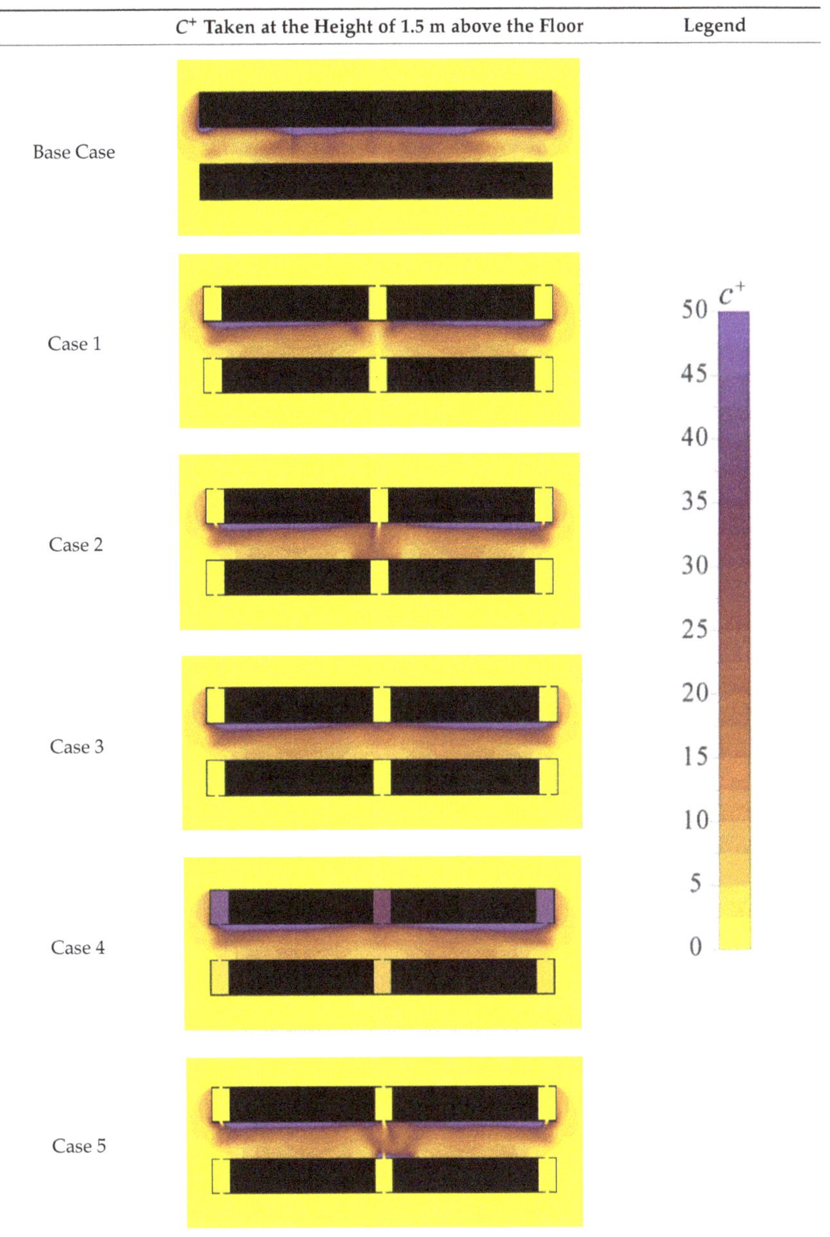

Table 4. Pollutant concentration at occupants' height on first floor.

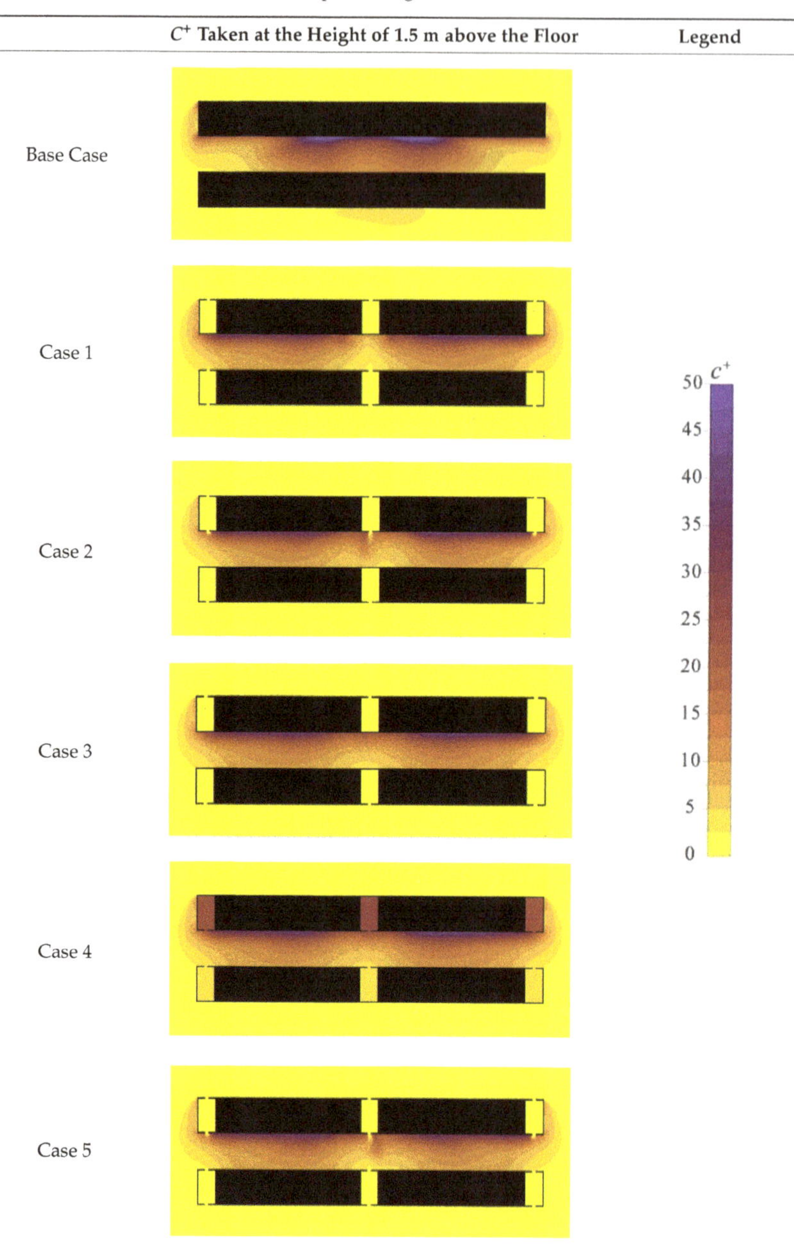

As was indicated previously, pollutant concentration decreases with height, and the same can be observed in the difference between c^+ values between UM2 and UU2. Table 5 shows the pollutant concentration on the second floor, at the height of 8.25 m above ground. Despite the high concentration of pollutants adjacent to Wall A in the base case, internal rooms do not record a significant pollution level. Although a slight increase in c^+ values are observed in the wake of the downstream building for the base case, facilitating air flow through the building causes dispersion of these pollutants. The pollution distribution pattern follows the pattern similar to that which exists on the lower floors, i.e., negligible c^+ values in the upstream building for all cases, except Case 4. Additionally, the rooms in the downstream building record low levels of pollutant concentration which is almost the same as lower floors. c^+ values of ~18 are seen in the spaces UU1, UU2, and UU3, while the upper rooms on the downstream building, DU1, DU2, and DU3 show a value of less than 5.

Table 5. Pollutant concentration at occupants' height on second floor.

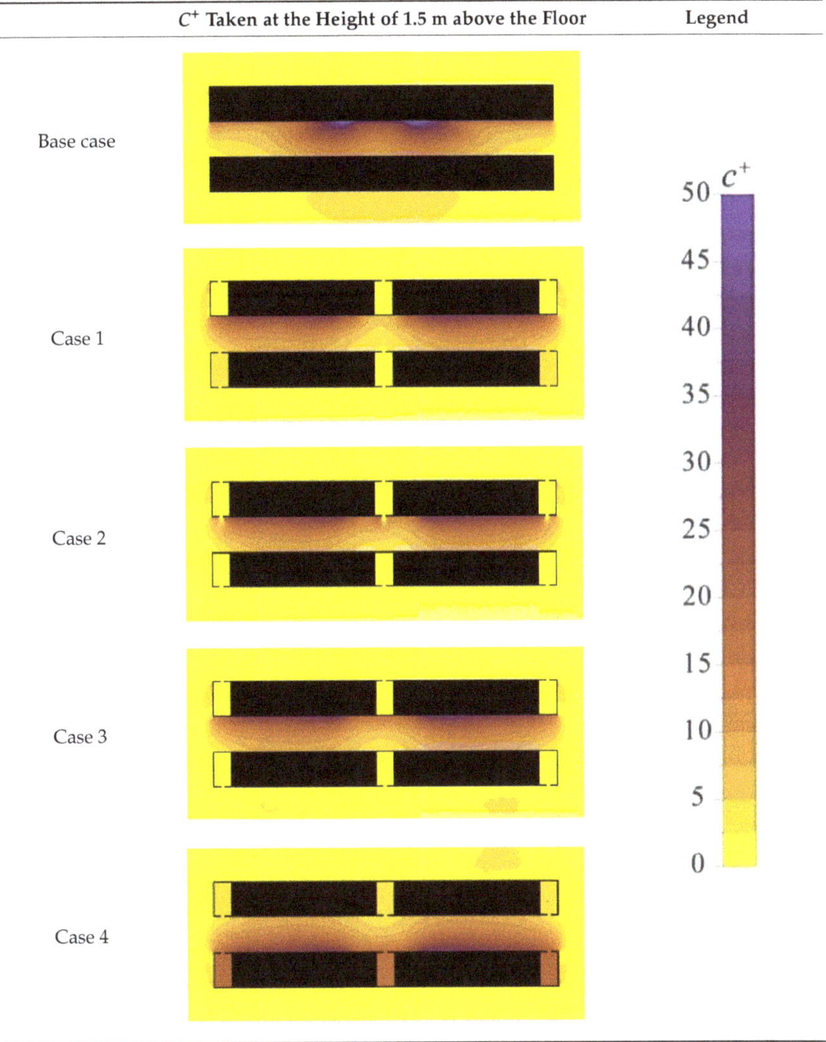

Table 5. *Cont.*

	C^+ Taken at the Height of 1.5 m above the Floor	Legend
Case 5	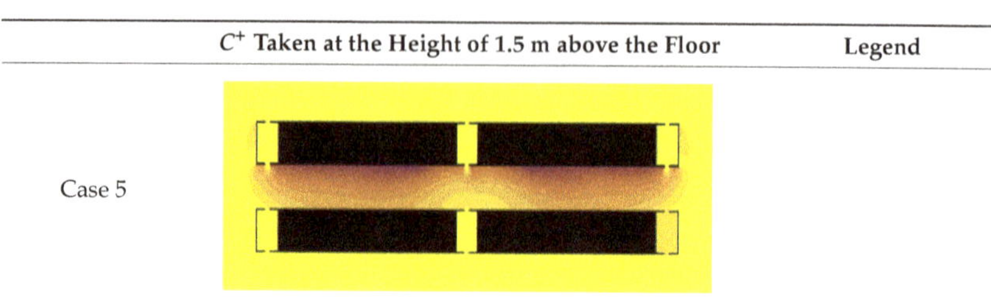	

A comparison of the average c^+ at the breathing plane is shown in Figure 5. Evidently, except for Case 4, all window opening strategies ensure that the pollution remains low in the indoor regions. The upstream building has a negligible level of pollutants penetrating the indoor spaces for most of the cases. The air flow direction ensures that the spaces are ventilated from the windward side. However, in the critical scenario (Case 4), air flow from the canyon side accumulates the pollutants inside the rooms. Concentration is highest on the ground floor, while there is a drop of about 30–45% on the first floor. Further reductions in the range of 20–30% are observed on the second floor. Additionally, the rooms near the edge of the canyon have a slightly higher pollutant concentration as compared to the centrally located ones. The average values differ by about 10%.

Low levels of infiltration are observed for all cases in the downstream building, arising due to recirculating air flow. Concentration in the range of 1–3 is recorded for all cases, and centrally located rooms are less polluted than the corner ones. Surprisingly, the upper floors are more susceptible to infiltration as compared to lower floors. In Case 1, for instance, DU1 has a 47% higher concentration than DL1. For Case 4, however, concentration in the range of 3–6 is observed, which is nearly a two-fold increase.

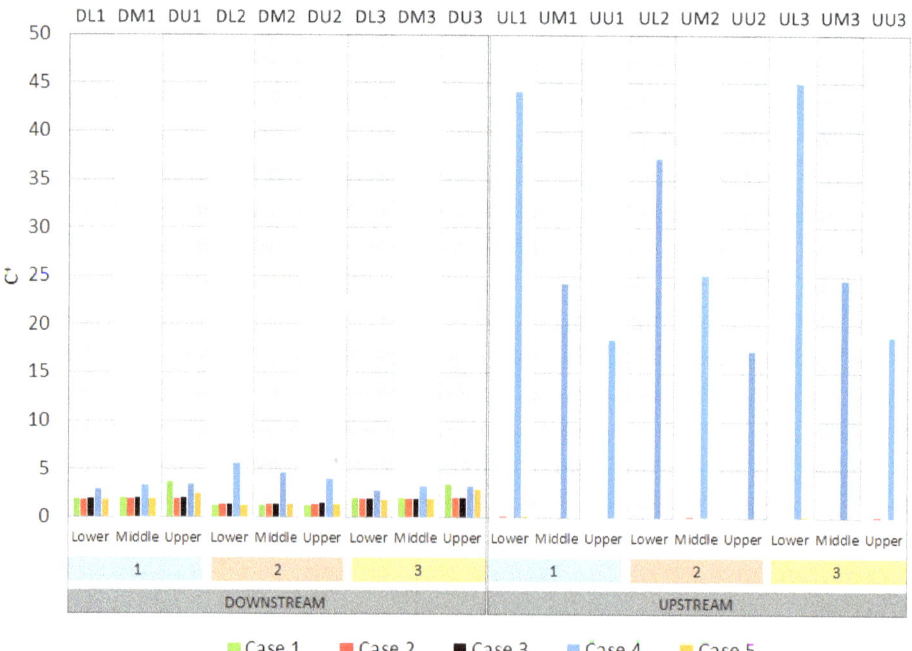

Figure 5. Comparison of average indoor c^+, measured at occupants' breathing plane.

4. Conclusions and Future Works

The study aimed to identify the transmission characteristics of outdoor pollutants into an indoor environment, using computational modelling in an idealised street canyon. For this purpose, a model was generated in Ansys® (Canonsburg, PA, Pennsylvania) R18.1 similar to the experimental study conducted by Karlsruhe Institute of Technology (Karlsruhe, Germany) [63] and numerical simulation by [44,64,65]. After verification and sensitivity analysis, the archetypal canyon model was modified to consider the indoor environment. The novelty of this study lay in this combination of the indoor and outdoor environment, which has not been tried earlier. Previous studies have either not modelled the indoor environment in a street canyon or have simplified the flow field and missed important parameters. Additionally, the impact of design and microenvironmental parameters on the transmission characteristics and flow field can be studied using this numerical model, including the effect of wind speed, direction, and thermal conditions. Most studies have carried out independent assessment of either indoor or outdoor pollutants, such as performance of HVAC systems or ventilation modes; however, the combined assessment and investigation of factors remains to be explored in detail, including the nature of transmission and finer nuances of air flow.

Investigations in the current study have been directed towards the impact of ventilation strategy on the transmission of outdoor pollutants into indoor spaces. Five window opening strategies were modelled and implemented on an urban canyon with an aspect ratio of 1. Rooms were located at the centre and the far end of the building. The results indicate that single-sided ventilation from the canyon side leads to the accumulation of outdoor pollutants for both the upstream and downstream buildings. Especially, pollution in the upstream building was 10–20 times higher than the downstream building. For other ventilation modes, the upstream building had negligible infiltration of pollutants as the airflow into the interior space was from the windward side—away from the polluting street. Whereas, the downstream building showed constant infiltration of pollutants, albeit in comparatively smaller quantity (in the range of 1–3). It can be concluded that living in the upstream building is usually a safer option than the downstream building; however, occupants in the upstream building must ensure that their spaces are cross ventilated from the windward side. Occupants in the downstream building are usually exposed to a constant pollution level regardless of the ventilation mode.

Future studies should explore the effect of aspect ratio, building shape, and configuration on the air quality and transmission into these spaces. The impact of varying outdoor conditions should be evaluated, which includes the wind speed and direction. The influence of the location, amount, and type of outdoor pollutant sources should also be evaluated. Furthermore, more focus should be given to the indoor environment and indoor pollutant source to help assess inter-unit transmission in urban canyons. The present model did not consider the impact of neighbouring buildings affecting the wind flow patterns around the urban canyon.

Author Contributions: Conceptualisation, M.M. and J.C.; methodology, M.M.; software, M.M.; validation, M.M. and J.C.; formal analysis, M.M.; investigation, M.M. and J.C.; resources, J.C.; data curation, M.M.; writing—original draft preparation, M.M.; writing—review and editing, J.C.; visualisation, M.M.; supervision, J.C.; project administration, J.C. Both authors have read and agreed to the published version of the manuscript.

Funding: This research received no external funding.

Institutional Review Board Statement: Not applicable.

Informed Consent Statement: Not applicable.

Conflicts of Interest: The authors declare no conflict of interest.

References

1. Ritchie, H. Indoor Air Pollution. Our World in Data, 16 November 2013. Available online: https://ourworldindata.org/indoor-air-pollution (accessed on 5 September 2020).
2. GBD 2017 Risk Factor Collaborators. Global, regional, and national comparative risk assessment of 84 behavioural, environmental and occupational, and metabolic risks or clusters of risks for 195 countries and territories, 1990–2017: A systematic analysis for the Global Burden of Disease Study 2017. *Lancet* **2018**, *392*, 1923–1994. [CrossRef]
3. WHO. *Review of Evidence on Health Aspects of Air Pollution—REVIHAAP Project*; World Health Organization: Geneva, Switzerland, 2013. Available online: https://www.euro.who.int/en/health-topics/environment-and-health/air-quality/publications/2013/review-of-evidence-on-health-aspects-of-air-pollution-revihaap-project-final-technical-report (accessed on 9 April 2020).
4. Effects of Air Pollution & Acid Rain on Wildlife. Available online: http://www.air-quality.org.uk/17.php (accessed on 9 April 2021).
5. Pope, C.A.; Ezzati, M.; Dockery, D.W. Tradeoffs between income, air pollution and life expectancy: Brief report on the US experience, 1980–2000. *Environ. Res.* **2015**, *142*, 591–593. [CrossRef]
6. Wu, X.; Nethery, R.C.; Sabath, B.; Braun, D.; Dominici, F.; James, C. Exposure to air pollution and COVID-19 mortality in the United States: A nationwide cross-sectional study. *medRxiv Preprint* **2020**. [CrossRef]
7. Coker, E.S.; Cavalli, L.; Fabrizi, E.; Guastella, G.; Lippo, E.; Parisi, M.L.; Pontarollo, N.; Rizzati, M.; Varacca, A.; Vergalli, S. The Effects of Air Pollution on COVID-19 Related Mortality in Northern Italy. *Environ. Resour. Econ.* **2020**, *76*, 611–634. [CrossRef]
8. Copat, C.; Cristaldi, A.; Fiore, M.; Grasso, A.; Zuccarello, P.; Signorelli, S.S.; Conti, G.O.; Ferrante, M. The role of air pollution (PM and NO_2) in COVID-19 spread and lethality: A systematic review. *Environ. Res.* **2020**, *191*, 110129. [CrossRef] [PubMed]
9. EPA. *Report to Congress on Indoor Air Quality, Volume II: Assessment and Control of Indoor Air Pollution*; EPA: Washington, DC, USA, 1989.
10. Menichini, E.; Iacovella, N.; Monfredini, F.; Turrio-Baldassarri, L. Relationships between indoor and outdoor air pollution by carcinogenic PAHs and PCBs. *Atmos. Environ.* **2007**, *41*, 9518–9529. [CrossRef]
11. Aftermath of Working from Home. University of Southampton. 2020. Available online: https://www.southampton.ac.uk/news/2020/07/long-term-implications-wfh.page (accessed on 30 September 2020).
12. Deloitte. The Impact of COVID-19 on Productivity and Wellbeing. 2020. Available online: https://www2.deloitte.com/uk/en/pages/consulting/articles/working-during-lockdown-impact-of-covid-19-on-productivity-and-wellbeing.html (accessed on 5 September 2020).
13. Cardiff University. *UK Productivity Could be Improved by a Permanent Shift towards Remote Working*; Cardiff University: Cardiff, UK, 2020. Available online: https://www.cardiff.ac.uk/news/view/2432442-uk-productivity-could-be-improved-by-a-permanent-shift-towards-remote-working,-research-shows (accessed on 30 September 2020).
14. Spataro, J. The Future of Work. Microsoft 365 Blog. 2020. Available online: https://www.microsoft.com/en-us/microsoft-365/blog/2020/07/08/future-work-good-challenging-unknown/ (accessed on 5 September 2020).
15. Tien, P.W.; Wei, S.; Liu, T.; Calautit, J.; Darkwa, J.; Wood, C. A Deep Learning Approach Towards the Detection and Recognition of Opening of Windows for Effective Management of Building Ventilation Heat Losses and Reducing Space Heating Demand. *Renew. Energy* **2021**, *177*, 603–625. [CrossRef]
16. Chang, W.R.; Cheng, C.L. Carbon monoxide transport in an enclosed room with sources from a water heater in the adjacent balcony. *Build. Environ.* **2008**, *43*, 861–870. [CrossRef]
17. Park, J.S.; Jee, N.-Y.; Jeong, J.-W. Effects of types of ventilation system on indoor particle concentrations in residential buildings. *Indoor Air* **2014**, *24*, 629–638. [CrossRef]
18. Chen, C.; Zhao, B.; Weschler, C.J. Indoor Exposure to "Outdoor PM_{10}": Assessing Its Influence on the Relationship Between PM_{10} and Short-term Mortality in U.S. Cities on JSTOR. *Epidemiology* **2012**, *23*, 870–878. [CrossRef]
19. Chan, A.T. Indoor-outdoor relationships of particulate matter and nitrogen oxides under different outdoor meteorological conditions. *Atmos. Environ.* **2002**, *36*, 1543–1551. [CrossRef]
20. Błaszczyk, E.; Rogula-Kozłowska, W.; Klejnowski, K.; Kubiesa, P.; Fulara, I.; Mielżyńska-Švach, D. Indoor air quality in urban and rural kindergartens: Short-term studies in Silesia, Poland. *Atmos. Health* **2017**, *10*, 1207–1220. [CrossRef]
21. Hassanvand, M.S.; Naddafi, K.; Faridi, S.; Arhami, M.; Nabizadeh, R.; Sowlat, M.H.; Pourpak, Z.; Rastkari, N.; Momeniha, F.; Kashani, H.; et al. Indoor/outdoor relationships of PM10, PM2.5, and PM1 mass concentrations and their water-soluble ions in a retirement home and a school dormitory. *Atmos. Environ.* **2014**, *82*, 375–382. [CrossRef]
22. Chen, A.; Gall, E.T.; Chang, V.W.C. Indoor and outdoor particulate matter in primary school classrooms with fan-assisted natural ventilation in Singapore. *Environ. Sci. Pollut. Res.* **2016**, *23*, 17613–17624. [CrossRef]
23. Zhao, J.; Birmili, W.; Wehner, B.; Daniels, A.; Weinhold, K.; Wang, L.; Merkel, M.; Kecorius, S.; Tuch, T.; Franck, U.; et al. Particle Mass Concentrations and Number Size Distributions in 40 Homes in Germany: Indoor-to-Outdoor Relationships, Diurnal and Seasonal Variation. *Aerosol Air Qual. Res.* **2020**, *20*, 576–589. [CrossRef]
24. Chiesa, M.; Urgnani, R.; Marzuoli, R.; Finco, A.; Gerosa, G. Site- and house-specific and meteorological factors influencing exchange of particles between outdoor and indoor domestic environments. *Build. Environ.* **2019**, *160*, 106181. [CrossRef]
25. Wang, F.; Meng, D.; Li, X.; Tan, J. Indoor-outdoor relationships of PM2.5 in four residential dwellings in winter in the Yangtze River Delta, China. *Environ. Pollut.* **2016**, *215*, 280–289. [CrossRef] [PubMed]

26. Rim, D.; Gall, E.T.; Kim, J.B.; Bae, G.N. Particulate matter in urban nursery schools: A case study of Seoul, Korea during winter months. *Build. Environ.* **2017**, *119*, 1–10. [CrossRef]
27. Niu, J.; Tung, T.C.W. On-site quantification of re-entry ratio of ventilation exhausts in multi-family residential buildings and implications. *Indoor Air* **2007**, *18*, 12–26. [CrossRef] [PubMed]
28. Mao, J.; Gao, N. The airborne transmission of infection between flats in high-rise residential buildings: A review. *Build. Environ.* **2015**, *94*, 516–531. [CrossRef]
29. Tippayawong, N.; Khuntong, P.; Nitatwichit, C.; Khunatorn, Y.; Tantakitti, C. Indoor/outdoor relationships of size-resolved particle concentrations in naturally ventilated school environments. *Build. Environ.* **2009**, *44*, 188–197. [CrossRef]
30. Lv, Y.; Wang, H.; Wei, S.; Zhang, L.; Zhao, Q. The Correlation between Indoor and Outdoor Particulate Matter of Different Building Types in Daqing, China. *Procedia Eng.* **2017**, *205*, 360–367. [CrossRef]
31. Gupta, A.; David Cheong, K.W. Physical characterization of particulate matter and ambient meteorological parameters at different indoor-outdoor locations in Singapore. *Build. Environ.* **2007**, *42*, 237–245. [CrossRef]
32. Chithra, V.S.; Shiva Nagendra, S.M. Impact of outdoor meteorology on indoor PM10, PM2.5 and PM1 concentrations in a naturally ventilated classroom. *Urban Clim.* **2014**, *10*, 77–91. [CrossRef]
33. Orza, J.A.G.; Cabello, M.; Lidón, V.; Martínez, J. Contribution of resuspension to particulate matter inmission levels in SE Spain. *J. Arid Environ.* **2011**, *75*, 545–554. [CrossRef]
34. He, L.; Hang, J.; Wang, X.; Lin, B.; Li, X.; Lan, G. Numerical investigations of flow and passive pollutant exposure in high-rise deep street canyons with various street aspect ratios and viaduct settings. *Sci. Total Environ.* **2017**, *584–585*, 189–206. [CrossRef]
35. Tominaga, Y.; Stathopoulos, T. Ten questions concerning modeling of near-field pollutant dispersion in the built environment. *Build. Environ.* **2016**, *105*, 390–402. [CrossRef]
36. Ai, Z.T.; Mak, C.M. *From Street Canyon Microclimate to Indoor Environmental Quality in Naturally Ventilated Urban Buildings: Issues and Possibilities for Improvement*; Elsevier: Amsterdam, The Netherlands, 2015; Volume 94, pp. 489–503.
37. Yuan, C. Improving Air Quality by Understanding the Relationship Between Air Pollutant Dispersion and Building Morphologies. In *Springer Briefs in Architectural Design and Technology*; Springer: Berlin/Heidelberg, Germany, 2018; pp. 117–140.
38. Salim, S.M.; Buccolieri, R.; Chan, A.; Di Sabatino, S. Numerical simulation of atmospheric pollutant dispersion in an urban street canyon: Comparison between RANS and LES. *J. Wind Eng. Ind. Aerodyn.* **2011**, *99*, 103–113. [CrossRef]
39. Hong, B.; Qin, H.; Lin, B. Prediction of wind environment and indoor/outdoor relationships for PM2.5 in different building-tree grouping patterns. *Atmosphere* **2018**, *9*, 39. [CrossRef]
40. Rai, P.K.; Panda, L.L.S. Dust capturing potential and air pollution tolerance index (APTI) of some road side tree vegetation in Aizawl, Mizoram, India: An Indo-Burma hot spot region. *Air Qual. Atmos. Health* **2014**, *7*, 93–101. [CrossRef]
41. Tong, Z.; Chen, Y.; Malkawi, A.; Adamkiewicz, G.; Spengler, J.D. Quantifying the impact of traffic-related air pollution on the indoor air quality of a naturally ventilated building. *Environ. Int.* **2016**, *89–90*, 138–146. [CrossRef]
42. Yang, F.; Kang, Y.; Gao, Y.; Zhong, K. Numerical simulations of the effect of outdoor pollutants on indoor air quality of buildings next to a street canyon. *Build. Environ.* **2015**, *87*, 10–22. [CrossRef]
43. Peng, Y.; Ma, X.; Zhao, F.; Liu, C.; Mei, S. Wind driven natural ventilation and pollutant dispersion in the dense street canyons: Wind Opening Percentage and its effects. *Procedia Eng.* **2017**, *205*, 415–422. [CrossRef]
44. Hong, B.; Qin, H.; Jiang, R.; Xu, M.; Niu, J. How Outdoor Trees Affect Indoor Particulate Matter Dispersion: CFD Simulations in a Naturally Ventilated Auditorium. *Int. J. Environ. Res. Public Health* **2018**, *15*, 2862. [CrossRef]
45. Wang, B.; Hang, J.; Buccolieri, R.; Yang, X.; Yang, H.; Quarta, F. Impact of indoor-outdoor temperature differences on dispersion of gaseous pollutant and particles in idealized street canyons with and without viaduct settings. In *Building Simulation*; Tsinghua University Press: Beijing, China, 2019; Volume 12, pp. 285–297. [CrossRef]
46. Hassan, A.M.; ELMokadem, A.A.; Megahed, N.A.; Abo Eleinen, O.M. Urban morphology as a passive strategy in promoting outdoor air quality. *J. Build. Eng.* **2020**, *29*, 101204. [CrossRef]
47. Gousseau, P.; Blocken, B.; Stathopoulos, T.; van Heijst, G.J.F. CFD simulation of near-field pollutant dispersion on a high-resolution grid: A case study by LES and RANS for a building group in downtown Montreal. *Atmos. Environ.* **2011**, *45*, 428–438. [CrossRef]
48. Huang, J.; Jones, P.; Zhang, A.; Hou, S.S.; Hang, J.; Spengler, J.D. Outdoor Airborne Transmission of Coronavirus Among Apartments in High-Density Cities. *Front. Built Environ.* **2021**, *7*, 48. [CrossRef]
49. Liu, X.; Peng, Z.; Liu, X.; Zhou, R. Dispersion Characteristics of Hazardous Gas and Exposure Risk Assessment in a Multiroom Building Environment. *Int. J. Environ. Res. Public Health* **2020**, *17*, 199. [CrossRef] [PubMed]
50. Yang, F.; Zhong, K.; Chen, Y.; Kang, Y. Simulations of the impacts of building height layout on air quality in natural-ventilated rooms around street canyons. *Environ. Sci. Pollut. Res.* **2017**, *24*, 23620–23635. [CrossRef]
51. Wang, J.; Huo, Q.; Zhang, T.; Wang, S.; Battaglia, F. Numerical investigation of gaseous pollutant cross-transmission for single-sided natural ventilation driven by buoyancy and wind. *Build. Environ.* **2020**, *172*, 106705. [CrossRef]
52. Mei, S.J.; Luo, Z.; Zhao, F.Y.; Wang, H.Q. Street canyon ventilation and airborne pollutant dispersion: 2-D versus 3-D CFD simulations. *Sustain. Cities Soc.* **2019**, *50*, 101700. [CrossRef]
53. Tominaga, Y.; Stathopoulos, T. CFD modeling of pollution dispersion in a street canyon: Comparison between LES and RANS. *J. Wind Eng. Ind. Aerodyn.* **2011**, *99*, 340–348. [CrossRef]
54. Wang, Y.; Zhao, T.; Cao, Z.; Zhai, C.; Wu, S.; Zhang, C.; Zhang, Q.; Lv, W. The influence of indoor thermal conditions on ventilation flow and pollutant dispersion in downstream industrial workshop. *Build. Environ.* **2021**, *187*, 107400. [CrossRef]

55. Gromke, C.; Ruck, B. Pollutant Concentrations in Street Canyons of Different Aspect Ratio with Avenues of Trees for Various Wind Directions. *Bound.-Layer Meteorol.* **2012**, *144*, 41–64. [CrossRef]
56. Milner, J.; Vardoulakis, S.; Chalabi, Z.; Wilkinson, P. Modelling inhalation exposure to combustion-related air pollutants in residential buildings: Application to health impact assessment. *Environ. Int.* **2011**, *37*, 268–279. [CrossRef] [PubMed]
57. Blocken, B. 50 years of Computational Wind Engineering: Past, present and future. *J. Wind Eng. Ind. Aerodyn.* **2014**, *129*, 69–102. [CrossRef]
58. Chaouat, B. The State of the Art of Hybrid RANS/LES Modeling for the Simulation of Turbulent Flows. *Flow Turbul. Combust.* **2017**, *99*, 279–327. [CrossRef]
59. Gromke, C.; Buccolieri, R.; Di Sabatino, S.; Ruck, B. Dispersion study in a street canyon with tree planting by means of wind tunnel and numerical investigations—Evaluation of CFD data with experimental data. *Atmos. Environ.* **2008**, *42*, 8640–8650. [CrossRef]
60. Yazid, A.W.M.; Sidik, N.A.C.; Salim, S.M.; Saqr, K.M. A review on the flow structure and pollutant dispersion in urban street canyons for urban planning strategies. *Simulation* **2014**, *90*, 892–916. [CrossRef]
61. Gan, C.J.; Salim, S.M. Numerical analysis of fluid-structure interaction between wind flow and trees. *Lect. Notes Eng. Comput. Sci.* **2014**, *2*, 1218–1223.
62. Jeanjean, A.P.R.; Hinchliffe, G.; McMullan, W.A.; Monks, P.S.; Leigh, R.J. A CFD study on the effectiveness of trees to disperse road traffic emissions at a city scale. *Atmos. Environ.* **2015**, *120*, 1–14. [CrossRef]
63. Karlsruhe Institute of Technology CODASC. Concentration Data of Street Canyons. Laboratory of Building- and Environmental Aerodynamics. 2008. Available online: https://www.windforschung.de/CODASC.htm (accessed on 13 October 2020).
64. Salim, S.M. Computational Study of Wind Flow and Pollutant Dispersion Near Tree Canopies. Ph.D. Thesis, University of Nottingham, Nottingham, UK, 2011.
65. Kang, G.; Kim, J.-J.J.; Kim, D.-J.J.; Choi, W.; Park, S.-J.J. Development of a computational fluid dynamics model with tree drag parameterizations: {Application} to pedestrian wind comfort in an urban area. *Build. Environ.* **2017**, *124*, 209–218. [CrossRef]
66. Buccolieri, R.; Salim, S.M.; Leo, L.S.; Di Sabatino, S.; Chan, A.; Ielpo, P.; de Gennaro, G.; Gromke, C. Analysis of local scale tree-atmosphere interaction on pollutant concentration in idealized street canyons and application to a real urban junction. *Atmos. Environ.* **2011**, *45*, 1702–1713. [CrossRef]

MDPI

St. Alban-Anlage 66

4052 Basel

Switzerland

Tel. +41 61 683 77 34

Fax +41 61 302 89 18

www.mdpi.com

Sustainability Editorial Office

E-mail: sustainability@mdpi.com

www.mdpi.com/journal/sustainability